Field and Laboratory Methods in Primatology
A Practical Guide

Field and Laboratory Methods in Primatology is a manual for students and researchers studying wild primates. Technological advances allow fieldworkers to collect a wide range of data, store samples for later analysis, and collect information remotely. These methods open up opportunities to gain new insights on previously studied populations and are the means of collecting data on species that have, until now, been difficult to study. However, information on the practicalities of using such methodology in the field has largely been lacking. Here, in this indispensable reference, experienced field- and laboratory-workers provide the first comprehensive practical guide to the wide variety of techniques available for the study of wild primates. Covering everything from pre-trip planning to laboratory analysis of endocrine and genetic samples, packed full of tips and emphasising practicalities and ethics throughout, it is a must-have for all field primatologists and others studying free-ranging animals.

JOANNA M. SETCHELL is a post-doctoral research fellow at the University of Surrey Roehampton, and the Centre International de Recherches Médicales in Gabon, where she studies mandrills. She has also conducted primatological fieldwork in Cameroon, Congo and Malaysia. Her research interests include primate reproductive strategies, sexual selection, behavioural endocrinology and the integration of field and laboratory studies of primates.

DEBORAH J. CURTIS is a post-doctoral research fellow at the University of Surrey Roehampton. Her research currently focuses on activity patterns in primates, primate reproductive endocrinology and plant–lemur co-evolution. Her work has included field studies on the behaviour and ecology of mongoose lemurs in Madagascar, research on lemur reproductive endocrinology, lemur diet and nutrition and the functional anatomy of the slow loris.

Field and Laboratory Methods in Primatology

A PRACTICAL GUIDE

EDITED BY

JOANNA M. SETCHELL
Centre for Research in Evolutionary Anthropology, School of Life and Sport Sciences, University of Surrey Roehampton, UK

AND

DEBORAH J. CURTIS
Centre for Research in Evolutionary Anthropology, School of Life and Sport Sciences, University of Surrey Roehampton, UK

CAMBRIDGE
UNIVERSITY PRESS

CAMBRIDGE UNIVERSITY PRESS
Cambridge, New York, Melbourne, Madrid, Cape Town, Singapore, São Paulo, Delhi

Cambridge University Press
The Edinburgh Building, Cambridge CB2 8RU, UK

Published in the United States of America by Cambridge University Press, New York

www.cambridge.org
Information on this title: www.cambridge.org/9780521820042

First published 2003

A catalogue record for this publication is available from the British Library

Library of Congress Cataloguing in Publication data
Field and laboratory methods in primatology: a practical guide / edited by Joanna M.
  Setchell and Deborah J. Curtis.
    p. cm.
  ISBN 0 521 82004 9 (hardback) – ISBN 0 521 52628 0 (paperback)
  1. Primates – Research – Methodology.   I. Setchell, Joanna M., 1973–
II. Curtis, Deborah J., 1964–

QL 737.P9F53 2003
599.8′072–dc21   2003043593

ISBN 978-0-521-82004-2 hardback
ISBN 978-0-521-52628-9 paperback

Transferred to digital printing 2008

To all those people in habitat countries who make fieldwork possible.

Contents

Contributors

Email addresses are included for corresponding authors only.

MARC ANCRENAZ
Kinabatangan Orang-utan Conservation Project, PO Box 3109,
Sandakan, 90734 Sabah, Malaysia
hutan1@tm.net.my

NICOLA ANTHONY
Department of Biological Sciences, University of New Orleans, New Orleans,
LA 70148, USA

SIMON K. BEARDER
Nocturnal Primate Research Group, Department of Anthropology, School of
Social Sciences and Law, Oxford-Brookes University, Oxford OX3 OBP, UK
skbearder@brookes.ac.uk

JAMES R. BELL
Centre for Research in Ecology and the Environment, School of Life and
Sport Sciences, University of Surrey Roehampton, West Hill, London SW15
3SN, UK

MICHAEL W. BRUFORD
School of Biosciences, Cardiff University, PO Box 915, Cathays Park,
Cardiff CF10 3TL, UK
brufordmw@cardiff.ac.uk

RICHARD T. CORLETT
Ecology and Biodiversity, University of Hong Kong, 5 Sassoon Road,
Hong Kong SAR, People's Republic of China

DEBORAH J. CURTIS
Centre for Research in Evolutionary Anthropology, School of Life and Sport
Sciences, University of Surrey Roehampton, West Hill, London SW15 3SN, UK
d.curtis@roehampton.ac.uk

BRIAN W. DARVELL
Dental Materials Science, Faculty of Dentistry, University of Hong Kong,
Hong Kong SAR, People's Republic of China

J. LAWRENCE DEW
Department of Anthropology, University of California, Davis, One Shields
Avenue, Davis, CA 95616, USA
jldew@ucdavis.edu

NATHANIEL J. DOMINY
Department of Ecology and Evolution, University of Chicago, 1101 East 57th
Street, Chicago, IL 60637-1573, USA

GREGORY A. ENGEL
Swedish Family Medicine Residency, Providence Campus, Seattle, WA 98122,
USA, and University of Washington National Primate Research Center,
Seattle, WA 98115, USA

HANS ERKERT
Institute of Zoology, University of Tübingen, Auf der Morgenstelle 28,
D-72076 Tübingen, Germany
hans.erkert@uni-tuebingen.de

HAFEJEE C. ESSACKJEE
Physiological Laboratory, University of Liverpool, Crown Street,
Liverpool L69 3BX, UK

ANNA T. C. FEISTNER
Research Department, Durrell Wildlife Conservation Trust, Les Augres
Manor, Trinity, Jersey JE3 5BP, Channel Islands

JÖRG GANZHORN
Animal Ecology and Conservation, Institute of Zoology and Zoological
Museum, Martin-Luther-King Platz 3, 20146 Hamburg, Germany
ganzhorn@zoologie.uni-hamburg.de

THOMAS GEISSMANN
Anthropological Institute and Museum, University of Zürich,
Winterthurerstr. 190, 8057 Zürich, Switzerland
thomas.geissmann@aim.unizh.ch

BENOÎT GOOSSENS
School of Biosciences, Cardiff University, PO Box 915, Cathays Park,
Cardiff CF10 3TL, UK

COLIN GROVES
School of Archaeology and Anthropology, Australian National University,
Canberra, ACT 0200, Australia
colin.groves@anu.edu.au

JOANNE HARDING
School of Archaeology and Anthropology, Australian National University,
Canberra, ACT 0200, Australia

MICHAEL HEISTERMANN
Department of Reproductive Biology, German Primate Center, Kellnerweg 4, 37077 Göttingen, Germany
mheiste@gwdg.de

J. KEITH HODGES
Department of Reproductive Biology, German Primate Center, Kellnerweg 4, 37077 Göttingen, Germany

PAUL E. HONESS
Department of Veterinary Services, University of Oxford, Oxford OX1 3PT, UK
and
Wildlife Conservation Research Unit, Department of Zoology, University of Oxford, Oxford OX1 3PS, UK
paul.honess@virgin.net

KAREL HUGHES
Centre for Research in Ecology and the Environment, School of Life and Sport Sciences, University of Surrey Roehampton, West Hill, London SW15 3SN, UK
k.hughes@roehampton.ac.uk

KATHRYN JEFFERY
School of Biosciences, Cardiff University, PO Box 915, Cathays Park, Cardiff CF10 3TL, UK

MIREILLE JOHNSON-BAWE
School of Biosciences, Cardiff University, PO Box 915, Cathays Park, Cardiff CF10 3TL, UK
and
Centre International de Recherches Médicales, B.P. 769 Franceville, Gabon

CLIFFORD J. JOLLY
Anthropology, New York University, Washington Square, New York, NY 10003, USA
clifford.jolly@nyu.edu

LISA JONES-ENGEL
University of Washington National Primate Research Center, Division of International Programs, Box 357330, Seattle, WA 98115, USA
jonesengel@bart.rprc.washington.edu

JULIA N. LLOYD
Nocturnal Primate Research Group, Department of Anthropology, School of Social Sciences and Law, Oxford-Brookes University, Oxford OX3 OBP, UK

PETER W. LUCAS
Department of Anatomy, University of Hong Kong, 5 Sassoon Road, Hong Kong SAR, People's Republic of China
pwlucas@hkucc.hku.hk

DAVID W. MACDONALD
Wildlife Conservation Research Unit, Department of Zoology, University of
Oxford, Oxford OX1 3PS, UK

ROBERT D. MARTIN
Academic Affairs, The Field Museum, 1400 S. Lake Shore Drive, Chicago,
IL 60605, USA
rdmartin@fieldmuseum.org

JULIAN MAYES
Centre for Research in Ecology and the Environment, School of Life and
Sport Sciences, University of Surrey Roehampton, West Hill, London SW15
3SN, UK
j.mayes@roehampton.ac.uk

ALEXANDRA E. MÜLLER
Anthropological Institute and Museum, University of Zürich,
Winterthurerstrasse 190, 8057 Zürich, Switzerland

MARC MYERS
Primate Conservation Inc., 1411 Shannock Rd, Charlestown, RI 02813, USA
mmyers@primate.org

K. A. I. NEKARIS
Nocturnal Primate Research Group, Department of Anthropology,
School of Social Sciences and Law, Oxford-Brookes University,
Oxford OX3 OBP, UK

DANIEL OSORIO
School of Biological Sciences, University of Sussex, Brighton BN1 9QG, UK

CLAIRE M. P. OZANNE
Centre for Research in Ecology and the Environment, School of Life and
Sport Sciences, University of Surrey Roehampton, West Hill, London SW15
3SN, UK
c.ozanne@roehampton.ac.uk

JANE PHILLIPS-CONROY
Anatomy and Neurobiology, Washington University School of Medicine,
Box 8108, 660 S. Euclid Avenue, St Louis, MO 63110, USA

JONATHAN F. PRINZ
Wageningen Centre for Food Sciences, PO Box 557, 6700 AN Wageningen,
The Netherlands

NIGEL REEVE
Centre for Research in Ecology and the Environment, School of Life and
Sport Sciences, University of Surrey Roehampton, West Hill,
London SW15 3SN, UK

PABLO RIBA-HERNANDEZ
School of Biology, University of Costa Rica, San Pedro, Apdo 10672-1000,
San Jose, Costa Rica

CAROLINE ROSS
Centre for Research in Evolutionary Anthropology, School of Life and Sport
Sciences, University of Surrey Roehampton, West Hill,
London SW15 3SN, UK
c.ross@roehampton.ac.uk

NOEL ROWE
Primate Conservation Inc., 1411 Shannock Rd, Charlestown, RI 02813, USA
nrowe@primate.org

MICHAEL A. SCHILLACI
Department of Social Sciences, University of Toronto at Scarborough, 1265
Military Trail, Ontario M1C1A4, Canada

JUTTA SCHMID
Experimental Ecology, University of Ulm, Albert Einstein Allee 11, D-89069
Ulm, Germany
jutta.schmid@biologie.uni-ulm.de

JOANNA M. SETCHELL
Centre for Research in Evolutionary Anthropology, School of Life and Sport
Sciences, University of Surrey Roehampton, West Hill,
London SW15 3SN, UK
j.setchell@roehampton.ac.uk

KATHRYN E. STONER
Institute of Ecology, Universidad Nacional Autónoma de México,
Departamento de Ecología de Recursos Naturales, Apartado Postal 27-3
(Xangari), Morelia, Michoacan, Mexico 58089

ELIZABETH A. WILLIAMSON
Scottish Primate Research Group, Department of Psychology, University of
Stirling, Stirling FK9 4LA, UK
e.a.williamson@stir.ac.uk

NAYUTA YAMASHITA
Department of Cell and Neurobiology, Keck School of Medicine, University
of Southern California, 1333 San Pablo Street BMT 401, Los Angeles,
CA 90089-9112, USA

Foreword

ROBERT D. MARTIN
Academic Affairs, The Field Museum, Chicago, Illinois, USA

In addition to their inherent value for primatology as such, results from field studies of primate behaviour and ecology have made numerous significant contributions to discussions of human evolution. In an entirely different direction, they have also become increasingly important for conservation biology. Primates typically inhabit tropical or subtropical forests, and their relatively high profile makes them prominent indicator species for the rapidly advancing process of deforestation that is a central problem for conservation generally. Reliable data on primate behaviour and ecology accordingly constitute one of the tools used in modern biology as increasing resources are devoted to the battle against extinction. Field studies have progressively expanded to cover the entire order Primates to varying degrees, yielding an impressive wealth of data.

Despite this veritable explosion in primate field studies, however, relatively little attention has been devoted to synthetic publications dealing with methodological issues. Apart from an early guide produced in Washington by the Subcommittee on Conservation of Natural Populations (1981), which focussed on ecological aspects and has now been out of print for some considerable time, there is no single published source for even a basic set of methods required for primate field studies. One is reminded of the lament made by Cain (1959) with respect to a similar failing that once afflicted taxonomy: 'Is it not extraordinary that young taxonomists are trained like performing monkeys, almost wholly by imitation, and that only in the rarest cases are they given any instruction in taxonomic theory?' The practical guide to field methods and certain allied laboratory procedures provided by the present edited volume, including contributions from many of the current generation of key players, is therefore sorely needed and most welcome. Future fieldworkers will now have a single source of invaluable practical guidance at their disposal. In this age of information flow and networking, it is becoming increasingly easy to track down basic documentation (as is obvious from the numerous websites helpfully cited at the ends of individual chapters),

Field and Laboratory Methods in Primatology: A Practical Guide, ed. Joanna M. Setchell and Deborah J. Curtis. Published by Cambridge University Press. © Cambridge University Press 2003.

but carefully distilled advice for budding fieldworkers has nevertheless been a regrettably rare commodity until now.

Primate field studies have come a long way over the past 70 years since anecdotal accounts first gave way to systematic data collection, a transition marked by the publication of Carpenter's (1934) classic field study of the behaviour and social relations of howler monkeys. This was followed by a series of detailed studies of the natural behaviour of individual primate species, emphasising Old World monkeys living in relatively open habitats (e.g. baboons, patas monkeys and vervets) and our closest relatives, the apes (notably chimpanzees). For many years, a notebook-and-pencil approach to fieldwork prevailed, but the emphasis was very much on careful compilation of detailed behavioural observations over extended periods.

Some 30 years passed before another major transition, reflecting the influence of innovative field studies of birds, was marked by the seminal publication of Crook & Gartlan (1966), in which the first attempt was made to generate a basic classification of primate social systems and to examine possible evolutionary developments in relation to ecological factors. This quickly led to the key concept of primate socioecology and instigated a new wave of primate field studies in which quantitative data were systematically collected for both behaviour and particular ecological variables (notably those relating to diet). The basic aim of generating an overall comparative framework for primate field studies was further consolidated by Eisenberg et al. (1972).

At this stage, reliable techniques for collection of unbiased quantitative data were recognised to be of crucial importance. Careful attention to objective, well-defined methods to guarantee the collection of reliable quantitative data has undoubtedly been one of the most important developments in the history of the subject. A fundamental publication by Altmann (1974) on behavioural sampling surely played a major part in transforming the approach taken to collection of data in the field. (A more recent guide to behavioural methods for use both in the field and in captivity was provided by Martin & Bateson, 1986.) Although it is now taken almost for granted, the shift towards focal animal sampling as the method of choice (because of its explicit aim of excluding observational bias) was one of the most important features of the post-Altmann era of primate field studies. In parallel, it was realised that careful quantification of ecological variables was also a prime need, and this led to clear recognition that a field study of any given primate species should cover a minimum of one annual cycle in order to take account of seasonal variation. Indeed, because there can be marked differences from year to year, primate field studies should ideally cover a period of several years if the role of ecological factors is to be assessed properly. Long-term field studies of individual primate species are, in any case, of particular value, as evolutionary arguments commonly require some indication of reproductive success, which is best measured over an individual's entire lifetime. Because primates are relatively long-lived mammals, the ideal requirements for long-term study are difficult to fulfil, but an increasing number of primate field studies are beginning to meet that very demanding standard.

Only a few years elapsed before quantitative data from several detailed, long-term studies of primate behaviour and ecology were brought together in an influential book edited by Clutton-Brock (1977). That book, in addition to presenting overviews from individual field studies, provided results from overall comparative analyses of data on group size and composition, ranging behaviour and diet that revealed certain general quantitative principles of primate ecology and social organisation (see also Clutton-Brock & Harvey, 1977; Martin, 1981). A decade later, another landmark edited volume further advanced the process of synthesis of quantitative information from a wide range of primate field studies (Smuts et al., 1987). Most recently, data from a very wide spectrum of field studies have been systematically presented in the three-volume work produced by Sussman (1999, 2000, in press).

In a parallel development, primate field studies were progressively transformed from fact-finding exercises to problem-oriented investigations, as was directly reflected in the title of another edited volume published soon after Clutton-Brock's (1977) compendium (Sussman, 1979). This marked a significant shift towards primate field studies that were specifically designed to investigate particular issues seen to be of theoretical importance (e.g. intra-specific variation in behaviour between study sites or distinctions between sympatric species). A number of single-author synthetic books have also been produced over the years, notably by Jolly (1985; the second edition of a book originally published in 1972), Fedigan (1982), Richard (1985) and Dunbar (1988).

For a variety of reasons, coverage of primate species both in primary field studies and in synthetic works has generally tended to be conspicuously uneven. For many years, there was a heavy emphasis on higher primates, predominantly on Old World monkeys and apes, while prosimians (lemurs, lorises and tarsiers) were largely neglected. The few prosimian species that were studied in the earlier detailed studies were monkey-like in the sense that they were relatively large-bodied and diurnal in habits. Following the initial concentration on species that either inhabit relatively open habitats (particularly savannah-living Old World cercopithecine monkeys) or are relatively closely related to humans (i.e. apes), increasing attempts were made to include a number of New World monkeys, which are all exclusively forest-living. Carpenter's (1934) classic study of howler monkeys in Panama was notable not only for its pioneering role but also because it exceptionally involved a forest-living New World monkey species. Despite the fact that they represent a quarter of the total number of extant primate species and markedly extend the adaptive and temporal range of primates, prosimians were for a long time much neglected as a group. One practical reason for this is that most prosimians are small-bodied, nocturnal and forest living and are hence relatively difficult to study. However, there was also a measure of deliberate neglect because of the myopic view that prosimians are of marginal importance, whereas in fact they yield a far richer foundation for comparative studies.

In the 1970s, one important insight derived from the first detailed studies of nocturnal prosimian species was that these 'solitary' primates actually have well-developed social networks involving encounters between individuals at night and nest-sharing by day (Charles-Dominique, 1977, 1978; Bearder & Martin, 1980; Bearder, 1987). Prosimians exhibit some special features that are completely lacking among higher primates, particularly with respect to activity pattern. As is noted by Erkert in his chapter on chronobiology (Chapter 17), most primates are either clearly diurnal (active from dawn to dusk) or clearly nocturnal (active from dusk to dawn). Higher primates are in fact typically diurnal, the only exception being provided by the owl monkeys (*Aotus*), whereas the majority of prosimians are nocturnal. Thus study of adaptations for nocturnal life among primates is confined largely to prosimians. Moreover, some prosimians (notably *Eulemur* spp.) show a highly unusual pattern of activity that includes both diurnal and nocturnal components. It was only relatively recently that this distinct activity pattern was clearly recognised and defined as cathemerality (Tattersall, 1988), and field studies are just beginning to reveal its significance (Curtis & Rasmussen, 2002). Studies of nocturnal prosimians also led to a number of methodological innovations, most notably that of radio-tracking, which is still used far more commonly for nocturnal primates than for diurnal species. Given the previous general neglect of prosimians in primate field studies, it is noteworthy that the first volume of Sussman's (1999) synthetic survey is devoted almost entirely to prosimian primates, reflecting both increased awareness and the marked growth in studies devoted to them in recent years.

One comparatively recent extension to the problem-oriented approach to primate field studies resided in explicit attention to sociobiological hypotheses. Various core concepts of sociobiology, most notably kin selection, inclusive fitness, reciprocal altruism and sexual selection, are clearly of direct relevance to primate behaviour. It is therefore understandable that an increasing number of field studies have been designed to test sociobiological hypotheses. However, it should be noted that it is exceedingly difficult to collect adequate data for convincing tests of sociobiological interpretations from long-lived, slow-breeding animals such as primates. As already noted, only a few primate studies have been continued long enough to yield data on lifetime reproductive success, which are ideally required to test many arguments about selective advantage of individual behaviour patterns.

Lest it be thought that the development of primate field studies has followed a simple linear trajectory of unmitigated progress, it should also be emphasised that some important aspects have fallen by the wayside, at least to some extent. One example is provided by studies designed to explore proximate causation, which typically require some form of field experimentation. Although this is a very promising approach, it has been relatively little used

(Kummer, 2002). A number of excellent studies have illustrated the particular value of playback experiments employing natural vocalisations (e.g. Seyfarth & Cheney, 1982; Semple, 1998), but the experimental approach otherwise remains a relative rarity in primate field studies.

The latest methodological landmark in primate field studies has come about through the availability of various relatively new techniques that can potentially be exploited. Some of these new techniques are technological in nature (e.g. coupling of radio-tracking with automatic activity recording (Chapters 10 and 17); Global Positioning System (GPS), Geographical Information Systems (GIS) and remote sensing (Chapter 4)), while others are attributable to a combination of new methods with non-invasive collection of samples (e.g. assaying of hormone levels (Chapter 19); genetic typing (Chapter 20)). In particular, the tripartite combination of field observations with hormone assays and genetic testing (notably for inference of paternity) has opened up new possibilities for testing long-established notions regarding the relationship between social behaviour and reproduction. An excellent example of an application of this tripartite approach is provided by a study of female cycles, mating behaviour and paternity in free-living Hanuman langurs (Heistermann *et al.*, 2001).

Because of the high standards now set for objective, bias-free data collection and the steadily growing spectrum of ancillary techniques that are available, primate field studies have become increasingly complex. For this reason, the need for thorough advance preparation has become particularly acute and it is becoming increasingly difficult for a single investigator to cope with the demands of a modern field study. In the future there will undoubtedly be an increasing trend towards field studies involving teamwork. In any event, effective advance preparation has now become essential, and the chapters in this book provide many useful indications to optimise preparedness. It cannot be emphasised enough that timely application for any relevant research approval and permits is a key part of any preparation. The basic design of any study naturally requires very careful thought and, as it is now also customary for a primate field study to address particular issues or hypotheses that have been identified in advance, the design of a study will be subject to fairly strict constraints from the outset. As part of the design, the question of achieving statistical independence in any data collected must be given careful consideration. As several authors in this volume stress, advance preparation should also include training in relevant skills, expert advice and prior familiarisation with particular techniques and equipment where appropriate.

The increasing need for careful planning of primate field studies is somewhat problematic because of the equally important need to maintain some degree of flexibility under unpredictable field conditions. The best-laid plans can be upset by unexpected conditions and developments in the field. It is all well and good to invest time and energy in advance preparation and training, but adaptability and the ability to come up with inspired improvisation are also called for in field studies. As noted by Curtis and Setchell (Introduction), for this and several other reasons, if at all possible it is best to conduct a pilot study at the proposed field study site prior to the main study, particularly if there is no prior history of comparable fieldwork at that site.

One issue of central importance for any modern field study highlighted in this volume is clear recognition of ethical responsibilities and appropriate decision-making by fieldworkers at an early stage. Awareness of ethical aspects has progressively increased over the years, partly because of rapidly growing concern about conservation issues but also because of mounting sensitivity to animal welfare in a very general sense. In the first place, proper compliance with regulations for permissions and permits is mandatory for ethical and not merely for practical reasons. Fieldworkers necessarily serve as 'ambassadors' and have a duty to promote understanding for conservation and welfare by their actions both at home and abroad. Applications for Convention on International Trade in Endangered Species (CITES) permits, in particular, may be time-consuming and occasionally vexatious (notably when the question-able topic of export/import of urine and faecal samples is involved), but it is imperative to show respect for international legislation that was introduced to control international trade in endangered species. Suitable prior train-ing in certain techniques (e.g. handling of animals and taking of blood sam-ples) is also important in this context, as it can significantly enhance animal welfare.

Another key area that involves several ethical aspects is whether or not animals should be captured (e.g. Chapters 7, 8, 10 and 18). In this respect, too, fieldworkers serve as ambassadors and it is therefore essential that any capture programme should be properly authorised and conducted with obvi-ous concern for animal welfare. If the decision is made to capture animals as part of a field study, every effort should be made to maximise the in-formation that can be obtained without prejudice to the subjects. While a captured animal is appropriately immobilised, it is possible to take standard measurements, check on health and reproductive condition and take various samples (e.g. dental casts, ectoparasites, plucked hairs, blood samples, X-rays (Chapters 7, 8 and 9)), thus maximising the value of the field study. Indeed, it could be argued that the fieldworker has an obligation to collect a maximum of biological information from captured animals, provided that physical in-tervention and risk are kept to a minimum. One productive possibility that requires retention of animals in captivity for a few days is measurement of basal metabolic rate (BMR). As is reported by Schmid (Chapter 18), it is now possible to measure BMR under field conditions using portable equipment. Here, a novel approach using a nestbox with a built-in thermal jacket as a metabolic chamber and sampling/assaying of air that is drawn out of the box rather than pumped through it (thus avoiding the need for hermetic sealing) offers advantages with respect to the level of disturbance imposed on study subjects (Genoud et al., 1997).

One of the greatest benefits of capturing animals is undoubtedly the potential for applying various kinds of marking to facilitate recognition of individuals, provided that such marking is conducted with a proper concern for animal welfare (Chapter 10). In particular, radio-tracking can be used, dramatically increasing the reliability of individual identification and the ease with which study animals can be located, potentially facilitating the process of habituation. However, it should be remembered that prominent questions of animal welfare arise in this context too. Careful consideration must be given to the size and method of attachment of the transmitter package, and timely recapture must be reliably feasible in order to remove the package in due course. Removal of transmitter packages from all animals by the end of a field study is absolutely essential, but regrettably this is a topic that is inadequately reported or discussed in the existing literature (Chapter 10).

Another crucial area in which ethical issues arise is that of taxonomy (Chapter 9). All primate fieldworkers should give due attention to taxonomy in designing and conducting their studies. In addition to contributing to the general database that permits continued refinement of primate taxonomy, correct identification of the taxon under study may directly affect the outcome. One example of this is provided by Richard's (1974) study of two sifaka populations in Madagascar, which were at that time identified as subspecies of *Propithecus verreauxi* (*P.v. verreauxi* and *P.v. coquereli*). That study was designed specifically to examine variability in behaviour and ecology within a single primate species. However, analysis of mitochondrial DNA sequences has now revealed that these two populations belong to two distinct clades of sifakas and that the degree of genetic divergence between *verreauxi* and *coquereli* suggests a species-level divergence (Pastorini *et al.*, 2001). Hence, a study that was explicitly designed to examine intra-specific variability may well have been an investigation of inter-specific differences instead.

As Groves and Harding (Chapter 9) discuss, taxonomic aspects are so fundamental to any fieldwork on primates, particularly if any contribution is made to conservation measures, that investigators arguably have a duty to collect data and specimens that will help to refine taxonomy. This, however, is precisely where a major ethical issue arises. Current museum-based taxonomic practice uses a strategy of collecting reference specimens, and type specimens play a fundamental role in formal recognition of species and subspecies. The question that must now be posed is whether it is ethically justifiable to continue to collect primate specimens (other than incidental finds) from the field specifically for the purpose of contributing to reference collections. This question is, of course, particularly acute where severely threatened populations are involved, but there are now many investigators who would simply rule out further collection of any primate specimens for museum reference purposes, even for species that are still relatively common. This is a matter that urgently requires informed debate among primatologists in order to establish generally acceptable guidelines. Formal recognition (including publication) of a new primate species or subspecies currently requires at least one reference specimen. In the absence of such a specimen, newly discovered primate taxa may remain in limbo. In principle, it should now be possible to design an acceptable reference base (e.g. combining photographs, measurements taken on the live animal, dental casts, field X-rays, plucked hairs and genetic typing) that no longer requires the collection and preparation of a cadaver. However, there is a pressing need for international agreement on some alternative system if we are to accept that collection of museum specimens is no longer mandatory for taxonomic recognition. Furthermore, if it should prove possible to design an acceptable system to replace collection of classical reference specimens, the need for effective broad-based sampling from live animals in primate field studies would become even greater.

In closing, it should be emphasised that some thought should be given to what happens when the investigator returns home after completing a long-term primate field study in a distant foreign country. Quite apart from its scientific benefits, a long period spent abroad for such a purpose has a particular value as a stimulating (if frequently challenging) life experience that broadens the investigator's horizons and permits an external perspective on the latter's own culture. However, it is not often realised that the investigator may suffer a fairly acute form of culture shock on returning home. This 're-entry syndrome' (a term coined by one of the respondents) was specifically reported by Hinde (1979) in a scarce publication based on replies to a questionnaire that he distributed to gauge the overseas experiences of recent graduates and others. Completed questionnaires concerning the first field experience lasting 5–24 months were received from 63 people in four groups (30 primatologists, 7 social anthropologists, 5 modern linguists and 21 participants in Voluntary Service Overseas (VSO)), most of whom replied to the questionnaire within two years after returning from the field.

Hinde's survey incidentally revealed some important basic points about primate field studies. Significantly, respondents generally emphasised the value of prior training in methods of data collection and of previous field experience, thus emphatically underlining the value of appropriate preparation and pilot work. In Hinde's words: 'Proper preparation is crucial. Problems must be formulated, data collection methods outlined and practised, and the fieldworker trained in the skills necessary for his project.' Prior language training was also identified as a valuable asset. Most of the primatologists were very positive about their field experiences, although they reported that the first three months or so tended to be quite problematic. It is noteworthy that half of the primatologists thought that they worked too hard because of their self-imposed work schedules (which may do much to explain the phenomenon of 'fieldworker's procrastination syndrome' recognised in the final chapter by Bearder et al.), and many of the primatologists and anthropologists thought that a break after three to six months would be a sensible measure. Academic isolation was a serious problem for many of the respondents and a key factor was the presence or absence of effective local support at the field study site. Hinde noted: 'It is desirable, but not usually possible, for research supervisors to visit research students in the field – if possible in the early phases of their study. If that is not possible, it is an asset to have a supervisor who knows the study area, and who keeps in as close touch as possible by correspondence.'

The most striking point to emerge from Hinde's (1979) survey, however, was the high frequency with which respondents reported feeling some form of disorientation or alienation after returning home. This was specifically mentioned by 23 primatologists (77%), by all of the anthropologists, by three modern linguists (60%) and by 13 VSO workers (62%). The 're-entry syndrome' thus identified involved culture shock in relation to 'civilised society' and the perception of its fast pace, difficulties with social interaction and difficulties in settling down to work (reported by half the primatologists). Clearly, readapting to the culture of origin after completion of a field study is also a serious issue that deserves attention as part of the experience of fieldwork. At the very least, recognition of the fact that this is a widespread phenomenon and hence a shared experience may make it easier for returning primatologists to adjust to this additional challenge.

As a final note, I would like to pay a personal tribute to all of the dedicated primate fieldworkers who have braved tough and demanding conditions – including the problems of 're-entry syndrome' – to build the impressive database that is now available. They have performed a major service to the academic community and, increasingly, to conservation biology. I pass on my very best wishes to the next generation of fieldworkers, who will surely be involved in studies of ever-increasing complexity, but who at least will have this comprehensive methodological survey to guide them.

REFERENCES

Altmann, S.A. (1974). Observational study of behaviour: sampling methods. *Behaviour* **49**, 227–65.

Bearder, S.K. (1987). Lorises, bushbabies, and tarsiers: diverse societies in solitary foragers. In *Primate Societies*, ed. B.B. Smuts, D. Cheney, R.M. Seyfarth, R. Wrangham & T. Struhsaker, pp. 11–24. Chicago: Chicago University Press.

Bearder, S.K. & Martin, R.D. (1980). The social organization of a nocturnal primate revealed by radio-tracking. In *A Handbook on Biotelemetry and Radio Tracking*, ed. C.J. Amlaner & D.W. Macdonald, pp. 633–48. Oxford: Pergamon Press.

Cain, A.J. (1959). The post-Linnaean development of taxonomy. *Proc. Linn. Soc. Lond.* **170**, 234–43.

Carpenter, C.R. (1934). A field study of the behavior and social relations of howling monkeys. *Comp. Psychol. Monogr.* **10**, 1–168.

Charles-Dominique, P. (1977). *Ecology and Behaviour of Nocturnal Primates*. (Translated by R.D. Martin.) London: Duckworth.

(1978). Solitary and gregarious prosimians: evolution of social structure in prosimians. In *Recent Advances in Primatology*, vol. 3, *Evolution*, ed. D.J. Chivers & K.A. Joysey, pp. 139–49. London: Academic Press.

Clutton-Brock, T.H. (ed.) (1977). *Primate Ecology*. London: Academic Press.

Clutton-Brock, T.H. & Harvey, P.H. (1977). Primate ecology and social organization. *J. Zool. Lond.* **183**, 1–39.

Crook, J.H. & Gartlan, J.S. (1966). On the evolution of primate societies. *Nature* **210**, 1200–3.

Curtis, D.J. & Rasmussen, M.A. (2002). Cathemerality in lemurs. *Evol. Anthrop.* **11**, Suppl. 1, 83–6.

Dunbar, R.I.M. (1988). *Primate Social Systems*. London: Croom Helm.

Eisenberg, J.F., Muckenhirn, N.A. & Rudran, R. (1972). The relation between ecology and social structure in primates. *Science* **176**, 863–74.

Fedigan, L.M. (1982). *Primate Paradigms: Sex Roles and Social Bonds*. Montreal: Eden Press.

Genoud, M., Martin, R.D. & Glaser, D. (1997). Rate of metabolism in the smallest simian primate, the pygmy marmoset (*Cebuella pygmaea*). *Am. J. Primatol.* **41**, 229–45.

Heistermann, M., Ziegler, T., van Schaik, C.P., Launhardt, K., Winkler, P. & Hodges, J.K. (2001). Loss of oestrus, concealed ovulation and paternity confusion in free-ranging Hanuman langurs. *Proc. Roy. Soc. Lond. B* **268**, 2445–51.

Hinde, R.A. (1979). Report on replies to a questionnaire concerning the experiences of recent graduates and others abroad. *Primate Eye*, **11**, Suppl., 1–18.

Jolly, A. (1985). *The Evolution of Primate Behavior*, 2nd edition. New York: Macmillan.

Kummer, H. (2002). Topics gained and lost in primate social behaviour. *Evol. Anthrop.* **11**, Suppl. 1, 73–4.

Martin, P. & Bateson, P. (1986). *Measuring Behaviour*. Cambridge: Cambridge University Press.

Martin, R.D. (1981). Field studies of primate behaviour. *Symp. Zool. Soc. Lond.* **46**, 287–336.

Pastorini, J., Forstner, M.R.J. & Martin, R.D. (2001). Phylogenetic history of sifakas (*Propithecus*: Lemuriformes) derived from mtDNA sequences. *Am. J. Primatol.* **53**, 1–17.

Richard, A.F. (1974). Intra-specific variation in the social organization and ecology of *Propithecus verreauxi*. *Folia Primatol.* **22**, 178–207.

(1985). *Primates in Nature*. New York: W.H. Freeman.

Semple, S. (1998). The function of Barbary macaque copulation calls. *Proc. Roy. Soc. Lond. B* **265**, 287–91.

Seyfarth, R.M. & Cheney, D.L. (1982). How monkeys see the world: a review of recent research on East African vervet monkeys. In *Primate Communication*, ed. C.T. Snowdon, C.H. Brown & M.R. Petersen, pp. 239–52. Cambridge: Cambridge University Press.

Smuts, B.B., Cheney, D., Seyfarth, R.M., Wrangham, R. & Struhsaker, T. (eds.) (1987). *Primate Societies*. Chicago: Chicago University Press.

Subcommittee on Conservation of Natural Populations (1981). *Techniques for the Study of Primate Population Ecology*. Washington, DC: National Academy Press.

Sussman, R.W. (ed.) (1979). *Primate Ecology: Problem Oriented Field Studies*. New York: John Wiley.

Sussman, R.W. (1999). *Primate Ecology and Social Structure*, vol. 1, *Lorises, Lemurs and Tarsiers*. Needham Heights, MA: Pearson Custom Publishing.

(2000). *Primate Ecology and Social Structure*, vol. 2, *New World Monkeys*. Needham Heights, MA: Pearson Custom Publishing.

(in press). *Primate Ecology and Social Structure*, vol. 3, *Old World Monkeys and Apes*. Needham Heights, MA: Pearson Custom Publishing.

Tattersall, I. (1988). Cathemeral activity in primates: a definition. *Folia Primatol.* **49**, 200–2.

Introduction

DEBORAH J. CURTIS AND JOANNA M. SETCHELL
Centre for Research in Evolutionary Anthropology, School of Life and Sport Sciences, University of Surrey Roehampton, London, UK

> The ideal program for the future study of animal behaviour would involve the coordination of field and laboratory investigation.
>
> Schaller (1965, p. 624)

In the early days of expeditions, a naturalist's field equipment consisted of little more than a gun, the means to preserve specimens, and luggage bearers. Later on, field primatologists relied on pencil and paper, binoculars, a compass and, if they were studying nocturnal species, a torch. More recently, a shift in emphasis towards integration of methods has led to collaboration between laboratory and field researchers working on wild primates. Technological advances have presented fieldworkers with the opportunity to collect more sophisticated data, store samples for later laboratory analysis, analyse samples in the field and collect information remotely. This has led to an increase in data concerning (for example): population and group structure; growth, development and physiology; and the application of non-invasive techniques for DNA analyses and hormonal assays. Ecological methods and techniques available for monitoring primate habitats have also improved, with the application of remote sensing, mapping (Global Positioning System, GPS), and data integration (Geographical Information Systems, GIS). These methods open up possibilities for collecting new information on previously studied populations, and a means for collecting data on species that, for example, cannot be habituated for behavioural observations. However, these technological advances come at a time when 64% of primate species are under threat and insufficient data are available to assess the status of a further 11% of species (2000 IUCN Redlist of threatened species). Knowledge of the behavioural ecology of a species is essential for conservation, and this makes an integrated approach combining laboratory and field techniques all the more important, particularly as some of the most endangered species are those about which least is known. However, although fieldworkers are generally well prepared in terms of basic behavioural and ecological methodology, there is currently no easy

Field and Laboratory Methods in Primatology: A Practical Guide, ed. Joanna M. Setchell and Deborah J. Curtis. Published by Cambridge University Press. © Cambridge University Press 2003.

source of information concerning the wide variety of additional data and samples that can be collected for subsequent laboratory analysis and methods must usually be compiled from the primary literature. We aim to fulfil this need with this volume.

In our introduction, we illustrate briefly the methodological paradigm shifts that have occurred in the history of primate field studies with a case study (the aye-aye). We then give our own advice to those preparing for the field, with a general overview of how to plan and conduct a field study, before detailing our aims in compiling this volume and presenting an overview of the chapters included.

METHODOLOGICAL PARADIGM SHIFTS IN FIELD PRIMATOLOGY: THE AYE-AYE

Several methodological paradigm shifts have occurred in the history of primate field studies (Strum & Fedigan, 2000; see also the Foreword to this volume). Until the early twentieth century, the driving force behind field studies was classification, through the collection and dissection of specimens, and comparative anatomy. Knowledge of animal behaviour in the wild increased much more slowly and, until the 1920s, comprised incidental observations by adventurers, seafarers, hunters, explorers and missionaries, some of whom were naturalists. This interest in behaviour, albeit anecdotal, constituted the first paradigm shift leading towards modern day behavioural and ecological studies. This and subsequent changes in methodological approach are well illustrated by studies of the aye-aye (*Daubentonia madagascariensis*; family Daubentonidae).

The aye-aye was discovered by Sonnerat (1782, cited in Owen, 1866) in Madagascar during a voyage to the East Indies and China. Sonnerat described it as bearing resemblance to a squirrel, but also having characters that allied it to lemurs and monkeys. Cuvier (1798, cited in Owen, 1866) classified the aye-aye as a rodent, and although Schreber (cited in Peters, 1865) placed it within the lemurs as early as 1803, it was not definitively recognised as such until 60 years later, with the publication of Owen's monograph on aye-aye anatomy in 1866.

Expeditions aimed almost uniquely at the collection of aye-aye specimens for anatomical study continued into the early twentieth century (for a review, see Lavauden, 1933). Pollen (1863) neatly summed up the reason for the lack of information on the natural history of the Malagasy vertebrates, and his explanation remains generally applicable today:

> It remains incomprehensible for those who do not know the obstacles faced by scientific explorations, that one knows so little of the natural history of this large country, even though a number of travellers have visited different parts of it. But, if one notes the numerous difficulties that these enterprising

travellers have taken in their stride, in traversing the vast expanses of this isle, be it through the influence of the tropical climate, the insalubrious nature of the country, the numerous rivers and lakes that made their progress difficult, be it due to the defective means of transport and the season during which they were there, one is no longer surprised. On the contrary, one is obliged to pay homage to these zealous travellers who have risked their lives for science and sacrificed their health for the collection of the limited number of specimens they managed to obtain...

> Translated from the original French.

Early descriptions of aye-aye behaviour stem from animals kept in captivity in La Réunion (Vinson, 1855) and Mauritius (Sandwith, 1859, cited in Owen, 1866). Sandwith (1859, cited in Owen, 1866) provides an excellent description, stating:

I...was much struck with the marvellous adaptation of the creature to its habits, shown by his acute hearing, which enables him to aptly distinguish different tones emitted from the wood by his gentle tapping; his evidently acute sense of smell, aiding him in his search; his secure footsteps on the slender branches, to which he firmly clung by his quadrumanous members; his strong rodent teeth, enabling him to tear through wood; and lastly, by the curious slender finger, unlike that of any other animal, and which he used alternately as a pleximeter, a probe, and a scoop.

Lamberton (1911) carried out the longest such study, observing a captive individual in Madagascar for over a year. Given its nocturnal habits, Lamberton concluded that the only way to study the aye-aye's natural behaviour was in captivity, and was ahead of his time in remarking that:

individuals taken to Europe are certainly in bad condition. Made feeble by the voyage, embittered by suffering, sick due to the change in climate and food, how to distinguish between what is natural and what is a consequence of the conditions in which the animal finds itself? Observations made on aye ayes held in captivity in their country of origin certainly approach reality far more...

> Translated from the original French.

The earliest descriptions of aye-aye behaviour in the wild come from missionaries in Madagascar, who compiled information by interviewing the Malagasy (e.g. Baron, 1882). Lavauden (1933) made the earliest qualitative behavioural observations of wild aye-ayes, including photographs of wild individuals, a nest and a larval tunnel exposed by an aye-aye. The first comprehensive description of aye-aye behaviour and ecology was published by Petter in 1962, but no quantitative studies were conducted in the wild until the late 1980s (Sterling, 1993; Ancrenaz et al., 1994). Since then no further field studies have been carried out, possibly because, despite its taxonomic uniqueness, Daubentonia is no longer regarded as a highly endangered species. In terms of the integration of laboratory and field techniques that we aim to encourage with this volume, only diet and nutrition have been investigated for the aye-aye, by combining field data with chemical analyses of food items (Sterling *et al.*, 1994).

PLANNING AND CONDUCTING A FIELD STUDY

Our knowledge of aye-aye biology could be greatly expanded by the application of some of the methods detailed in this volume. However, field trips to distant habitat countries like Madagascar are expensive, and may well be once-in-a-lifetime opportunities for both the researcher and, unfortunately, the study population. It is therefore imperative to prepare well and to maximise the information collected. In this section we provide some guidance on how to deal with some of the non-scientific aspects of fieldwork. However, this is by no means to be regarded as covering all eventualities, as it is based mainly on our personal experience and on conversations with colleagues. Many of the following chapters give specific advice on various problems.

PREPARATION

Before you leave for the field, ensure that you know what your objectives are. The stated aims of the project must be fulfilled by your planned data collection. It is likely that these aims will change over time, usually because you have been too ambitious at the outset, but you need to start somewhere. Think about how you will analyse your data before you go into the field. Make sure that you are familiar with your study species by referring to previous work, contacting others who have worked on the same animals, and, if possible, spending time at a zoo, observing them to give you a feel for the species' behaviour and general characteristics. You can also try out behavioural sampling methods at the zoo although you might need to change things once you are in the field if, for example, your contact time with the animals is low or visibility is poor. If possible, test your ecological methods before you go and, if your study requires particular techniques (e.g. collection and preparation of botanical samples, blood sampling), take expert advice and get the necessary training. Make sure that you are familiar with all your equipment before you leave. If your study will involve the capture of study animals, training and preparation are essential. Capture is a risky procedure, and you should take the opportunity to collect as much information as you can about each captured individual (e.g. morphometrics, health status and biological samples). Even if these data, or other samples collected, are not directly related to your current research questions, they may constitute a future research project, or provide opportunities for collaboration.

Gather as much information on potential study sites as you can, read up on previous studies carried out at the same or similar study sites and contact colleagues who have worked in these areas. Inform yourself of particular problems that may pertain to the study area(s) (e.g. difficult terrain, climate, health hazards, antagonism from local people, political stability) and do not hesitate to ask for advice from those with more experience. Inoculations, applying for permits, and setting up collaborations in your destination country all take time. You may also need to book flights some time in advance if you will be travelling during busy periods. Insurance, at the very least medical cover, is expensive but essential if something goes wrong.

TERRAIN AND CLIMATE

Ensure that you are fit enough to cope with difficult terrain, and that you can be evacuated within a reasonable time-scale, should you injure yourself and need hospital treatment. This will apply generally, but particularly in remote areas, and being sensible will greatly reduce the probability of an accident. For example, employ local assistants to climb trees, should this be necessary, as they will usually be much better at it than you are. Ensure that you have adequate protection from the elements, such as warm clothing, clothing to protect you from the sun, and rainwear.

HEALTH HAZARDS

Assess the health hazards that you may face and take the necessary precautions. Get a medical check-up before you leave, have all the recommended vaccinations and take the necessary medication with you. It is also advisable to seek the advice of a health professional once you reach the habitat country (especially if you will be carrying out a long-term study), as a local doctor will know more about local diseases and the precautions or prophylaxes that may be necessary. For example, many countries produce their own malarial prophylaxes, which are perfectly adequate in most cases, as well as being cheaper. Furthermore, you are less likely to contribute to the growing resistance of the parasite to drugs that are too expensive for local people to afford. As is the case the world over, see a doctor before you take antibiotics (if possible), and finish the course – again, to avoid contributing to the ever-increasing problem of antibiotic-resistant bacteria.

If you will be handling animals and/or will be working at some distance from medical services then rabies vaccinations are a good idea. However, if you are bitten by an animal that may be rabid, then you must seek medical attention immediately, as the vaccination does not give you full protection. It is also a good idea to have a selection of syringes and needles with you, as well as the necessary equipment for perfusions, to be used if you are admitted to hospital. You will have been inoculated against hepatitis A and B as a standard procedure, but the human immunodeficiency virus (HIV) is rife the world over and, if you do not have your own equipment with you, then you run the risk of infection from recycled syringes.

Take large amounts of standard first-aid kit items with you (e.g. plasters, bandages, painkillers, etc.). This is not just for your own use; people may come to you for help before they go to the nearest health professional, particularly if you are working in a remote area. In Madagascar, D.J.C. was asked to help in cases of serious ear and eye infections, septic wounds, sexually transmitted diseases, malaria, minor cuts and bruises and headaches, as well as sick zebu cattle. In most cases, all you can do is send your 'patients' to the local doctor, in others, a painkiller or plaster will do the world of good and you will have made a friend. Think about taking a first-aid course before you leave, and, finally, if you are employing local assistants, think about acting as their health insurance (we believe that this is only reasonable) and pay for any medical and dental treatment. In addition to moral considerations, it is to your own advantage to do so, as healthy employees are more inclined to work.

Precautions against dangerous animals require common sense, vigilance and, for example, thick leather garters as a protection against venomous snakes, and long trousers and long-sleeved shirts as protection against insects. Look after your mental health, ensure you take regular breaks from the field and do not push yourself beyond your limits. Remember, you will be working hard, but you are not supposed to return from the field a mental and physical wreck – again, use your common sense. Little things can help you through bad times – D.J.C. found the shortwave radio useful, as well as the luxury of an inexpensive reading lamp in the tent, powered by a car battery charged using solar panels; J.M.S. kept a diary.

POLITICAL STABILITY

Check your government's website for detailed information on the political stability of the country and area in which you intend to work, and follow its advice. If you will be working in a highly volatile country, then ensure you have your escape route planned in advance, should you need to leave in a hurry. Even if a country is stable, it is always a good idea to register your presence with the officials at the local embassy or consulate of your country of origin and let them know your itinerary.

THE PILOT PHASE

If you can afford it, then a pilot phase is invaluable for finding a study site, setting up collaborations, arranging permits and testing methodology. If you are travelling around to find an appropriate study site, then make contacts with people working and living in the area to become acquainted with the advantages and disadvantages of working there. Spend some time at the site to assess the logistics of provisioning. If you will be working in the forest, assess visibility, density of the forest, and, above all, try to find and follow your study animals to test your methods. D.J.C. visited three potential study sites during her pilot phase to study the cathemeral mongoose lemur. One region was out of the question as the density of the study animal was low and the forest highly degraded. The second had the advantage that it was at a field station and therefore transport and provisioning would have been relatively easy to deal with. However, the density of mongoose lemurs was low. In one area of forest near the field station where visibility was good, the flora was mostly introduced. A second area where visibility was good contained no study animals. In the third area where the animals did occur, the forest was dense, visibility low and the canopy very high. The third region, where the study was eventually carried out, was chosen because the density of study animals was extraordinarily high, the forest was open enough that no transects needed to be cut, the canopy was not too high and visibility was good. The downside was that the site was fairly remote, making provisioning complicated, and transport and access difficult, particularly during the rainy season.

Use your pilot phase to test sample collection and storage methods. For example, the practicalities of collecting faeces, or food specimens, from small canopy monkeys may only become apparent when sitting below subject animals in a 30 metre canopy. A few representative samples taken back to the laboratory at this stage will allow you to test preservation methods and may save a great deal of trouble at a later stage.

WORKING WITH OTHER PEOPLE

Studying primates inevitably involves interactions with other people, be they other researchers, local populations or government officials. Respect the country you are in. Bureaucracy can be a nuisance, but rules are there for a reason, and you would not break the law in your own country. Good relations with your host country from top officials down to workers and villagers are very important both for your study and for researchers coming after you. Working with in-country collaborators is to be encouraged (not doing so could be regarded as 'scientific colonialism'), and will help with language, cultural understanding and bureaucracy.

At some point you will probably encounter antagonism from local people and much of this can be prevented by respecting local customs (e.g. dress and food prohibitions). Ensure that people living in your study area are aware of what you are doing and why you are doing it. You may need to visit the headmen of surrounding villages, and perhaps give a talk to the villagers – do not rely on officials to do the job for you. Your paperwork might be in order, but that will not be enough. Make some kind of contribution to the local economy, i.e. do not try to set up a self-contained study site, but employ some of the local population (not just strangers to the area) and buy local produce. This will also help to spread the word about what you are doing in the area. Be aware of payment scales, and avoid overpaying and underpaying for supplies and assistance.

If you are in a situation where you need to employ assistants (e.g. you are not working on a pre-existing field station), then we recommend drawing up some kind of contract. This can prevent many problems and also makes it clear who is the boss. You and your local workers are not a team (this observation stems from bitter experience!), and any working situation usually needs some kind of hierarchy in order to function efficiently.

WHEN YOU GET BACK

Once you are back from the field, data analysis will quickly take over your life, but make sure that you get a medical check-up. Some parasites can remain undetected in your body for years, but cause serious harm to your health. Watch out for strange symptoms and always inform your doctor that you have been working in the field. In addition to your physical health, be aware of the 're-entry syndome' described in the Foreword to this volume.

AIMS OF THIS VOLUME

The impetus for editing this volume stems from personal experience in the field, and that of many of our collaborators and colleagues. The increasingly competitive nature of scientific research, in combination with space limitations in scientific journals, means that it is often difficult to come by information on methodological problems, troubleshooting and, in particular, negative results. Journal articles tend to present a clear, logical development of ideas, hiding the methodological problems encountered along the way. Furthermore, as many governments increasingly insist on returns from research in terms of either technological or biomedical advances that are beneficial to economic growth, it is becoming more and more difficult to obtain funding for studies rooted in behaviour and ecology. The combination of field and laboratory techniques may well make a study more interesting to the biomedical industry, for example.

We have brought together expert authors to cover methods at the field/laboratory interface, in order to acquaint the reader with techniques used in the study of primates and non-primates, to guide in the selection of appropriate methods and to produce a volume that can act as a reference book in the field. We aim to encourage fields of research that are currently underexploited in primatological and zoological fieldwork, and also to address the (often neglected) broader cultural and legal implications of fieldwork. Although the book is aimed primarily at primatologists, many methods will also be applicable to those studying other groups of animals. We do not include an introduction to the primates, primate behavioural ecology, behavioural methodology, or statistics, as these subjects are amply and ably covered by other authors (e.g. Siegel & Castellan, 1988; Martin & Bateson, 1993; Lehner, 1998; Fleagle, 1999; Sussmann, 1999, 2000, in press; Strier, 2000). Neither do we intend to provide a comprehensive review of each topic, as this would fill several volumes. As noted by a number of the contributors, researchers will still need to refer to the specialist literature for additional information, but the chapters will provide them with an excellent starting point. The focus is on the technical and practical aspects of each method or area, rather than on research results, and chapters include the following information, where appropriate:

- Point-by-point instructions for sample collection, processing and preservation for later analysis.
- The time and equipment required for work in the field and laboratory.
- Methodology and techniques, including the difficulties of various methods and how to overcome them.
- The ideal situation/equipment and what will do if necessary.
- Exactly what data the samples will provide to the fieldworker.
- Ethical considerations.
- Comparative costs.
- Selected useful references.

Where authors mention the use of commercially available equipment and supplies, this does not constitute endorsement either by the editors or the publisher. Similarly, we have endeavoured to ensure that the URLs for external websites referred to in this book are correct and active at the time of going to press. However, neither we nor the publisher has any responsibility for the websites and we can make no guarantee that a site will remain live or that the content is or will remain appropriate.

OVERVIEW OF CHAPTERS

It is now almost impossible to study non-human primates without becoming aware of contact (and conflict) between them and humans, whether via hunting, tourism, pet ownership, food competition, crop-raiding or research. Chapter 1 (Jones-Engel et al.) discusses the practicalities of studying such interactions between human and non-human primates (Ethnoprimatology). Jones-Engel et al. cover the design of surveys, conducting interviews and sampling both human and non-human primate subjects, with a particular emphasis on cultural fluency and the ethics of working with people in habitat countries. This is followed by Williamson and Feistner (Chapter 2), who discuss the advantages and disadvantages of methods for habituating primates. Chapters 3–5 concern the methods used to describe a habitat. Ganzhorn (Chapter 3) discusses vegetation as the biotic matrix for the evolution of life history traits, detailing methods for forest classification, habitat structure and phenology. Hughes (Chapter 4) then addresses the potential applications of three modern technologies in primate research: the GPS, GIS and remote sensing. These have, as yet, rarely been applied in field studies of primates, and Hughes provides an introduction to each technology, describes the decisions to be made before applying them and stresses the importance of thorough preparatory work and the necessity of 'ground-truthing'. Weather and climate have important influences on ecosystems and Mayes (Chapter 5) describes the types of local weather information that can be collected in field studies. He demonstrates how weather can vary spatially and temporally and gives methods for measuring local atmospheric conditions. Ross and Reeve (Chapter 6) then introduce some of the key issues to consider in primate population studies and discuss survey and census methods for estimating the population and population density of a species.

Many field studies involve capturing primates, for example to mark or radio-collar study animals, or to collect morphological or physiological data. In Chapter 7, Jolly *et al.* cover the practicalities of trapping primates of all shapes and sizes, while Ancrenaz *et al.* (Chapter 8) present the veterinary aspects of handling primates, including anaesthesia, health evaluation and methods for collecting biological samples. Both chapters emphasise the necessity of collecting as much information as possible when animals are captured, owing to the risks involved in the procedure. Following on from capture methods, Groves and Harding (Chapter 9) introduce morphology, morphometrics and taxonomy, with details of how to preserve dead specimens and measure both dead primates and anaesthetised live animals. Honess and Macdonald (Chapter 10) then discuss the advantages and disadvantages of various methods for marking study animals and the uses and practicalities of radio-tracking primates, satellite tracking and biotelemetry.

The next four chapters are concerned with food habits and nutritional ecology. In Chapter 11, Dew introduces the study of feeding ecology and seed dispersal, detailing methods for the observation of feeding behaviours, the collection and classification of feeding remains and faecal contents analysis. This is followed by two contributions by Lucas *et al.* (Chapters 12 and 13) covering dietary analysis. Chapter 12 concentrates on the physical properties of food items as they might influence feeding behaviour, including methods for measuring food geometry (size and shape), colour, texture and mechanical properties. Chapter 13 then deals with the chemical properties of food, including sample preservation and extraction and chemical tests for food components. In both contributions Lucas *et al.* emphasise the value of *in situ* analysis, in a 'field laboratory', but they also include information on laboratory analyses if this is not possible. Food availability is an important factor in primate diet and methods of estimating fruit and leaf availability to frugivorous and folivorous primates are covered in Chapter 3, while Ozanne and Bell (Chapter 14) cover methods of assessing (and identifying) the insect prey available to insectivorous species.

The next two chapters give practical advice on equipment and methods available for recording study animals. Geissmann (Chapter 15) discusses how to record primate vocalisations. This is important in terms of archival data, sound analysis to determine the properties of vocalisations and also for playback experiments. In Chapter 16, Rowe and Myers discuss still and video photography of primates, with advice on choosing, using, protecting, and travelling with photographic equipment.

Chapters 17 to 19 concern primate physiology. Erkert (Chapter 17) introduces the subject of chronobiology, the study of biological rhythms, and cautions all primate fieldworkers to be aware of the implications of circadian rhythms, zeitgebers and masking effects, and our own anthropocentric measurement of time for studies of physiology and behaviour. He also describes methods for measuring luminance and recording activity patterns in the field. Schmid (Chapter 18) introduces methods for the study of energy expenditure and body temperature in wild animals, discussing various techniques and the advantages and disadvantages of each. Hodges and Heistermann (Chapter 19) then discuss methods for monitoring hormonal changes in free-ranging primates using non-invasive samples (faeces and urine). More non-invasive techniques are described in Chapter 20 (Goossens *et al.*), which covers the collection, storage and analysis of non-invasive genetic material and includes advice for those able to obtain more traditional blood or tissue samples. Finally, in a more light-hearted style, Bearder *et al.* (Chapter 21) close the book with an A–Z of tips to make fieldwork safer and more comfortable.

CONCLUDING COMMENTS

Although most research on primates relies almost entirely on observation, the methods that can be useful in the field are clearly limited only by the investigator's ingenuity...

Cheney *et al.* (1987, p. 8)

This is a book about methods, which fieldworkers are continually developing and improving through practice, trial and error. Both primatology and primate conservation are dependent upon people who are prepared to go into the field, put in hard work and get dirty. Fieldwork can be demanding, and exhausting, but it is also highly rewarding (see any fieldworker's diary). Although we are advocating the use of technology in this book, we are encouraging its integration with standard field techniques, and not suggesting in any way that it can replace traditional behavioural and ecological data collection. Field conditions have not changed dramatically since the early days (Fig. I.1), although our tools have: pencil and paper have replaced guns (Greif & Schmutz, 1995). It is still only by the researcher travelling on foot (like the early naturalists) that new populations, and even new species, are discovered.

Fig. I.1. *Above*: Hans Bluntschli in his field camp, Madagascar 1931–1932 (from the archives of the Anthropological Institute and Museum, University of Zurich). *Below*: Field camp at Anjamena, Madagascar 1994–1995 (Curtis & Zaramody, 1998).

If this book proves useful to fieldworkers, acts to stimulate research and understanding of primates in their natural state, and through that increased knowledge can make some small contribution to primate conservation, then we will have achieved our aim.

ACKNOWLEDGEMENTS

We are indebted to the contributors to this book for their patience, enthusiasm and encouragement during the editorial process, to Tracey Sanderson at Cambridge University Press for guiding us through the process of publication, to the staff of Life and Sport Sciences at University of Surrey Roehampton for their encouragement, to Steve Wagg for advice on approaching publishers, to Richard Thomas for the redrawing of illustrations, and editorial advice, and to Hans Schmutz for information on Hans Bluntschli. D.J.C. is grateful to Urs Thalmann, with whom she discussed a similar volume several years ago. J.M.S. thanks her parents and Benoît Goossens for their support during her own fieldwork, and during the production of this volume.

REFERENCES

Ancrenaz, M., Lackman-Ancrenaz, I. & Mundy, N. (1994). Field observations of aye-aye (*Daubentonia madagascariensis*) in Madagascar. *Folia Primatol.* **62**, 22–36.

Baron, L. (1882). Notes on the habits of the aye-aye of Madagascar in its native state. *Proc. Royal Soc. London* **XLIII**, 639–40.

Cheney, D.L., Seyfarth, R.M., Smuts, B.B. & Wrangham, R.W. (1987). The study of primate societies. In *Primate Societies*, ed. B.B. Smuts, D.L. Cheney, R.M. Seyfarth, R.W. Wrangham & T.T. Struhsaker. Chicago and London: University of Chicago Press.

Curtis, D.J. & Zaramody, A. (1998). Group size, home range use and seasonal variation in the ecology of *Eulemur mongoz*. *Int. J. Primatol.* **19**, 811–35.

Fleagle, J.G. (1999). *Primate Adaptation and Evolution*. San Diego: Academic Press.

Greif, R. & Schmutz, H.-K. (1995). Hans Bluntschli als Morphologe. *Gesnerus* **52**, 133–57.

Lamberton, C. (1911). Contribution à l'étude des moeurs du aye-aye. *Bull. Acad. Malg.* **VIII**, 129–140.

Lavauden, L. (1933). Le aye-aye. *La Terre et la Vie Paris* **3**, 77–9.

Lehner, P.N. (1998). *Handbook of Ethological Methods*, 2nd edition. Cambridge: Cambridge University Press.

Martin, P. & Bateson, P. (1993). *Measuring Behaviour: An Introductory Guide*, 2nd edition. Cambridge: Cambridge University Press.

Owen, R. (1866). On the aye-aye. *Chiromys*, Cuvier. *Trans. Zool. Soc. Lond.* **68**, 33–101.

Peters, W. (1865). Ueber die Säugethiergattung *Chiromys*. *Abh. Akad. Wiss. Berlin*, 79–100.

Petter, J.J. (1962). Recherches sur l'écologie et l'éthologie des Lémuriens malgaches. *Mém. Mus. Nat. Hist. Nat. Sér. A, Zool.* **27**, 115–20.

Pollen, F. (1863). Enumération des animaux vertébrés de l'île de Madagascar. R. Amsterdam, *Nederl. Tijdschr. Dierk.* **1**, 277–345.

Schaller, G. B. (1965). Field procedures. In *Primate Behavior: Field Studies of Monkeys and Apes*, ed. I. De Vore, pp. 623–9. New York: Holt, Rinehart and Winston.

Siegel, S. & Castellan, N. J. (1988). *Nonparametric Statistics for the Behavioural Sciences*, 2nd edition. New York: McGraw-Hill.

Sterling, E. J. (1993). Behavioral ecology of the aye-aye (*Daubentonia madagascariensis*) on Nosy Mangabe, Madagascar. Ph.D. thesis, University of Yale.

Sterling, E. J., Dierenfeld, E. S., Ashbourne, C. J. & Feistner, A. T. C. (1994). Dietary intake, food composition and nutrient intake in wild and captive populations of *Daubentonia madagascariensis*. *Folia Primatol.* **62**, 115–24.

Strier, K. (2000). *Primate Behavioural Ecology*. Boston, MA: Allyn & Bacon.

Strum, S. C. & Fedigan, L. M. (2000). Changing views of primate society: a situated North American perspective. In *Primate Encounters: Models of Science, Gender and Society*, ed. S. C. Strum & L. M. Fedigan, pp. 3–49. Chicago: University of Chicago Press.

Sussmann, R. W. (1999). *Primate Ecology and Social Structure*, vol. 1, *Lorises, Lemurs and Tarsiers*. Needham Heights, MA: Pearson Custom Publishing.

(2000). *Primate Ecology and Social Structure*, vol. 2, *New World Monkeys*. Needham Heights, MA: Pearson Custom Publishing.

(in press). *Primate Ecology and Social Structure*, vol. 3, *Old World Monkeys and Apes*. Needham Heights, MA: Pearson Custom Publishing.

Vinson, A. (1855). Observations sur un Aye-aye gardé deux mois à la Réunion. *C. R. Acad. Sc.* **XLI**, 638–40.

Further reading and useful Internet sites

The Royal Geographical Society (UK)'s expedition planning handbook and other publications (some of which can be downloaded from their website <www.rgs.org>) are useful further reading for the practicalities of planning fieldwork.

The 2000 IUCN Redlist of threatened species can be found at <www.redlist.org>.

1 • Human–nonhuman primate interactions: an ethnoprimatological approach

LISA JONES-ENGEL[1], MICHAEL A. SCHILLACI[2] AND GREGORY A. ENGEL[1,3]

[1]University of Washington National Primate Research Center, Seattle, USA
[2]Department of Social Sciences, University of Toronto at Scarborough, Ontario, USA
[3]Swedish Family Medicine Residency, Providence Campus, Seattle, USA

INTRODUCTION

Over the past decades, economic, political and social forces in the developing world have brought about deforestation on a massive scale, depleting the remaining natural habitats of wild nonhuman primates (NHPs). Squeezed into ever smaller domains surrounded by human society, NHPs are coming into increasingly regular contact with humans. Poaching and habitat destruction are recognised dangers to NHP populations in the wild. In contrast, the potentially devastating threat posed by human-to-NHP disease transmission in wild NHP populations is under-appreciated and not well studied (see also Chapter 8). We believe that effective programmes for the conservation of wild NHP populations must acknowledge the interrelation of habitat destruction, bushmeat hunting and human-to-NHP disease transmission.

Pathogens endemic to humans have the capacity to devastate NHP populations. This phenomenon has been observed repeatedly in laboratory settings, where epidemics of endemic human diseases such as influenza, tuberculosis, chicken pox and measles can cause mortality rates greater than 90% among NHPs, including animals newly captured from the wild (Padovan & Cantrell, 1986; Mansfield & King, 1998). If endemic human pathogens can cause such profound destruction among captive NHPs, it follows that we should explore the threat that human contact poses to wild NHPs.

Though competition and commensality between humans and NHPs are not new phenomena, few have studied the complex ways in which human societies influence the survival of NHPs in the wild (but see Wallis & Lee, 1999; Beck et al., 2001; Fuentes & Wolfe, 2002). Our research on interspecific pathogen transmission seeks to describe how diseases are transmitted both from humans to NHPs and from NHPs to humans. Humans and NHPs come into contact in a variety of contexts: hunting, logging, ecotourism,

Field and Laboratory Methods in Primatology: A Practical Guide, ed. Joanna M. Setchell and Deborah J. Curtis. Published by Cambridge University Press. © Cambridge University Press 2003.

pet ownership, competition for food, and research. The emerging field of Ethnoprimatology focuses on the interconnections between human culture and NHP populations. Ethnoprimatological research is important as a means of understanding how human activities and diseases affect NHPs, and also how NHPs (in particular, as a vector for primate-borne diseases) can affect human populations.

This chapter deals with some technical aspects of pursuing research on human–NHP interactions. The recommendations contained herein are based on our experiences in investigating inter-specific disease transmission between human and NHPs (macaques, *Macaca* spp.) on the Indonesian islands of Sulawesi and Bali (Jones-Engel *et al.*, 2001; Engel *et al.*, 2002). We recognise that each research project will encounter unique challenges and obstacles. Here we have tried to set out some general considerations for those contemplating research on NHPs and the human communities with which they come into contact.

MULTIDISCIPLINARY NATURE OF THE STUDY

It is important to recognise that humans and NHPs interact in complex ways. As a result, the study of interspecies interactions is facilitated by a multidisciplinary approach. Our research on bidirectional disease transmission has benefited from the contributions of team members who possess expertise in a variety of fields: primatology, anthropology, human medicine, veterinary medicine, epidemiology and virology. These diverse perspectives have proved helpful at all stages of the work, from planning to data collection to data analysis. Regardless of the study subject – be it crop-raiding, bushmeat hunting or conservation – it is advisable to avail yourself of a broad range of research perspectives.

WORKING WITH IN-COUNTRY/HABITAT COUNTRY COLLABORATORS

Successful research involving human and NHP interactions in habitat countries will almost certainly rely heavily on a strong collaborative relationship between the researcher and his/her in-country collaborator(s). Often these relationships are established through professional or academic contacts. However the initial contact is made, researchers should consider discussing all issues regarding the research with their collaborators prior to departure for the field. These issues can include timing of the field research, acquisition of necessary permits, in-country transport, recruitment of field assistants and arranging access to research subjects. We strongly recommend that compensation and acknowledgement of collaborators be agreed to prior to arrival in the habitat country. Academic recognition (i.e. co-authorship on manuscripts) can be very important to in-country collaborators and the need for financial compensation often looms large in any working collaboration. A researcher going into the field ideally should know beforehand what constitutes appropriate compensation. There are dangers in compensating in-country collaborators on the same monetary scale that the researcher would use at home – over-paying collaborators (by in-country standards) may create animosities at his or her institution, or make it difficult for future researchers to work in that area.

One of the greatest challenges of doing research in another culture is to maintain a focus on the research agenda. It sometimes occurs that collabora-tors or field assistants have agendas that run contrary to the stated objectives of the research project or that may even be harmful to NHP populations. It is important to identify potential actions that could compromise human or NHP populations.

CULTURAL FLUENCY

To conduct research effectively it is necessary to be adequately familiar with local cultures and customs. Cultural fluency must extend beyond linguistic competence to include a basic understanding of the subtext of the culture in which one is working. The researcher must not only be able to work with local collaborators and research subjects but he or she must also interact ef-fectively with local and governmental officials. This can be difficult to achieve, especially in multi-ethnic societies where the primary collaborator's ethnicity may differ from that of research subjects. In such cases it is advisable to hire field assistants or go-betweens who are of the ethnicity of the area where the research is being conducted and who are the most familiar with the local traditions, customs and power structure.

DESIGNING SURVEYS

The literature on designing ethnographic surveys is extensive, and survey con-tent will be dictated by the area of interest. For our research on inter-species pathogen transmission, we gather data on hunting practices, crop protection and bushmeat consumption as well as pet ownership and trade practices in order to characterise the contexts of human–NHP interactions. These data in-clude: frequency of hunting; methods used to hunt; and the disposition of bushmeat, i.e. how are the animals prepared, is the meat sold to others or consumed by the hunter and his family, are any portions of the animal used for medicinal or ritual applications, etc. Our human subjects are also inter-viewed about the frequency of crop-raiding that they experience and methods used to protect their crops. NHP pet owners are interviewed about various as-pects of NHP pet ownership: how was their pet acquired; who cares for the pet; has the pet ever been ill; how is the pet housed; does the pet have contact with other animals (domestic or wild); has the owner had previous pet NHPs, and if so what happened to them, etc. Additionally, we collect demographic and health information from our human subjects relating to prior NHP bites and scratches.

In general, care should be taken to make survey questions simple, direct and specific. Avoid leading questions. A frequently encountered problem is inaccurate responses from subjects due to cultural attitudes or poorly designed survey questions. Some subjects try to guess what the 'right' answer is in order to please the interviewer. In other situations the subject's responses may be constrained by cultural or legal considerations. For example, a subject who is questioned about their bushmeat hunting practices may be less than forthcoming if bushmeat hunting is illegal in their country. One way to circumvent this tendency is to design the survey so that the most sensitive questions are asked in several different ways. It is also important to consider the context of the interview. Subjects who are interviewed in the presence of others, be they peers or higher or lower ranking individuals, may answer differently from how they would if they were speaking only with the interviewer. Surveys should be checked for cultural relevance by in-country collaborators and tested before being used in the field.

Here, we would like to emphasise that, when going into the field with a survey instrument that has been meticulously researched and prepared, the researcher should be prepared to modify the survey (i.e. take the original disc version with you). For example, many informants may be embarrassed by particular questions, or certain questions may have little or no cultural relevance and therefore be difficult to answer. The survey should be as short as possible yet still allow you to access the information necessary for your study. The construction and implementation of an ethnographic survey is another area where cultural fluency is critical. It took three field seasons before our standard ethnographic survey reached its final streamlined form. Optimally, the primary researcher should be responsible for administering the surveys. This is especially important in early versions of the survey as shades of meaning, inconsistencies, misunderstandings and concepts that simply do not cross cultural boundaries are worked out. Strongly consider a pilot survey to clarify meanings and work out 'bugs'. One of our early surveys required respondents to indicate whether they 'agreed strongly, agreed, didn't know, disagreed or disagreed strongly' with the statement: 'monkeys can make you sick'. This very Western notion of stratified answers to a question was met with bewilderment by most of the respondents.

If you intend to undertake a quantitative analysis of the survey data we suggest using a program such as Epi Info or Sudaan, which are designed for epidemiology and public health survey research (see List of suppliers at the end of this chapter).

CONDUCTING INTERVIEWS

It is not at all uncommon for NHP hunting or NHP pet ownership to be 'illegal' in source countries. Because of this we often noticed that interviewees were hesitant to provide information about some of their practices. We found that we were able to improve the efficiency of our data gathering by investigating and involving the leading figures in the village political structure. Upon arriving in a community we sought out the leaders in the village and explained our research to them. These leaders frequently led us to hunters, pet owners and farmers who came into contact with crop-raiding NHPs. Their knowledge of their communities both saved us valuable time in locating research subjects and broke down barriers to communication.

When interviewing people about their practices/beliefs towards NHPs it is also important to remember that the way that researchers present themselves is important. A conservative, respectful and more formal type of clothing should be worn when researchers go to interview subjects at their homes or farms.

A small honorarium as thanks for the time that subjects spend answering survey questions should be considered. The amount should be gauged on the typical daily wage for that given local economy. Additionally, having on hand a generous supply of tobacco, betel nut, pens, pencils, small notepads, sweets for the children, sample packets of perfume, lotion, lipstick, nail polish and a Polaroid camera/film to document the occasion, have a value among the subjects that far exceeds their cost. When we collected blood samples from pet owners we typically doubled our standard honorarium and also gave them a week's supply of vitamins. Monetary gifts are potentially problematic. Our experience has been that in some communities receiving money is awkward for recipients who, though they desire such gifts, are at pains to avoid the external appearance of being paid by the researcher whom they regard in one sense as a guest. Our solution to this has been to offer honoraria and to comment that the money is being provided for the owner's children or for monkey food. This practice has been well received.

ETHICAL CONSIDERATIONS OF GIFT-GIVING

The practice of giving gifts is rooted in the practical realities of doing field research in developing countries. Without such gifts it is extremely difficult to gain access to people, NHPs and information. There is also the impulse on the part of researchers to engender goodwill towards themselves and towards the research. This is especially true when they plan to return to the community. However, giving gifts is not without ethical dilemmas. When we give cigarettes as gifts we are both promoting tobacco use and, in a sense, legitimising it, no matter what our feelings are about the product. Women find sex-specific gifts such as lipstick or perfume samplers desirable because they are luxuries to which they might not otherwise have access. Yet dispensing such gifts promotes and legitimises the Western ideals of beauty and consumerism.

The researcher who believes that he or she can enter a community and do research without in some way changing that community is self-deluded. As foreigners it is impossible for us to know all the myriad ways in which our research and very presence impacts on these communities. Our approach is to maximise our positive impacts and attempt to mitigate the negative ones. Ultimately, if we were not convinced that our research benefited the animals, environments and communities where we work, we would not continue.

SAMPLING PROTOCOLS

We collect both survey data and biological samples from our subjects. Human subjects are chosen on the basis of the kinds of interaction they have with NHPs. Blood samples are collected from pet owners, hunters and/or people who report that they eat NHPs. Control subjects, those reporting no contact with NHPs, are also identified and sampled. Additionally, demographic and health information relating to any prior exposures to bite or scratch wounds by NHPs are elicited.

Only adults (over the age of 18 years) are asked to donate blood samples. After the procedure is explained in detail, written consent is obtained and 6 ml of blood are collected by venipuncture from the individual's antecubital vein. This blood is processed, stored, transported and analysed in the same manner as the NHP blood (see below).

Detailed protocols for handling, anaesthetising, sampling and giving NHPs health examinations are presented elsewhere in this volume (Chapter 8). When sampling wild NHPs we use trapping protocols similar to those outlined in Chapter 7. In situations where trapping is not appropriate (such as the Monkey Forests of Bali), we use air-powered dart guns and blowpipes to anaesthetise free-ranging NHPs. All of our pet NHP subjects are hand injected in order to minimise stress for the pet and owner. We have immobilised more than 280 NHPs during three field seasons using Telazol® (tiletamine/zolazepam). We prefer Telazol® to ketamine because of its amnesic and antianxiety properties, its wide margin of safety, the smooth induction and recovery that it promotes and the small injection volume needed.

We collect a variety of samples from our NHP subjects: whole blood, serum, genital and rectal swabs for bacterial and viral culture, faecal samples and gastric samples. We also take radiographs (Schillaci *et al.*, 2001), morphometric data (Chapter 9) and dermatoglyphic data. The entire sampling process typically requires less than 30 minutes. A portion of whole blood is centrifuged in the field to extract serum, and both serum and whole-blood samples are distributed to smaller tubes for safety and ease of storage and transport. Whole blood, serum, swabs and gastric samples are frozen and stored at $-20\,°C$ in a portable 12 V freezer (weighing 11 kg, Engel 15®).

SAMPLE ANALYSES

We use dot immunobinding assays, indirect fluorescence or enzyme-linked immunobinding assays for analysis of blood samples collected from humans and NHPs for the presence of antibodies to the following viruses: measles, polio, rubella, mumps, adenovirus, influenza A and B, parainfluenza 1, 2 and 3, human T-cell lymphotrophic virus, *Herpes virus hominis* types 1 and 2, respiratory syncytial virus, rotavirus, hepatitis A, B, and E, Epstein–Barr virus, dengue, Japanese encephalitis, simian varicella-zoster, cytomegalovirus, simian retrovirus, simian T-cell lymphotrophic virus, simian foamy virus, simian immunodeficiency virus and herpes B. We also carry out a comprehensive faecal analysis including a physical description, biochemical analysis, ova and parasite identification and bacterial culture. Additionally, viral swabs collected from NHP conjunctival, oral and genital regions are analysed for the presence of herpes viruses. Gastric aspirates are collected and analysed for the presence of mycobacterium as are nasal swabs for the presence of *Staphylococcus* spp. Results of the analyses of biological samples, combined with ethnographic and epidemiological data, allow us to characterise bidirectional pathogen transmission between humans and NHPs.

PERMITS

Research that entails both human and NHP subjects typically means that there are now at least twice the number of governmental agencies that must be negotiated both at the federal and local levels. Never assume that permits granted at one level of bureaucracy will automatically be valid at another. In countries where the provincial governments are geographically removed from the central governments, researchers should be prepared to pursue additional permits for their research. Much of the negotiation for these permits is often better handled by the in-country collaborator, as they are typically more familiar with the 'dealings' of their governmental agencies. We cannot stress enough that the permitting process be initiated well in advance of the intended start date for the field research, this applies to both the process in the habitat country and any procedures that are required from the researcher's own university. Most Western-based universities require that any research on animals, whether it be in the laboratory or in the field, be reviewed and approved by a committee established by the university. Additionally, research that involves humans must typically be reviewed and evaluated by the university's Human Research and Review Committee or equivalent body (e.g. Ethics Committee). The review processes for research involving humans and NHPs can be time consuming and arduous – plan accordingly.

Research that involves the collection of biological samples from humans and/or NHPs may be subject to international regulations regarding the import and export of such samples. Any researchers undertaking this type of research and data collection should familiarise themselves with local, national and international regulations.

TIMING

When considering field studies on human and NHP interactions several factors must go into the researcher's planning, including the season, climate, holidays and the political and economic stability of their field site's country.

For example, in Bali, Indonesia, we discovered that, while the months of May to August worked very well for our schedules, this period corresponded to the height of the tourist season. We encountered local resistance to our research, with guides, shopkeepers and guards expressing concern that sampling during the tourist season might impact negatively on tourist revenues. Consultation with local administrative and governmental officials, in conjunction with informing local villagers about our research and the potential positive and negative consequences of our project, was essential for belaying some of these concerns.

The timing of local holidays can have a significant impact on field research. In Muslim countries, Ramadan, the Muslim fasting month, can profoundly affect the ability of a researcher to move around the country and interact with people. Similarly, access to governmental agencies, and the permits they provide, can be severely restricted during national or religious holidays.

When considering scheduling of interviews with farmers, researchers must take into account the planting season. Harvest times typically require intense and extended efforts by the farmers. During these periods the latter may be less inclined to take the time to sit and talk or answer survey questions. Similarly, if you are interested in documenting crop-raiding by NHPs it is imperative that you identify the seasons that are associated with these behaviours. In Northern Sulawesi, hunting pressure on NHPs is most severe in December, because the 'Christmas monkey' is a preferred feast dish among the Minahasa. When planning a field season consult with your in-country collaborators about expected local weather conditions. Interviewing and collecting samples during the rainy season can be extremely challenging, if not impossible.

TRANSPORTATION AND EQUIPMENT

The importation and transportation of research equipment to a research site can be a logistical nightmare. Equipment-intensive research requires provision for international transport of the equipment, for example make arrangements with airlines well in advance of departure and make sure that any valuable capital items that are to be brought into the host country have the appropriate permits and declarations. Make arrangements with in-country collaborators to meet them immediately upon arrival. No matter how thorough your permits and documents may be it is not wise to spend time standing around in the airport or customs surrounded by a mound of expensive imported equipment.

If your work requires that you move about the country you can either rely on local transportation, which is generally inexpensive, or hire personal transport. The latter option is always more expensive, but often more convenient, especially when one is transporting a great deal of equipment or going to remote and difficult-to-access areas. However, bear in mind that arriving in a village in an expensive automobile may immediately set up expectations from the villagers.

CAVEATS

Researchers who come into direct or even indirect contact with NHPs should be cognizant of the potential for pathogen transmission both from humans to NHPs and from NHPs to humans. Recognising the risk of inter-specific disease transmission is even more crucial when one is collecting biological samples. Any researcher who embarks on a study that will involve collecting biological samples from NHPs and humans must be meticulous in avoiding the transmission of pathogens to their subjects. It is also the responsibility of the researcher to avoid or contain the risk of exposure, both to themselves and to their human and NHP subjects.

Here we would like to emphasise that meticulous attention to technique is essential for safe sample collection and good technique is attained through training. Researchers should be well versed in all sample collection techniques before venturing out into the field. Field conditions are often suboptimal and can compound the difficulty of collecting samples. Weather conditions and poor lighting are common obstacles. Crowds pose a particular difficulty. In our experience, the very presence of foreign scientists attracts curiosity, and sedating and sampling animals often causes a sensation. It was not uncommon to have a group of 15–100 people crowding around us as we worked. Of course, spectators are more than inconvenient; they pose an infectious disease threat to the animals. In circumstances such as these it is important to maintain your composure, and request some member of the community in a position of authority to control the crowd.

CONCLUSION

In short, field research on human and NHP interactions needs to factor flexibility into the schedule. Not only is the researcher forced to account for unpredictability in locating and sampling NHPs, but they must also account for the cultural, ritual and occupational commitments of their human subjects.

ACKNOWLEDGEMENTS

This research has benefited from the intellectual, moral and often physical support of a number of people representing a variety of disciplines and perspectives: J. Allan, J. Cheek, D. Cohn, A. Cooke, D. Cox, A. Fuentes, J. Froehlich, R. Grant, R. Heberling, J. Heidrich, R. Kyes, U. Paputungan, A. Putra, A. Rompis, K. Schwerin, J. Setchell and A. Sobel. Funding for this research has been provided by American Society of Primatologists, Chicago Zoological Society,

Sigma Xi Grants-in-Aid, Unconventional Concepts Inc., Sandia National Labs and the University of New Mexico School of Graduate Studies.

REFERENCES

Beck, B.B., Stoinski, T.S., Hutchins, M., Maple, T.L., Norton, B., Rowan, A., Stevens, E.F. & Arluke, A. (2001). *Great Apes and Humans. The Ethics of Coexistence.* Washington, DC: Smithsonian Institution Press.

Engel, G.A., Jones-Engel, L., Schillaci, M.A., Suaryana, K.G., Putra, A., Fuentes, A. & Henkel, R. (2002). Human exposure to herpesvirus B-seropositive macaques, Bali, Indonesia. *Emerg. Infect. Dis* 8 [online] URL: <www.cdc.gov/ncidod/EID/vol8no8/01-0467.htm>.

Fuentes, A. & Wolfe, L.D. (2002). *Primates: Face to Face.* Cambridge: Cambridge University Press.

Jones-Engel, L., Engel, G.A., Schillaci, M.A., Babo, R. & Froehlich, J. (2001). Detection of antibodies to selected human pathogens among wild and pet macaques (*Macaca tonkeana*) in Sulawesi, Indonesia. *Am. J. Primatol.* 54, 171–8.

Mansfield, K. & King, N. (1998). Viral diseases. In *Nonhuman Primates in Biomedical Research: Diseases*, ed. B.T. Bennett, C.R. Abee & R. Henrickson, pp. 1–57. London: Academic Press.

Padovan, D. & Cantrell, C.A. (1986). Varicella-like herpesvirus infections of nonhuman primates. *Lab. Animal. Sci.* 36, 7–13.

Schillaci, M.A., Jones-Engel, L., Heidrich, J.E., Miller, G.P. & Froehlich, J.W. (2001). A field methodology for lateral cranial radiography of nonhuman primates. *Am. J. Phys. Anthropol.* 116, 278–84.

Wallis, J. & Lee, D.R. (1999). Primate conservation: the prevention of disease transmission. *Int. J. Primatol.* 20, 803–26.

List of suppliers

Epi Info is available as shareware from the Internet: <www.cdc.gov/epiinfo/ei2000.htm>.

Sudaan can be ordered from <www.rti.org/sudaan/home.cfm>, and Engel 15® from <www.engel-usa.com>.

2 • Habituating primates: processes, techniques, variables and ethics

ELIZABETH A. WILLIAMSON[1] AND ANNA T. C. FEISTNER[2]
[1]*Scottish Primate Research Group, Department of Psychology, University of Stirling, Stirling, UK*
[2]*Durrell Wildlife Conservation Trust, Trinity, Jersey, Channel Islands*

INTRODUCTION

Field biologists adopted the term habituation from physiology, as the relatively persistent waning of a response as a result of repeated stimulation that is not followed by any kind of reinforcement (Thorpe, 1963). Repeated neutral contacts between non-human primates (hereafter called primates in this chapter) and humans can lead to a reduction in fear, and ultimately to the ignoring of an observer. The techniques and processes involved have only rarely been described (e.g. Schaller, 1963; Kummer, 1995), as habituation has generally been viewed as a means to an end (Tutin & Fernandez, 1991). The few studies that have quantified primate behaviour in relation to habituators describe the process with African great apes (Grieser Johns, 1996; van Krunkelsven *et al.*, 1999; Blom *et al.*, 2001). As we become increasingly aware of the potential effects of observer presence on primate behaviour, and especially the potential risks of close proximity with humans, it behoves us to measure as much as possible about the habituation process.

Many behavioural responses are taxon specific, and these should be taken into account when one is trying to habituate human-naïve wild primates. Between us we have had experience with a wide range of wild primates, ranging in size from marmosets (*Callithrix* spp.) to gorillas (*Gorilla* spp.), from South America, Africa, Madagascar and Asia, which, together with discussions with colleagues, we have used to make this chapter as broadly applicable as possible. Our emphasis is on what to take into account when considering primate habituation, particularly in relation to preparation, observer behaviour, and assessing the risks that habituation to humans may pose for increasingly endangered primates. Since habituating African great apes probably presents the greatest challenge, the aspects that we illustrate with reference to apes will in most cases be applicable to designing an approach with other primates.

There are limits to what studies of unhabituated primates can achieve: it is difficult to observe at close range, so subtle or cryptic behaviour such as facial expressions and soft vocalisations may be missed, and even individual

Field and Laboratory Methods in Primatology: A Practical Guide, ed. Joanna M. Setchell and Deborah J. Curtis. Published by Cambridge University Press. © Cambridge University Press 2003.

identification may be difficult, resulting in analyses based only on age–sex classes. Primates disturbed by the presence of observers will show altered patterns of behaviour and it may not be possible to follow groups on the move. Habituation allows an observer to approach closely, because subjects no longer flee and do not seem to respond to observers, allowing us to become familiar with individuals, and fine-level behaviours such as subtle social interactions or food processing can be observed. When the researcher is following habituated animals, behaviour can be sampled consistently.

METHODS

PREPARATION

Knowing your primate

To prepare for a study involving primate habituation, first read as much about your target species and its environment as possible. Familiarise yourself with what others have done and with your species' basic behavioural repertoire. Secondly, try to gain exposure to the species itself, perhaps by visiting a zoo to observe the animal directly. The more naturalistic the physical and social environment of the captive primate, the better its potential as a model for wild conspecifics. Spending time observing captive primates will increase familiarity with their basic locomotor patterns and postures, facial expressions, and other behaviours, including foraging and food processing, grooming, play, reproductive and agonistic interactions. Pay particular attention to vocalisations, since these may be a means of locating wild primates and will help you to interpret behaviour in the absence of clear observation. Concentrate on alarm and display behaviours, as these are likely to be initial responses to human presence. Observing captive primates will also provide an opportunity to learn how to distinguish individuals and will give you a head start in recognising primates in the wild.

Equipment

In the early stages, little equipment is needed: binoculars, a notebook, a compass, maps and a Global Positioning System (Chapter 4). Binoculars and cameras should be used with care; pointing a big 'eye' towards primates can make them uneasy.

Using a trail system

Trails can facilitate movement within the study area and make it easier to follow primate groups, but this depends greatly on vegetation type and even more on the area to be covered, which is likely to be determined by home range size. For example, in Taï, Côte d'Ivoire, researchers studying chimpanzees (*Pan troglodytes verus*) worked alone, and did not cut transects or trails but learnt to orient themselves in the forest with a compass. We took the same approach with lowland gorillas (*Gorilla gorilla gorilla*) at Lopé, Gabon, where an additional factor was fast growing Marantaceae vines that would have required labour-intensive trail maintenance. Learning the lie of the forest, using features such as streams and vegetation types, without trail cutting keeps disturbance to a minimum. In African forests with high elephant density, it is also possible to use established networks of elephant trails.

Some researchers studying monkeys with home ranges that are smaller than those of apes have found a trail system to be invaluable. Kaplin judged that establishing a trail system was crucial in keeping up with blue monkeys (*Cercopithecus mitis*) and therefore to habituating them, and that 'building it as we went was key. It was created to facilitate animal follows, and improved as we came to know the animals and began to predict where we could find them. The trail system allowed us to move more quickly through the forest, and to open access to places where the monkeys had travelled previously' (B.A. Kaplin, personal communication, 2002). Other researchers have established trails on a grid system prior to, or during, the early part of studies (e.g. at Bezamahafaly, Madagascar) or have worked in areas that have previously been logged, so an old trail or grid system may exist (e.g. at Lopé and at Kirindy, Madagascar).

FINDING YOUR PRIMATE

Frequent contacts with the same individuals are necessary to achieve habituation, so a daily objective is usually to locate a known group or individual. Some researchers have trapped and radio-tagged primates as an aid in locating them. This option has to be considered carefully for several reasons, including ethical ones. In addition, depending on the species and procedures used, catching and marking may actually set back the habituation process (Sterling, 1993; Chapters 7 and 10). Assuming that catching and marking is not used, primates need to be located by other means. Searching at random in the hope of encountering your primate is unlikely to be a successful strategy, especially when the animals are in small and cryptic groups or occur at low densities (e.g. lowland gorillas usually occur at about one group per 10 km²). It is extremely helpful to have prior knowledge about the behavioural ecology of your species, but when this is not possible, some practical decisions and a number of clues may help in locating primates and ensuring regular contact.

Auditory clues

Often the first indication of primate presence will be an auditory clue: vocalisations such as long-calls (e.g. chimpanzees, *Pan troglodytes*; gibbons, *Hylobates* spp.; howler monkeys, *Alouatta* spp.; arboreal guenons, *Cercopithecus* spp.; ruffed lemurs, *Varecia variegata*; indri, *Indri indri*), alarm calls or warning barks. Taking a compass bearing and heading in that direction may facilitate your approach. Alternatively, the sound of movement in the vegetation as animals travel may alert you to their presence. Groups of arboreal guenons and colobines may be quite noisy as branches and leaves move with their passage; species such as colobus and sifaka (*Propithecus verreauxi*) make a thump as they land on larger branches and trunks; or you may hear the sound of food processing or plant parts dropping to the ground. Indirect clues to the previous presence of primates in the area include ape nests and a variety of feeding signs.

Waiting at key sites

An alternative to searching for your study primates is to wait for them to come to you by identifying key sites where it may be possible to encounter them regularly. Being at an animal's sleeping site before it wakes up and leaves means that you can locate your animal early in the day and become part of the surroundings the animal sees on waking up. Most primates sleep above ground, so aim to be close to nests, sleeping-cliffs, sleeping-trees or close to tree holes before dawn (before dusk for nocturnal primates). Predictable food resources such as figs, salt licks, water sources and seasonally available foods with limited distribution are also good places to locate primates.

It is possible to establish key sites through provisioning, which is the use of artificial feeding as a positive incentive to tolerate human presence, first used 50 years ago in the habituation of Japanese snow monkeys (*Macaca fuscata*) at Koshima, Japan (Watanabe, 2001). If food is provided at limited sites, as at the feeding station for chimpanzees at Gombe in Tanzania, then the animals will be attracted to that site. It may be easier to observe and habituate them there, rather than to try to locate them in the forest. They can then gradually be followed as they leave the site. However, provisioning carries with it the disadvantage of modifying natural behaviour patterns. At Gombe and Mahale, Tanzania, changes in chimpanzee activity budgets, aggression, and territoriality were recorded (Wrangham, 1974; Power, 1986). Provisioning also requires considerable resources, otherwise only a few individuals may become habituated. For example, the dominant 'alpha' males monopolise bananas at Gombe unless a large quantity is supplied (D. A. Collins, personal communication, 2002). Moreover, provisioning can affect non-target species: at Gombe baboons (*Papio cynocephalus*) were habituated as a side effect of the provisioning for chimpanzees (D. A. Collins, personal communication, 2002).

APPROACHING YOUR ANIMALS

How contact is established is one of most important elements of successful habituation, and various factors should be taken into account. Above all, avoid surprise; sudden contacts that frighten the animals will always be a negative experience. Boesch-Achermann & Boesch's (1994) basic method was 'being seen by the chimpanzees as often as possible and always in a peaceful, discreet attitude'. Thus it is desirable to be clearly visible to the primates, initially at a distance greater than that which invokes alarm and flight. Obviously the practicalities of this will depend on habitat variables (discussed below).

Habituation can be achieved only when the primate sees you, but try to choose the moment. Once you make your presence known, the animals are likely to leave, so it will be necessary to sacrifice observations in order to achieve progress. Many primates rapidly identify human faces, therefore the same observers should contact the animals, at least during the initial stages of habituation. It is also helpful to be consistent in your appearance (wear the same clothing, hat, rucksack, etc.). Habituators should behave calmly and attempt to 'reassure' animals by remaining still and/or mimicking natural behaviours, such as grooming or feeding. A crouching position is less intimidating and bonobos (*Pan paniscus*), for example, responded more positively when an observer was sitting rather than walking (van Krunkelsven *et al.*, 1999).

Signalling your presence

With apes it is often useful to adopt a signal that communicates your presence and which they learn to associate with your approach. This signal should be a specific noise, which becomes identified with a non-threatening neutral presence. 'Belch' or 'tongue-clacking' vocalisations are used with gorillas.

Keeping your distance

Generalisations are difficult to make, but you should err on the side of caution, while trying to maintain good visibility between primate and observer. Appropriate distances will depend on the environment, especially vegetation density, visibility, and species concerned.

What not to do

Certain behaviours should be avoided, such as making loud noises, sudden gestures, or surreptitious movements, although tolerance will depend on species and habitat. Most primates do not like to be followed (which may resemble predator tracking), but seem to respond better to arriving upon an observer. Moreover 'pushing' primates ahead of you by following them may cause them to incur unnecessary energetic costs. Primates do not like to be crept up on, so avoid hiding and help them to keep track of your location by vocalising. Avoid pointing with telephoto lenses, sticks or guns.

Primates are also sensitive to the number of people present, so it is better to habituate with only one or two observers, preferably the same people. Even habituated primates may alter their behaviour with increased numbers of observers; for example, stumptail macaques (*Macaca arctoides*) reacted differently to a team of observers than to the presence of one or two (Rasmussen, 1991) and Sulawesi crested black macaques (*Macaca nigra*) were more likely to flee or climb trees when visited by larger groups of people (Kinnaird & O'Brien, 1996). Mongoose lemurs (*Eulemur mongoz*) exhibited signs of agitation

Table 2.1. *Typical reactions of primates to observer presence during the habituation process*

Behaviour	Definition
Flight	Rapid, often noisy, panicked departure coupled with alarm or fear vocalisations, but no display
Avoidance	Groups are relatively calm, silent and disappear quickly without displaying; they 'melt' into the forest
Curiosity	Responses range from brief monitoring (surveillance), to moving to acquire a clearer view of an observer, to approaching the observer
Display	Vocalisations and species-typical displays (e.g. chest-beats, branch bouncing, yawns) are directed at the observer
Ignore	

if unknown observers were present or more than one to two people followed them (D. J. Curtis, personal communication, 2002).

DATA RECORDING

Systematic records are useful for assessing progress towards habituation, so we recommend recording the following: (1) time at which search started, (2) time at which animals were located, (3) how animals were located, (4) observer–primate distance (often useful to measure perpendicular distance and height, e.g. guenons at 40 m and at 15–20 m height), (5) activity of animals when contacted, (6) animal's behaviour in response to observer, (7) time at end of contact, and (8) how contact ended. As a minimum, both cumulative contact time (after your presence has been detected), observation time and location of subjects can be readily collected at all field sites.

Habituation can occur only when the same individuals are contacted on a regular basis, so it will be important to know whether you are seeing the same individuals. In addition to sex and age class, try to record also shape of face, ears and nose pattern (for apes), pelage pattern (e.g. some lemurs) and any scars, notches in ears, missing fingers or toes, broken canines, bent/broken tails.

Changes in behaviour during habituation are rarely monitored, but provide valuable information for quantifying the process and measuring the changing impact of observer presence. Choose simple categories of behaviour that can be quantified (Table 2.1).

KNOWING WHEN YOUR PRIMATE IS HABITUATED

Ideally your subjects will be aware of your presence yet learn to ignore you. Over the course of habituation both primate and observer behaviour is likely to change. As familiarity with the study species and its habitat increases, your ability to move around the study area, pick up clues to primate location and identify individuals should improve, reducing search time. Moreover, as the primates become habituated to your presence, there should be an increase in the duration of average contact time, and decreases in flight, avoidance and display behaviours. Work with both lowland gorillas and Geoffroy's tamarins (*Saguinus geoffroyi*) has shown that day ranges decreased with habituation (Rasmussen, 1998; Blom *et al.*, 2001). The ratio of habituation time to observation time should change, but be patient and do not have high expectations: mean contact time with lowland gorillas increased from about 7 minutes to only 20 minutes over three years (Blom *et al.*, 2001). Consistency of observers, or possible 'hand-over' periods may be helpful if different people are likely to be involved in habituation, although this should be avoided during the early phases. Rasmussen (1991) reported that observers who were at ease, and not intimidated (by stumptail macaques), evoked a different response than neophyte observers. It is worth sounding a note of caution here: can we ever assume that a study animal behaves as it would if no observer were present? Rasmussen (1991) found that habituated stumptail macaques still adjusted their travel in the presence of observers after almost daily contact for 14 years.

Habituation is an ongoing process, especially as group composition can change during the study period owing to immigration, emigration, births and deaths. The time taken to habituate depends on a large range of factors, which may be taxon specific, habitat based, or linked to the primates' previous experience of humans. A rough guide to habituation time is presented in Table 2.2.

Most nocturnal primates (bushbabies, *Galago* spp.; sportive lemurs, *Lepilemur* spp.; woolly lemurs *Avahi* spp.; and dwarf lemurs, *Cheirogaleus* spp.; but not aye-ayes) do not appear to need habituating at all, they simply continue their activity after noticing the presence of an observer. Diurnal and cathemeral lemurs that have not been hunted and are accustomed to humans passing through the forest can be habituated in a matter of a few days to less than a month (Andrews & Birkinshaw, 1998; Curtis, 1998). Human-naïve lemurs approach observers and may even touch them (these lemurs need habituating to the point of ignoring rather than approaching the observer). It took about three months to be able to follow arboreal mangabeys (*Cercocebus albigena*) without disturbing them or causing them to flee, but six months before there was no noticeable effect of observer presence on their behaviour (R. Kormos, personal communication, 2002). Similarly, it took three months before a group of patas monkeys (*Erythrocebus patas*) could be followed all day, and six months before animals could be reliably observed from 50 m (Chism & Rowell, 1988). Hamadryas baboons (*Papio hamadryas*) could be approached to 60 m during daily marches within a year, but to be able to walk among them took about two years (Kummer, 1995). Only after a year of repeated contacts did lowland gorillas' reactions change from

Table 2.2. *Approximate times to habituate non-provisioned primates that have not been hunted*

Time frame	Primate taxa	Source
<1 hour–<1 week	Bushbabies (*Galago* spp.), most nocturnal lemurs	S. Bearder, A. Mueller, U. Thalmann, personal communication, 2002
∼ 1 month	Most diurnal and cathemeral lemurs, orangutans (*Pongo pygmaeus*)	A. T. C. Feistner, personal observation; Rodman, 1979
2–5 months	Aye-aye (*Daubentonia madagascariensis*), pottos (*Perodicticus potto*), marmosets, tamarins, savannah baboons (*Papio* spp., except hamadryas)	Rylands, 1986; Barton & Whiten, 1993; Sterling, 1993; Cowlishaw, 1997; Passamani, 1998; Rasmussen, 1998; E. Pimley, personal communication, 2002; Y. Warren, personal communication, 2002
3–8 months	Colobus (*Colobus satanus*), guenons (*Cercopithecus l'hoesti, C. mitis*), mangabeys (*Cercocebus albigena*), muriqui (*Brachyteles arachnoids*)	Strier, 1999; R. Ham, B. A. Kaplin, M. J. S. Harrison, personal communication, 2002
6+ months	Patas (*Erythrocebus patas*), siamang (*Hylobates syndactylus*)	Chivers, 1974; Chism & Rowell, 1988
1–2 years	Hamadryas baboons (*Papio hamadryas*), mountain gorilla (*Gorilla beringei beringei*)	Schaller, 1963; Kummer, 1995; Butynski, 2001
2–5 years	Bonobo (*Pan paniscus*), chimpanzee (*Pan troglodytes*), lowland gorilla (*Gorilla gorilla gorilla*)	Boesch-Achermann & Boesch, 1994; Susman, 1984; Blom *et al.*, 2001

aggression to ignore (Blom *et al.*, 2001), yet titi monkeys (*Callicebus personatus personatus*) were habituated without systematic effort in about 12 weeks while following marmosets (*Callithrix geoffroyi*) (E. Price, personal communication, 2002).

FACTORS AFFECTING PRIMATE HABITUATION

Many factors influence the success of habituation: some can be determined by the observer, such as contact distances and observer behaviour, while there are others which you cannot control, such as species-specific reactions, habitat variables and previous experience with humans, especially hunting. Many of these factors interact, but here we attempt to evaluate some of them.

SPECIES-SPECIFIC FACTORS

Species differences

In general, species that are opportunistic, such as macaques (*Macaca* spp.) and baboons (*Papio* spp.), are relatively easy to habituate because they are extrovert and adapt readily to changing circumstances. Unhabituated individuals may even approach observers with minor mobbing, eyebrow threats or lip smacking. Primates that live in stable groups also tend to be easier because when you encounter one individual you have usually found the whole group. Moreover, if one macaque or baboon approaches with no negative consequences, others may learn from watching that 'non' interaction. These species also tend to be much more interested in exploring novel foods and so respond quickly to provisioning. In contrast, more conservative chimpanzees travel in small groups and their fission–fusion society means that, if you follow two or three one day, you may encounter a different subgroup next day (D. A. Collins, personal communication, 2002).

Diet is an important factor. Arboreal folivores such as colobus tend to be easier to habituate than arboreal frugivores such as guenons. This is likely to be due to the former having smaller home ranges, and their food being available in larger patches so that more individuals can feed together and increased time is spent resting.

Home range size is also important: primates with large home ranges are more difficult to encounter consistently, partially explaining differences between mountain and lowland gorilla habituation, and perhaps the difficulties of habituating mandrills (*Mandrillus sphinx*) and drills (*Mandrillus leucophaeus*).

Even when sympatric, species can vary in their ease of habituation. For example, white-throated capuchins (*Cebus capuchinus*) have habituated to people, whereas squirrel monkeys (*Saimiri oerstedi*) in the same area of Costa Rica have not (Boinski & Sirot, 1997). In addition, monkeys that sometimes occur in polyspecific groups may react differently depending on the circumstances. Kormos reports 'habituated mangabeys would flee without even looking if they were with unhabituated guenons who alarm called. On the other hand, they were much bolder in a polyspecific group' (R. Kormos, personal communication, 2002).

Finally, evolutionary history can influence ease of habituation. It may be that Madagascan prosimians are remarkably easy to habituate as a result of the reduced predator assemblage in Madagascar, combined with the very recent arrival of humans, only some 2000 years ago.

Sex and age differences

Responses differ according to age and sex of an individual primate. In multi-male multi-female or harem groups, adult males tend to be larger, more aggressive and play patrolling or sentinel roles in groups. Thus they are more overt and often more exposed to observer contact than are females, which may well be carrying or protecting young. Male baboons, macaques, chimpanzees and gorillas respond to observers with displays and curiosity, while juveniles generally respond with more curiosity than do adults.

HABITAT FACTORS: VISIBILITY

A clear view of the observer is one of the most important factors in habituation. Primates (such as baboons) living in more open habitats tend to be easier to habituate than those (such as mandrills) living in dense forest, both within and between taxa. In open habitats, such as savannah areas, primates can see you from a distance and you can gradually move closer. However, open habitats may still have poor visibility and patas monkeys, for example, are notoriously difficult to habituate (Chism & Rowell, 1988).

In the Virunga volcanoes, low vegetation and uneven topography provided ideal conditions where mountain gorillas (*Gorilla beringei beringei*) could be watched from the opposite side of a ravine. In contrast, visibility in lowland forest is poor, and lowland gorillas are usually obscured even within 10 m of an observer, unless they climb trees (Williamson, 1988). Sudden contacts are difficult to avoid in dense forest, and probably hinder habituation by frightening the animals.

Issues of visibility could explain differences between terrestrial and arboreal forest primates. Arboreal guenons are generally easier to habituate than terrestrial ones. For example, terrestrial l'Hoest's guenons (*Cercopithecus l'hoesti*) are more wary when followed on the ground; they flee and become completely still. In contrast, arboreal blue monkeys are apparently less concerned by observer presence when high in the trees (B. A. Kaplin, personal communication, 2002). Semi-terrestrial pigtail macaques (*Macaca nemestrina*) are more difficult to habituate than arboreal long-tailed macaques (*Macaca fascicularis*). Terrestrial sun-tailed guenons (*Cercopithecus solatus*) seem almost impossible to habituate owing to their large home ranges and cryptic habits.

ETHICAL ISSUES

One should seriously evaluate whether or not to attempt habituation if doing so may have a significant long-term effect on normal activity and behaviour patterns, and in particular if habituation may increase disease risks by changing the nature of contact between humans and non-human primates.

RISKS TO OBSERVERS

Risks during habituation

During the habituation process, aggression initially increases, peaks and then diminishes to low levels (Blom *et al.*, 2001); thus, in the early stages, a habituator may be subject to intimidating displays. These are more likely to be unpleasant than dangerous; however, people have received canine puncture wounds during mountain gorilla habituation.

Risks as a result of habituation (overhabituation)

A primary aim of habituation is for the observer to be a neutral element in the habitat. However, primates such as chimpanzees, baboons and macaques are highly intelligent, complex, socially manipulative animals, and overfamiliarity with the observer may change the observer from being a 'piece of the furniture' to a social tool, available for inclusion in their social relations. Being 'used' by your study animals is rarely mentioned in the literature, although it probably happens regularly (A. T. C. Feistner & E. A. Williamson, unpublished data). Well-habituated mountain gorillas sometimes redirect aggression towards a more vulnerable observer rather than a conspecific, and baboons and macaques may use observers in a display or to avoid aggression from others. Even seemingly harmless interactions such as a juvenile playing with a shoelace can quickly lead to incite-screaming, leaving the observer in trouble.

Loss of fear of humans can lead to negative interactions with local people: baboons at Gombe attack and steal from villagers (D. A. Collins, personal communication, 2002); habituated mountain gorillas in Uganda feed in fields outside a national park and have attacked people in fields (Butynski, 2001). Loss of fear through over-habituation and/or poorly managed contacts poses severe problems for local human populations, ecotourism programmes and primate conservation.

RISKS TO PRIMATES

Disease transmission between humans and non-human primates

Close proximity with researchers is likely to increase risks of disease transmission (Chapters 1 and 8). Illnesses to which animals have never been exposed are potentially the most dangerous. Homsy (1999) reviewed epidemiological data and disease transmission between people and apes. Studies of captive apes show that they have a definite susceptibility to human diseases, but not the same resistance as humans. As a result of this shared susceptibility, certain human pathogens, both respiratory (measles, herpes, pneumonia) and enteric diseases (polio, salmonella), can affect apes, and deaths are known to have occurred in wild populations (Butynski, 2001). Risks are greatest to terrestrial primates, which tend to come into closer proximity to people. Quarantine for researchers and minimum distance rules, such as those implemented at

Karisoke Research Centre, Rwanda, should be in place for all populations of primates. A suite of common-sense observer behaviours can reduce the risks: avoiding physical contact, maintaining minimum distances, careful management of waste products, avoiding excretion in the forest and not working if feeling unwell.

Stress
Stress provoked during the habituation process could potentially reduce reproductive success or result in immunosuppression and increase susceptibility to disease (Woodford *et al.*, 2002).

Generalisation to other humans
Since habituation is basically the loss of fear of humans, we may make animals easily approachable by hunters and poachers, and render them extremely vulnerable in dangerous situations. Thus the future of a study population and the ability to sustain protection should be contemplated carefully before the researcher decides to habituate.

OTHER IMPACTS ON PRIMATE BEHAVIOUR
The presence of an observer can have a direct effect on the activity of habituated primates, and several indirect effects by changing the target species' interactions with its conspecifics, predators, prey, and other species in its environment.

Protection from natural predators
Carnivores such as hyena (*Hyaena hyaena*), lions (*Panthera leo*), leopards (*Panthera pardus*), ocelots (*Leopardas pardalis*), and fossa (*Cryptoprocta ferox*) are unlikely to approach when a human observer is present, so habituated groups may inadvertently be protected from predation.

Behaviour with conspecifics
Regular contact with humans may cause changes in primate behaviour such as hindering reproduction, altering inter-group dynamics and impeding transfer between groups. Habituated and unhabituated primates respond differently to people, so a human presence may well alter the nature of interactions between neighbouring groups. The presence of an observer thus unintentionally benefits a habituated group, for example a naïve group may flee, enabling the study group to take over a favoured food resource (Rasmussen, 1991).

Impacts on activity
Observer presence can have negative impacts on behaviour: study groups have larger day ranges if they try to flee or avoid human contact during the initial

stages of habituation. They also expend more energy. Activity budgets may change, as feeding and other activities will be disrupted while the animals move away from habituators. In some cases primates appear to alter their activity in the presence of an observer, for example chimpanzees 'sit down and wait for us to catch up if we're slow and also disappear into deep thickets to lose us' (J. Setchell, personal communication, 2001).

CONCLUSIONS

In this chapter we have tried to explore the factors that affect habituation of primates to human observers. The key to success is persistent, regular and frequent neutral contact with the same individuals. In most cases habituation requires considerable time investment – from months to years, depending on species and environment (with Madagascan prosimians the exception). We suggest that anyone considering habituating primates for research purposes should: (1) think carefully about the pros and cons, especially the potential risks; (2) familiarise themselves with their study species prior to going to the field; (3) engage and use their common sense; and (4) record, measure and evaluate the habituation process. Habituation is usually hard work but the rewards, both personal and academic, are great: being in the company of wild primates who have accepted you into their environment and allow you an insight into their daily lives is a privilege.

ACKNOWLEDGEMENTS

We wish to thank James R. Anderson, D. Anthony Collins, Rebecca Kormos, Michael J. S. Harrison, Beth A. Kaplin, Eluned C. Price, Urs Thalmann and Ymke Warren for sharing their experiences during the preparation of this chapter.

REFERENCES

Andrews, J. R. & Birkinshaw, C. R. (1998). A comparison between daytime and nighttime diet, activity and feeding height of the black lemur, *Eulemur macaco* (Primates: Lemuridae), in Lokobe Forest, Madagascar. *Folia Primatol.* 69, Suppl. 1, 175–82.

Barton, R. A. & Whiten, A. (1993). Feeding competition among female olive baboons, *Papio anubis. Anim. Behav.* 46, 777–89.

Blom, A., Cipolletta, C., Brunsting, A. R. H. & Prins, H. H. T. (2001). Behavioural responses of gorillas to habituation in the Dzanga-Ndoki National Park, Central African Republic. In *Ecological and economic impacts of gorilla-based tourism in Dzanga-Sangha, Central African Republic*, A. Blom, pp. 91–124. Ph.D. thesis: University of Wageningen.

Boesch-Achermann, H. & Boesch, C. (1994). The Tai Chimpanzee Project in Côte d'Ivoire, West Africa. *Pan Africa News* 1, 5–7.

Boinski, S. & Sirot, L. (1997). Uncertain conservation status of squirrel monkeys in Costa Rica, *Saimiri oerstedi oerstedi* and *Saimiri oerstedi citrinellus*. *Folia Primatol.* **68**, 181–93.

Butynski, T. M. (2001). Africa's great apes. In *Great Apes & Humans: The Ethics of Coexistence*, ed. B. B. Beck, T. S. Stoinski, M. Hutchins, T. L. Maple, B. Norton, A. Rowan, E. F. Stevens & A. Arluke, pp. 3–56. Washington, DC: Smithsonian Institution Press.

Chism, J. & Rowell, T. H. (1988). The natural history of patas monkeys. In *A Primate Radiation: Evolutionary Biology of the African Guenons*, ed. A. Gautier-Hion, F. Bourlière, J.-P. Gautier & J. Kingdon, pp. 412–38. Cambridge: Cambridge University Press.

Chivers, D. J. (1974). The Siamang in Malaya: a field study of a primate in tropical rain forest. *Contrib. Primatol.* **4**, 1–335.

Cowlishaw, G. (1997). Trade-offs between foraging and predation risk determine habitat use in a desert baboon population. *Anim. Behav.* **53**, 667–86.

Curtis, D. J. (1998). Group size, home range use, and seasonal variation in the ecology of *Eulemur mongoz*. *Int. J. Primatol.* **19**, 811–35.

Grieser Johns, B. (1996). Responses of chimpanzees to habituation and tourism in the Kibale Forest, Uganda. *Biol. Conserv.* **78**, 257–62.

Homsy, J. (1999). *Ape Tourism and Human Diseases: How Close Should We Get?* Report to the International Gorilla Conservation Program. Website: <www.wildlifeinformation.org/subdirectories_fer_search/sampleEL/greatape.pdf>.

Kinnaird, M. F. & O'Brien, T. G. (1996). Ecotourism in the Tangkoko DuaSudara Nature Reserve: Opening Pandora's box? *Oryx* **30**, 65–73.

Kummer, H. (1995). *In Quest of the Sacred Baboon*. Princeton, NJ: Princeton University Press.

Passamani, M. (1998). Activity budget of Geoffroy's marmoset (*Callithrix geoffroyi*) in an Atlantic forest in southeastern Brazil. *Am. J. Primatol.* **46**, 333–40.

Power, M. (1986). The foraging adaptation of chimpanzees, and the recent behaviors of the provisioned apes in Gombe and Mahale National Parks, Tanzania. *J. Hum. Evol.* **1**, 251–65.

Rasmussen, D. R. (1991). Observer influence on range use of *Macaca arctoides* after 14 years of observation? *Lab. Prim. Newsl.* **30**, 6–11.

(1998). Changes in range use of Geoffroy's tamarins (*Saguinus geoffroyi*) associated with habituation to observers. *Folia Primatol.* **69**, 153–9.

Rodman, P. S. (1979). Individual activity patterns and the solitary nature of orangutans. In *Perspectives On Human Evolution*, vol. 5, *The Great Apes*, ed. D. A. Hamburg & E. R. McCown, pp. 235–55, Menlo Park, CA: Benjamin/Cummings Publishing Co.

Rylands, A. B. (1986). Ranging behaviour and habitat preference of a wild marmoset group, *Callithrix humeralifer* (Callitrichidae, Primates). *J. Zool. Lond.* **210**, 489–514.

Schaller, G. B. (1963). *The Mountain Gorilla: Ecology & Behavior*. Chicago: University of Chicago Press.

Sterling, E. J. (1993). Behavioral ecology of the aye-aye (*Daubentonia madagascariensis*) on Nosy Mangabe, Madagascar. Ph.D. thesis: University of Yale.

Strier, K. B. (1999). *Faces in the Forest: The Endangered Muriqui Monkeys of Brazil*. Cambridge, MA: Harvard University Press.

Susman, R. L. (1984). The locomotor behavior of *Pan paniscus* in the Lomako Forest. In *The Pygmy Chimpanzee. Evolutionary Biology and Behavior*, ed. R. L. Susman, pp. 369–93. New York: Plenum Press.

Thorpe, W. H. (1963). *Learning and Instinct in Animals*. London: Methuen.

Tutin, C. E. G. & Fernandez, M. (1991). Responses of wild chimpanzees and gorillas to the arrival of primatologists: behaviour observed during habituation. In *Primate Responses to Environmental Change*, ed. H. O. Box, pp. 187–97. London: Chapman & Hall.

van Krunkelsven, E., Dupain, J., van Elsacker, L. & Verheyen, R. (1999). Habituation of bonobos (*Pan paniscus*): first reactions to the presence of observers and the evolution of response over time. *Folia Primatol.* **70**, 365–8.

Watanabe, K. (2001). A review of 50 years of research on the Japanese monkeys of Koshima: status and dominance. In *Primate Origins Of Human Cognition And Behavior*, ed. T. Matsuzawa, pp. 405–17. Tokyo: Springer-Verlag.

Williamson, E. A. (1988). Behavioural ecology of Western Lowland Gorillas in Gabon. Ph.D. thesis: University of Stirling.

Woodford, M., Butynski, T. M. & Karesh, W. (2002). Habituating the great apes: the disease risks. *Oryx* **36**, 153–60.

Wrangham, R. W. (1974). Artificial feeding of chimpanzees and baboons in their natural habit. *Anim. Behav.* **22**, 83–93.

3 • Habitat description and phenology

JÖRG U. GANZHORN

Animal Ecology and Conservation, Institute of Zoology and Zoological Museum, Hamburg, Germany

INTRODUCTION

Habitat characteristics represent the matrix for the evolution of morphological, physiological and behavioural adaptations and life history traits of animals. These include bottom-up factors (such as distribution and abundance of food) and top-down constraints (such as protection from predators). The role of different habitat components in the survival and reproduction of individuals is of prime interest and is relevant for evolutionary understanding as well as for primate conservation. So far, this has been studied by linking habitat characteristics with phenotypic traits of animals, assuming that these traits represent adaptation to a given habitat constraint. However, we should bear in mind that the characteristics we observe today may represent adaptations to constraints that acted in the past (the Epaminondas effect, named after the little boy who always did the right thing for the previous situation; or, in evolutionary terminology, animals are tracking fitness optima by trying to climb the sides of shifting adaptive peaks (Cody, 1974)). However, testing hypotheses based on these 'ghosts of past constraints' has so far been impossible.

The present chapter focuses on the description of the vegetation of forest habitats as the biotic matrix for the evolution of life history traits.

BEFORE YOU START

Any statistical analysis requires quantitative measurements of habitat and animal characteristics. Most habitats are extremely complex and there are many features that can be measured, spending endless time and money, without measuring the right variable. The following issues that apply to all research projects should be considered before one starts:

- What is the question? Don't just measure habitat features simply because somebody else did it (e.g. Do you need to sample herbaceous vegetation and bushes for primate species that rarely leave the canopy? How relevant

Field and Laboratory Methods in Primatology: A Practical Guide, ed. Joanna M. Setchell and Deborah J. Curtis. Published by Cambridge University Press. © Cambridge University Press 2003.

is a 10 m × 100 m^2 plot of forest to a group of chimpanzees (*Pan* spp.) that range over several square kilometres?).

- What is the scale of the study?
- What is the unit of analysis?
- Which habitat variable do you need to measure to answer your question?
- How accurately do you need to measure variables?
- Which statistical test will be applied?
- Are your samples independent?
- What sample size is required for statistical analysis?

Primate body mass ranges from 0.03 to 100 kg, home ranges from 0.01 to 400 km^2, and habitats from snow-covered mountains to deserts and evergreen tropical rain forests. There is therefore no universal sampling method appropriate to answering all questions. All methods can be adjusted to satisfy the needs and special circumstances of the study in question and the development of new methods just to do something differently should be avoided where possible, as studies gain from the number of other studies with which they can be compared. All samples should be geo-referenced using a Global Positioning System (GPS) (Chapter 4). Even though this may not be of immediate relevance to the present study, it will add tremendous potential for future studies to monitor environmental or other long-term changes. As a general rule, statistical comparisons are facilitated by sampling more smaller units rather than by taking fewer large ones.

FOREST CLASSIFICATION

Forests can be classified on the basis of their overall structure, phenology, leaf shape, climate, geology/soil characteristics and history of human utilisation. Structural classifications are, for example, closed- or open-canopy forests, mixed forest with open or closed understorey, woodland or wooded grassland. Phenological, climatic, soil/water relationships and leaf characterisations are included in descriptions such as wet, moist or dry forest, gallery or mountain forest, seasonally inundated forest, peat or swamp forest, evergreen or deciduous forest, broad-leaved or spiny forest (Whitmore, 1990). The history of human utilisation is included in terms such as primary or secondary forest. These classifications are usually rather subjective and based on the overall impression of the fieldworker, yet they give a reasonable impression of the physiognomy of different forests.

Floristic characterisation of forests is often beyond the skills of zoologists. Field guides are available for woody plants (Gentry, 1993; Keller, 1996; APG, 1998; Schatz, 2001), but plant species identification usually requires collaboration with botanists in the field or at least collection of herbarium specimens (Chapter 11). To prepare a specimen, a twig with leaves (the inclusion of flowers and fruits is best for identification) is pressed between newspapers (or any absorbent paper) until it is dry. Simple plant presses can be constructed with two boards. Plant material must be fully dry before it is transferred to a solid cardboard, otherwise leaves etc. will come off the twig. A simple drying box can be constructed, as suggested by Hladik (1977), for processing food samples. Arrangements for help from experts should be made prior to sample collection.

A rough but useful forest classification can be based on families of plants or simply trees. Within communities, families (or species) can be assigned different degrees of dominance. On the basis of the percentage of individuals in the community, a family (or species) is called 'eudominant' if it accounts for >10% of the individuals. Other categories include 'dominant' (5–10%) and 'subordinate' (2–5%). Another standard classification for vegetation cover has been developed by Braun-Blanquet (1967) as:

+:   <1%
1:   1–5%
2:   6–25%
3:   26–50%
4:   51–75%
5:   76–100%

Other measures of diversity are summarised in most textbooks (e.g. Brower et al., 1990; Krebs, 1998). The most comprehensible form of documentation of floristic characteristics is a simple table or graph that shows the relative abundance of families or species.

HABITAT STRUCTURE

There are three basic types of method used to describe the density, frequency, coverage and biomass of plants in forest habitats: plot methods, transects and point-quarter sampling (Brower et al., 1990).

PLOT METHODS

In plot methods, all individual plants within a given area are counted, measured and identified. A tree should be considered to be in the plot if the centre of the trunk is within the limits of the plot. Rectangular plots with sides in a ratio of 1:2 are recommended. The size of the plot depends on the vegetation to be sampled. Recommended sizes are: 0.71 m × 1.41 m (2 m^2) for herbaceous vegetation; 2.24 m × 4.47 m (10 m^2) for bushes, shrubs and trees up to 3–4 m tall; 7.07 m × 14.14 m (100 m^2) for trees taller than 3–4 m. For tropical forests with large trees, the size of the sampling area must be extended. The standard measurement for trees is the diameter at breast height (DBH), related to the circumference (C) as DBH = C/3.14, and measured at a height of 1.3 m. DBH of buttressed and stilt-rooted trees should be measured above the

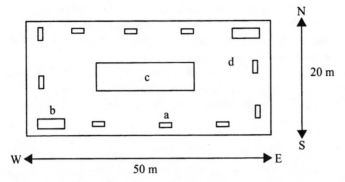

Fig. 3.1. Modified Whittaker plot consisting of nested subplots (after Stohlgren & Chong, 1997.) a, b, c, and d indicate subplots of different size.

buttresses/stilts, or by estimating the diameter of the trunk without the buttresses. Forestry suppliers provide 'DBH tapes' that give tree diameters when measuring circumference. Different studies use different lower size limits and size categories for measuring trees. This is due to different forest structure and structural requirements of primates in different forests. However, a lower size limit of 10 cm DBH and 5 or 10 cm increments would allow comparisons with most other descriptions of vegetation.

For tropical forests, modified Whittaker plots are proposed for multi-scale vegetation sampling (Fig. 3.1). Nested subplots of different sizes within a larger plot allow the development of species–area curves and estimation of the number of species in a larger unsampled area. This method has been applied by Smithsonian's Programme for Monitoring and Assessment of Biodiversity in Peru (Stohlgren & Chong, 1997). For primate habitats, all trees ≥10 cm DBH should be identified and measured within the largest plot – smaller sized trees (<10 cm DBH) are rarely considered. This might make inclusion of the small-sized nested subplots unnecessary. However, the principle of the sampling design remains valid and can be adjusted to include different sized trees (Davies, 1987). For example, trees ≥10 cm DBH can be recorded in subplots of 100 m² that are nested within larger plots in which only trees ≥ 30 cm DBH are registered.

As a rule of thumb, about 5–10 plots are needed per stratum to characterise a vegetation formation if within-stratum variation is low. If within-stratum variation is high, more sites are needed (Stohlgren & Chong, 1997).

Other plots consist of series of uniformly sized square or rectangular plots in which trees are identified and measured (e.g. Ganzhorn, 1989; Schwarzkopf & Rylands, 1989). The only difference to the methods described above is that these plots are not nested and are all of the same size. This has been the most commonly used method in primatological studies.

Advantages: Plot methods are most efficient for all kinds of density-related features. Plot sizes can be adjusted to the size of microhabitats relevant for the questions to be studied, such as analyses of used versus unused habitats (Southwood & Henderson, 2000).

Cautionary note: Plots of different shapes are likely to give different density estimates even if they cover the same surface (e.g. a 2 m × 50 m plot is likely to give a higher density estimate of trees than a 10 m × 10 m plot).

TRANSECT SAMPLES

A belt transect is a long rectangular plot. In primatological studies, 0.1 ha (10 m × 100 m²) transects have proved useful as they are of a manageable size and provide a good idea of the structure and floristic composition of the forest. This belt can be subdivided into intervals (e.g. 10 m × 10 m) to facilitate recordings and check for species–area relationships. However, single subplots cannot be used as independent measures for statistical analyses because they are likely to be autocorrelated. Transects have also been used to analyse Kloss gibbon (*Hylobates klossii*) ranging patterns (Whitten, 1982).

Transects seem more appropriate than plots in heterogeneous habitats as they sample a greater variety of microhabitats. In principle, they can also describe environmental gradients, although these are difficult to analyse in conjunction with primate ranging patterns. Analyses are facilitated if clearly defined categories of vegetation are used.

The most complete transect descriptions that are available for comparison are 'Gentry transects'. They measure 2 m × 50 m and were designed to sample floral diversity in many different countries (Gentry, 1995; Phillips & Miller, 2002). Large-scale plots (50 ha) are being installed by the Smithsonian Institution at various sites for comparative descriptions of different forests and long-term monitoring.

Advantages: Transects are less susceptible to patchy distributions of certain plant species than are plot samples. In tropical forests, sample sites are rarely selected at random, but rather by accessibility. This bias is reduced by sampling along transects. While the choice of the starting point of the transect is unlikely to be random, subsequent sites are likely to provide a more representative picture of the habitat than are plots.

Cautionary note: Subsequent subplots aligned along continuous transects suffer from autocorrelations, so cannot be treated as independent samples. This bias can be reduced by spacing subplots along transects.

POINT-QUARTER SAMPLING

The point-quarter method (= point-centred quadrants) is a plot-less method. For this, a number of randomly determined points are selected. These points

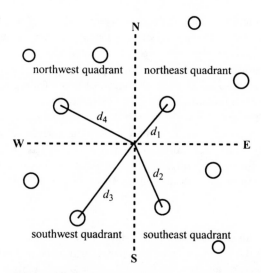

Fig. 3.2. Point-centred quadrants or point-quarter sampling (after Brower *et al.*, 1990)

can be distributed either along a transect line or throughout the habitat to be described. Each point represents the centre of four compass directions that divide the sampling point into four quarters. In each quarter, the distance (d_1 to d_4) from the centre of the nearest plant to the sampling point is measured (Fig. 3.2). Plants to be measured must be defined (e.g. only trees $\geq$ 10 cm DBH are considered). Since it is difficult to measure from the centre of a large tree, the radius of the tree trunk is added to the distance from the tree to the central sampling point.

Densities can be calculated as follows: Compute the mean of d_1 to d_4 per sampling plot ($= d$). This can be used as the unit for further analyses. The mean area in which a single plant occurs is equal to the mean distance squared. Thus, theoretically, each plant covers an area d^2. The density of plants per unit area (A) is then: A/d^2 (example: mean distance of trees $\geq$ 10 cm DBH to the sampling point = 5.24 m. Density per hectare (1 ha = 100 m $\times$ 100 m = 10 000 m^2) = 10 000/5.24^2 = 364 trees/ha). This method has been used to contrast primary forest with various kinds of secondary vegetation formations or to monitor small-scale variation within primary forest (Ganzhorn *et al.*, 1997; Ganzhorn, 2002) and can be modified to measure the density of single species or any other structural habitat component.

Advantage: The point-quarter method is very quick and yields large numbers of independent datapoints that can be used for statistical analyses. It avoids the tedious work of setting up plots and deciding whether trees along the boundaries of plots are in or out of the plot. The statistical results seem to indicate differences in the structure between habitats appropriately.

Cautionary note: Under field conditions, and given that primatologists normally have interests other than measuring plants, it is almost impossible to achieve a random set of sampling points. The tendencies are either to select a sapling that can be marked with flagging tape as the centre of the four quadrats, or to select points at some distance from the nearest trees. If the same person identifies the sampling point, then these biases do not average out. Though the statistical results seem to indicate differences in the structure between habitats appropriately, the calculated values of tree densities can deviate substantially from the actual density, especially at small sample sizes (<50 sampling points). Thus statistical differences indicate relative rather than absolute differences.

EQUIPMENT

Plots and transects can be measured with a tape or hip chains. They can be marked temporarily with flagging tape or permanently with numbered aluminium tags. If flagging tape is used it should be biodegradable and must be removed after completion of the study. Corners of plots should be georeferenced with a GPS, although to keep a given direction on transects a compass may be easier to handle in the forest. Directions can be maintained by using three poles arranged in a straight line. Once they have been established, the first pole can be removed and added again after the last. This results in more accurate compass bearings than using trees as landmarks.

PLOTS OR TRANSECTS?

In principle, plots and transects should be set up at random (Southwood & Henderson, 2000). Random 'choice' can be approximated by putting a labelled grid over a map of the study area and drawing a predefined number of grid cells. The predefined corner of a given cell can be used as the starting point for vegetation analyses. Random numbers can be generated by most computers or by taking the last two digits of telephone numbers. Realistically, randomness is never achieved in studies of primate natural history.

If boundaries for different vegetation communities and their extension are known, plot methods may provide more information, such as patchiness or species–area relationships in a homogeneous habitat. Results of the plots can then be extrapolated to the whole surface. Vegetation units could be swamp forest as compared to forest on solid ground, primary forest as compared to secondary forest, or areas used versus areas not used. In this case, plots should be located at randomly selected sites within the vegetation formation (stratified random sampling; e.g. Oates *et al.*, 1990; Smith *et al.*, 1997). However, primate home ranges are usually not known precisely before the study begins, 'uniform' vegetation types may not be known or delineated, it may not be possible to extend the existing trail system (e.g. in National Parks), or new trails may have to be installed where it is possible logistically and not where random numbers fall. In these cases transects seem to be a better compromise between data requirements, statistical rigour and time spent on vegetation issues.

CANOPY COVER

Canopy cover can be measured by taking photographs from ground level and measuring the percentage of open sky. However, the results of this method can be analysed only after films have been developed. Given solar panels, portable computers and digital cameras, it will be possible to measure canopy cover at many sites almost on-line, even at field stations with very basic infrastructure. Estimates of canopy cover based on photographic documentation must be calibrated to account for distortions. Estimates of the percentage of the sky covered can be obtained with any Geographical Information System (GIS) program. Aerial photographs allow classification of canopy cover, but it is essential that these classifications are 'ground-truthed' (Chapter 4).

A low budget alternative consists of recording at fixed intervals along a transect (e.g. every 1 m) whether or not the open sky is visible through a vertically held pipe. The advantage of this method over the photographic measurements is that canopy cover can be estimated at different heights or for different strata.

VERTICAL AND HORIZONTAL STRUCTURE

The line intercept technique is another transect method used to describe forest structure. Here, contacts of vegetation with a vertical line are recorded from 100 points spaced at regular intervals along a defined transect (Gautier et al., 1994; Fig. 3.3a). The length of the baseline should be about 10 times the height of the canopy. A fine example of this method is provided by Messmer et al. (2000; Fig. 3.3b). They measured the heights at which all plants came in contact and species encountered above each of 100 census points per transect with an 8 m pole marked at 20 cm intervals and held vertically. Heights of contact above 8 m were estimated. For dense foliage, the lower and upper points of the foliage mass were recorded. Alternatively, readings can be taken using a telelens focussed vertically at defined intervals. This method allows very detailed analyses of forest structure that can be presented in illustrative graphs. The relative densities of vegetation from different transects can be compared with respect to their horizontal or vertical orientation using Kolmogorov–Smirnov tests (Siegel & Castellan, 1988). However, this method does not sample an area (it can be adapted to sample unit areas by sampling a surface rather than along a straight line; but there are no comparative databases in primatology).

The lateral density of vegetation can be measured by positioning a square or longitudinal chequerboard at different heights in the forest. A reasonable

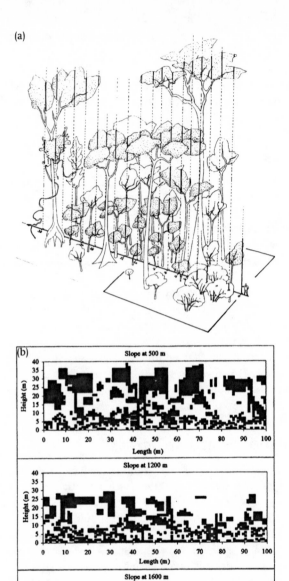

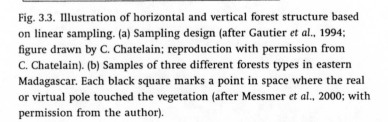

Fig. 3.3. Illustration of horizontal and vertical forest structure based
on linear sampling. (a) Sampling design (after Gautier *et al.*, 1994;
figure drawn by C. Chatelain; reproduction with permission from
C. Chatelain). (b) Samples of three different forests types in eastern
Madagascar. Each black square marks a point in space where the real
or virtual pole touched the vegetation (after Messmer *et al.*, 2000; with
permission from the author).

size is about 50 cm × 50 cm with 10 cm × 10 cm black and white squares to measure vegetation density on a small scale. For larger distances a board of 25 cm × 200 cm can be subdivided into quadrats of 25 cm. It can then be recorded how many (what percentage) of the different squares can be seen from specified distances. This method is not applied routinely in primatological studies, but might be useful to evaluate 'thickness of vegetation' (e.g. as protection from predators or as potential cover for predators).

POPULATION DISPERSION/PATCHINESS

Dispersion patterns of plants can be assessed by dividing the variance of tree densities by the mean density per plot (variance-to-mean ratio). Values <1 indicate uniform or overdispersed distribution, values = 1 indicate random distributions and values >1 indicate clumped (underdispersed, clustered) distributions. This method is labour intensive, as inventories of many different plots must be made. The variance-to-mean ratio is sensitive to population and plot size, requiring some thought on the scale of the study that is biologically relevant. Brower *et al.* (1990) provide an introduction to measurements and indices for population dispersions and Ganzhorn (1993) provides primate examples. Plotless (distance) methods for analysing spatial patterns are described by Ludwig & Reynolds (1988).

SINGLE-TREE-RELATED STRUCTURAL DATA

Some useful data that might be collected for a single tree are listed in Table 3.1. The variables measured will always be a compromise between their possible relevance for primates and the limited time available.

PHENOLOGY

Phenological measurements are taken to get an idea of seasonal changes in productivity of the forest and food abundance for primates. The plant parts measured are young and mature leaves, flowers, and unripe and ripe fruits; binoculars will be necessary for large trees. Data on seasonal variation in the production of exudates can be useful for specialised species (lemurs, galagos, tamarins; Heymann & Smith, 1999), but standardised quantitative measurements of gum production are problematic.

Ideally, phenological records should characterise (1) seasonal changes in the productivity of the forest and (2) seasonal changes in food availability for primates. The former can be achieved by random or stratified sampling. The latter is much more difficult because it requires detailed knowledge of the diet of the study species before the phenological sampling is initiated. Also, some seasonally very important resources may be scarce and may even be located outside the normal range of the study group. These so-called keystone resources (e.g. Gautier-Hion & Michaloud, 1989) are often identified only during

Table 3.1. *Variables measured to characterise trees as elements of forest structure*

Variable	Details	Source
DBH	Measuring all trees ≥10 cm diameter at breast height is standard; DBH is related to many important tree characteristics that are difficult to measure (tree height, biomass, age, crop size)	Chapman *et al.*, 1992
Crown width	Crown width (crown diameter) can be measured as the mean of the distances between the opposite extremes of the crown measured along two perpendicular axes. Like DBH, crown diameter can be related to biomass and food patch size. Crown diameters can be used instead of DBH when trees are highly buttressed	Terborgh, 1983; Wright, 1999
Crown volume	Crowns can take many different shapes that are important to quantify biomass. However, accurate measurement of crown shape is beyond the scope of primatological studies, and the effort required far exceeds possible benefits	Chapman *et al.*, 1992
Tree height	Tree height has sometimes been measured but seems to be of limited value in primatological studies. Height can be measured with a calibrated telelens, a range finder (clinometer) or by measuring the distance at which the top of the crown is seen at an angle of 45° from the observer. In the last case, tree height is calculated as: distance between the observer and the tree plus the height of the observer. This procedure looks nice in textbooks until you try it in the forest	Whitten, 1982; Ganzhorn, 1989
Bole height	This is the height of the first branch. It provides an idea about the open space below the canopy that can be used by vertical clingers and leapers	Whitten, 1982; Ganzhorn, 1989
Solar exposure of crowns	Solar exposure influences tree productivity and food quality but there are no data yet that link primate behaviour to this variable in an evolutionarily significant sense	Ganzhorn, 1995
Support for locomotion	Very detailed information on the aspects and size of supports for primates might be required for the analysis of locomotion	Warren, 1997

DBH, diameter at breast height.

the course of the study and may not have been included in the initial phenological sample.

Other problems are that: it is sometimes difficult to distinguish between young and mature leaves; flowers and fruits can be very small, green and difficult to detect; ripeness of fruits is difficult to define; and, if tree species are dioecious, half of the trees (males) will never produce fruit.

PHENOLOGICAL TRANSECTS

Phenological transects are supposed to describe fruit, flower and leaf availability of the forest based on random samples. Phenological data can be used in conjunction with the abundance and biomass of the different tree species that are present at a site. To achieve close matches between phenological records and vegetation descriptions, it is best to take random samples from the same plots or transects that are used to describe the floristic and structural characteristics of the forest.

Since leaf and fruit production vary widely within and between tree species and sites, additional phenological samples can be taken along transects stretching over several kilometres. Phenological trails or plots therefore cover larger areas or span larger distances than the vegetation samples used to describe structural aspects of the habitat.

If vegetation formations can be classified and delineated, phenological samples that might allow extrapolation to the whole area of interest can also be recorded within plots. Phenological patterns of tropical forests may not follow environmental seasonality as tightly as in temperate climates (Wright, 1999) and, ideally, data should be recorded over several years to account for year-to-year variation in individual trees, individual species and on the community level (e.g. Leigh et al., 1996; Struhsaker 1997).

The application of different techniques to record phenological data is reviewed by Chapman and co-workers (Chapman et al., 1992, 1994). A review of phenological studies in ecological contexts is provided by van Schaik et al. (1993). Productivity is summarised for South American sites by Kay et al. (1997).

Phenological records and analyses

Trees for phenology should be labelled using numbered aluminium tags and nails available from forestry suppliers. Aluminium does not rust, does not cause as much damage to trees as other materials and aluminium nails do not destroy saws, which may be relevant for work in timber plantations. Phenological data are recorded every two weeks (or twice per month). Abundance of plant parts can be rated on relative scales from 0 to 4 or 0 to 10, although this is subject to observer bias. Alternatively, abundances can be estimated on a logarithmic scale by actually counting items, for example fruits, in a defined sector of the crown and extrapolating this figure to the total crown volume. This results in approximate absolute figures (0 to 1 fruit = 0; about 10 fruits = 1; about 100 fruits = 2, etc.).

Productivity can be recorded and analysed as simple presence/absence data. Alternatively, production can be weighted by the DBH or basal area of the tree, by crown volume and by the measure of abundance.

Surprisingly, the different measurements and different ways of weighting productivity yield highly correlated results (Chapman *et al.*, 1994). Therefore, the method of simply adding the number of trees with a given plant part (fruits) is satisfying. Using this method, each tree can be considered as an independent unit. This facilitates statistical analysis, which becomes more complicated once productivity is based on weighted data.

FRUIT TRAILS AND FRUIT TRAPS

Fruit trails consist of individuals of selected tree species that are of special interest (e.g. known food tree species). Tree species are sampled in replicate (about five individuals per species; the more the better), allowing the study of intra-specific variation in fruit production. For the sake of easy access, and since the trees are not supposed to represent a random sample, the trees can be chosen along existing trails.

Fruit traps consist of a plastic bag suspended from a frame which can be made of poles or bent wire. The collecting area of the surface must be measured (e.g. 0.08–0.25 m²; Hladik, 1980; Chapman *et al.*, 1994). The whole structure should be raised off the ground so that the bag is not perforated by rodents or eaten by ants and making small holes in the plastic allows water to drain out (care should be taken that small seeds are not flushed out).

Fruit traps have been used in Peru (Terborgh, 1983), Uganda (Chapman *et al.*, 1994) and Madagascar (Hladik, 1980; Wright, 1997). They are set at random or at predefined intervals, for example 1 m off the side of existing trails. This method is supposed to represent random sampling of fruit availability and is equivalent to phenological transects. While data based on phenological transects and fruit trails are significantly correlated, fruit trap data did not correlate with the results of either method (Chapman *et al.*, 1994).

Fruit traps can also be set under the crown of individual trees of special interest to measure variability of fruit production or consumption rates within and between tree species (Böhning-Gaese *et al.*, 1999). Fruits, seeds and leaves as well as caterpillar faeces (e.g. Hladik, 1980) should be collected, identified, counted and weighed at regular intervals (e.g. once per week).

A method similar to fruit traps consists of sweeping a trail and recording fruit and litter fall on the trail at regular intervals. This method covers a larger area but fruit and seed removal from the trails by secondary consumers is exaggerated (Butynski, 1990).

Advantages: Fruit traps represent the most objective measurement for fruit and litter fall. They can be standardised and allow comparisons between sites that are independent from observer bias.

Disadvantages: Fruit traps sample a very small surface, despite the substantial investment required to construct and set them. Only fruit and litter fall is recorded. This means that fruits available but consumed or remaining on the tree are not measured. Fallen fruits can also start to rot or can be eaten by rodents and insects in the trap.

CONCLUSIONS

Having described all these wonderful methods, it should be kept in mind that new technologies are available that have the potential to catapult the analyses of animal–habitat relationships to new standards of resolution. For example, near-infrared reflectance spectra, in combination with other remote sensing methods, permit the measurement of habitat productivity at very fine scales over much larger areas than will ever be possible on the ground (Chapter 4). Since animals do not care about Latin binomials but are rather more interested in the chemical composition and the availability of nutritional resources (Chapter 13), these remote-sensing techniques have an extremely high potential that has not yet found its way into primatology.

SELECTED REFERENCES FOR GENERAL ASPECTS (BOLD) AND OTHER REFERENCES CITED

APG (1998). An ordinal classification for the families of flowering plants. *Ann. Miss. Bot. Gard.* **85**, 531–53.

Böhning-Gaese, K., Gaese, B.H. & Rabemanantsoa, S.B. (1999). Importance of primary and secondary seed dispersal in the Malagasy tree *Commiphora guillaumini*. *Ecology* **80**, 821–32.

Braun-Blanquet, J. (1967). *Pflanzensoziologie*, 3rd edition Wien: Springer-Verlag.

Brower, J.E., Zar, J.H. & von Ende, C.N. (1990). *Field and Laboratory Methods for General Ecology.* Dubuque, IA: William C. Brown Publishers.

Butynski, T.M. (1990). Comparative ecology of blue monkeys (*Cercopithecus mitis*) in high- and low-density subpopulations. *Ecol. Monogr.* **60**, 1–26.

Chapman, C.A., Chapman, L.J., Wrangham, R., Hunt, K., Gebo, D. & Gardner, L. (1992). Estimators of fruit abundance of tropical trees. *Biotropica* **24**, 527–31.

Chapman, C.A., Wrangham, R. & Chapman, L.J. (1994). Indices of habitat-wide fruit abundance in tropical forests. *Biotropica* **26**, 160–71.

Cody, M.L. (1974). Optimization in ecology. *Science* **183**, 1156–64.

Davies, A.G. (1987). *The Gola Forest Reserves, Sierra Leone.* Gland: IUCN.

Ganzhorn, J.U. (1989). Niche separation of the seven lemur species in the eastern rainforest of Madagascar. *Oecologia* **79**, 279–86.

   (1993). Flexibility and constraints of *Lepilemur* ecology. In *Lemur Social Systems and their Ecological Basis*, ed. P.M. Kappeler & J.U. Ganzhorn, pp. 153–65. New York: Plenum Press.

(1995). Low level forest disturbance effects on primary production, leaf chemistry, and lemur populations. *Ecology* **76**, 2084–96.

(2002). Distribution of folivorous lemurs in relation to seasonally varying food resources: effects of qualitative and quantitative food characteristics. *Oecologia* **131**, 427–35.

Ganzhorn, J. U., Malcomber, S., Andrianantoanina, O. & Goodman, S. M. (1997). Habitat characteristics and lemur species richness in Madagascar. *Biotropica* **29**, 331–43.

Gautier, L., Chatelain, C. & Spichiger, R. (1994). Presentation of a relevé method for vegetation studies based on high resolution satellite imagery. In *Proceedings of the XIIIth Plenary Meeting AETFAT, Malawi*, vol. 2, ed. J. H. Seyani & A. C. Chikuni, pp. 1339–50.

Gautier-Hion, A. & Michaloud, G. (1989). Are figs always keystone resources for tropical frugivorous vertebrates? A test in Gabon. *Ecology* **70**, 1826–33.

Gentry, A. (1993). *A Field Guide to the Families and Genera of Woody Plants of Northwest South America*. Washington, DC: Conservation International.

(1995). Diversity and floristic composition of neotropical dry forests. In *Seasonally Dry Tropical Forests*, ed. S. H. Bullock, H. A. Mooney & E. Medina, pp. 146–94. Cambridge: Cambridge University Press.

Heymann, E. W. & Smith, A. C. (1999). When to feed on gums: temporal patterns of gummivory in wild tamarins, *Saguinus mystax* and *Saguinus fuscicollis* (Callitrichidae). *Zoo Biol.* **18**, 459–71.

Hladik, A. (1980). The dry forest of the west coast of Madagascar: climate, phenology, and food available for prosimians. In *Nocturnal Malagasy Primates*, ed. P. Charles-Dominique, H. M. Cooper, A. Hladik, C. M. Hladik, E. Pages, G. F. Pariente, A. Petter-Rousseaux, J.-J. Petter & A. Schilling, pp. 3–40. New York: Academic Press.

Hladik, C. M. (1977). Field methods for processing food samples. In *Primate Ecology*, ed. T. H. Clutton-Brock, pp. 591–601. London: Academic Press.

Kay, R. F., Madden, R. H., van Schaik, C. & Higdon, D. (1997). Primate species richness is determined by plant productivity: implications for conservation. *Proc. Natl. Acad. Sci., USA* **94**, 13023–7.

Keller, R. (1996). *Identification of Tropical Woody Plants in the Absence of Flowers and Fruits*. Basel: Birkhäuser Verlag.

Krebs, C. J. (1998). *Ecological Methodology*. Menlo Park, CA: Addison Wesley Longman.

Leigh, E. G., Rand, A. S. & Windsor, D. M. (eds.). (1996). *The Ecology of a Tropical Forest: Seasonal Rhythms and Long-term Changes*. Washington, DC: Smithsonian Institution Press.

Ludwig, J. A. & Reynolds, J. F. (1988). *Statistical Ecology*. New York: John Wiley & Sons.

Messmer, N., Rakotomalaza, P. J. & Gautier, L. (2000). Structure and floristic composition of the Parc National de Marojejy, Madagascar. In *A Floral and Faunal Inventory of the Parc National de Marojejy, Madagascar: With Reference to*

Elevational Variation, ed. S. M. Goodman, pp. 41–104. Chicago: Field Museum of Natural History.

Oates, J. F., Whitesides, G. H., Davies, A. G., Waterman, P. G., Green, S. M., Dasilva, G. L. & Mole, S. (1990). Determinants of variation in tropical forest primate biomass: new evidence from West Africa. *Ecology* **71**, 328–43.

Phillips, O. L & Miller, J. S. (2002). *Global patterns of Plant Diversity: Alwyn H. Gentry's Forest Transect Data Set*, Monographs in Systematic Botany no. 89. St Louis, MO: Missouri Botanical Garden Press.

Schatz, G. E. (2001). *Generic Tree Flora of Madagascar*. Kew, St Louis, MO: Royal Botanical Garden and Missouri Botanical Garden.

Schwarzkopf, L. & Rylands, A. B. (1989). Primate species richness in relation to habitat structure in Amazonian rainforest fragments. *Biol. Cons.* **48**, 1–12.

Siegel, S. & Castellan, N. J. (1988). *Nonparametric Statistics for the Behavioral Sciences*. New York: McGraw-Hill.

Smith, A. P., Horning, N. & Moore, D. (1997). Regional biodiversity planning and lemur conservation with GIS in western Madagascar. *Cons. Biol.* **11**, 498–512.

Southwood, T. R. E. & Henderson, P. A. (2000). *Ecological Methods*, 3rd edition Oxford: Blackwell Science.

Stohlgren, T. & Chong, G. (1997). Assessment of biological diversity and long-term monitoring plan for the Lower Urubamba Region. In *Biodiversity Assessment and Monitoring of the Lower Urubamba Region, Peru*, vol. 1, *SI/Monitoring and Assessment of Biodiversity Program*, ed. F. Dallmeier & A. Alonso, pp. 41–4. Washington, DC: Smithsonian Institution.

Struhsaker, T. T. (1997). *Ecology of an African Rain Forest*. Gainesville, FL: University Press of Florida.

Terborgh, J. (1983). *Five New World Primates* (USA). Princeton, NJ: Princeton University Press.

van Schaik, C. P., Terborgh, J. W. & Wright, S. J. (1993). The phenology of tropical forests: adaptive significance and consequences for primary consumers. *Annu. Rev. Ecol. Syst.* **24**, 353–77.

Warren, R. D. (1997). Habitat use and support preference of two free-ranging saltatory lemurs (*Lepilemur edwardsi* and *Avahi occidentalis*). *J. Zool., Lond.* **241**, 325–41.

Whitmore, T. C. (1990). *An Introduction to Tropical Rain Forests*. Oxford: Oxford University Press.

Whitten, A. J. (1982). A numerical analysis of tropical rain forest using floristic and structural data and its application to an analysis of gibbon ranging behaviour. *J. Ecol.* **70**, 249–71.

Wright, P. C. (1997). Behavioral and ecological comparisons of Neotropical and Malagasy primates. In *New World Primates: Ecology, Evolution, and Behavior*, ed. W. G. Kinzey, pp. 127–41. New York: Aldine de Gruiter.

   (1999). Lemur traits and Madagascar ecology: coping with an island environment. *Yearbk Phys. Anthropol.* **42**, 31–72.

Further reading

Kent, M. & Coker, P. (1992). *Vegetation Description and Analysis*. Boca Raton, FL: CRC Press.
Mueller-Dombois, D. & Ellenberg, H. (1974). *Aims and Methods of Vegetation Ecology*. New York: John Wiley & Sons.
Underwood, A.J. (1997). *Experiments in Ecology. Their Logical Design and Interpretation using Analysis of Variance*. Cambridge: Cambridge University Press.

Useful Internet site

Gentry plots at the Smithsonian Institution: <www.ctfs.si.edu>.

4 • The Global Positioning System, Geographical Information Systems and Remote Sensing

KAREL HUGHES

Centre for Research in Ecology and the Environment, School of Life and Sport Sciences, University of Surrey Roehampton, London, UK

INTRODUCTION

Geographical Information Systems (GIS) and the Global Positioning System (GPS) are powerful tools that enable ecologists to acquire, store, analyse and display spatial ecological data (Dominy & Duncan, 2001), whereas remote sensing can provide information unobtainable by traditional methods. Although there has been an exponential rise in the use of this technology-based synergism for spatial analyses, mapping and modelling in the ecological and plant sciences (Du Puy & Moat, 1998; Dale, 1999; Wadsworth & Treweek, 1999), there have been few, fully integrative applications by primatologists (but see Table 4.1). Applications relevant to primatologists include: vegetation classification and mapping; forest fragmentation assessment; biomass estimation; population viability analyses; animal censusing; habitat sensitivity/vulnerability mapping and analysis; location and delimitation of protected areas; radio-tracking/foraging patterns and cost–distance analyses; and nesting pattern analysis.

In remote areas, where species and habitat information is limited, predictive modelling within a GIS can be a useful precursor to fieldwork (Lenton *et al.*, 2000), and integrating remotely sensed imagery can enhance such models. This is particularly useful in areas such as Central Africa where political instability and boundary conflicts may significantly influence fieldwork decisions: satellite sensors can safely gather data across national boundaries. The use of remotely sensed imagery is particularly appropriate if suitable maps are not available.

Reliable geo-referenced information is the basis of all spatial and synergistic analyses and is the essential prerequisite to understanding spatiotemporal relationships between any species and its environment. Critical to all applications is the need to collect field data that is located in geographical space. Establishing positional correspondence between features in the real world and a remotely sensed image is essential for calibration. In the absence of reliable maps, this is dependent on locating ground features using GPS technology.

Field and Laboratory Methods in Primatology: A Practical Guide, ed. Joanna M. Setchell and Deborah J. Curtis. Published by Cambridge University Press. © Cambridge University Press 2003.

Table 4.1. *Remote sensing and GIS applications of potential interest to primatologists: focus on vegetation and habitat mapping*

Application	Sensor	Comments	Reference
Lemur conservation	Landsat TM NOAA–AVHRR	Madagascar: integrative remote sensing and GIS to predict lemur species distributions	\<www171.pair.com/indri\>
Gorilla conservation	Landsat TM Landsat MSS (with radar)	Central Africa: use of fully integrated RS/GIS data to map and monitor rain forest across political borders	\<www.gis.com/trends/E\> \<RDAS.html\> \<www.informatics.org/gorilla/\>
Continental scale forest mapping and classified mosaics	JERS-1 SAR	South America and Africa: development of operational monitoring techniques using fine resolution (12.5–100 m) radar imagery for baseline land cover mapping	\<lcluc.gsfc.nasa.gov/products/ pdfs/PrgRep-Chapman2001.pdf\>
Landscape ecology and diversity	Landsat TM	Guatemala: investigation of potential of TM-based land cover maps with limited field data	Reybenayas & Pope, 1995
Biophysical properties of tropical forest	NOAA–AVHRR	Assessment of ground data requirements for regional-scale remote sensing of tropical forest biophysical properties	Atkinson et al., 2000
Tropical forest biomass	NOAA–AVHRR	Recommend use of middle infrared (MIR) for remote sensing of tropical forests	Boyd et al., 1999
Vegetation classification and LAI	Landsat TM	Prediction of LAI using RS and GIS data	Wulder, 1998
Biodiversity and lemur conservation	Landsat TM	Madagascar: satellite imagery and GIS	Smith et al., 1997

Application	Sensor	Description	Reference
Phenology	Landsat TM	Brazil: temporal analysis of foliage changes	Bohlman et al., 1998
Mapping rain forest	SPOT (HRV)	Congo: assessment of mapping accuracy using high resolution satellite data (results indicate high order of accuracy)	Thenkabail, 1999
Discrimination of vegetation and mapping	Landsat TM	Madagascar and Central Africa (Rwanda)	Reybenayas & Pope, 1995; Smith et al., 1997
Animal tracking	Review	Discusses the potential role of RS and GIS in biogeography	Gillespie, 2001
Mammals and habitat transformation	Landsat TM	Mexico: RS/GIS modelling of impact of landscape changes on mammal assemblages	Velazquez et al., 2001
Forest bioindicators	CASI	Infer physiological conditions from hyperspectral data	Sampson et al., 2000
Forest mapping and monitoring	JERS-1 SAR Interferometry (InSAR)	Overview (radar): tree-bole volume, biomass, canopy height	Balzter, 2001
Forest structure	ERASME (airborne radar)	French Guiana: biomass, tree density, basal area, height attributes	Riera et al., 1994
Phenologically based, regional-scale mapping	IRS WiFS	Himalayas: tropical forest correlation with ground data established	Joshi et al., 2001
Rain forest canopy structure	Overview	Cameroon example discussed	Bongers, 2001
Vegetation indices	AVIS	British Columbia forest: spectral and spatial sensitivity analysis	Teillet et al., 1997
LAI	Landsat TM EOS-MODIS	Temperate forests: vegetation indices	Turner et al., 1999

(cont.)

Table 4.1. *(cont.)*

Application	Sensor	Comments	Reference
Stand height estimation	Radar airborne scatterometer	France: very satisfactory results (relative error 6%)	Martinez et al., 2000
Forest biomass	JERS-1 SAR	Venezuelan pine plantation: relatively small errors achieved	Castel et al., 2002
Forest fragmentation	Landsat TM (with radar)	Bahian Atlantic forest, Brazil: potential for conservation and modelling	Saatchi et al., 2001
Vertical forest structure	LIDAR (airborne)	Overview: habitat mapping and forest wildlife management	Dubayah & Drake, 2000
Structural characteristics	LIDAR (airborne)	Costa Rica: biomass estimates in tropical wet forest	Drake et al., 2002
Canopy structure, height	SLICER (airborne)	Closed canopy broadleaved forests: methods and validation	Harding et al., 2001
Ecological modelling	Overview	Combining ecological process models and remotely sensed data	Plummer, 2000
Mapping stand-level canopy phenology	Landsat TM	Brazil: temporal changes in tropical forest	Bohlman et al., 1998
Forest condition and productivity	CASI (airborne)	Maple forest Ontario: variance in physiological condition	Sampson et al., 2001
Monitoring forest change	Multi-temporal JERS–SAR and ERS-2 SAR and SPOT HRV	Sumatra Island, Indonesia: deforestation. Radar and optical data	Takeuchi et al., 2000
Vegetation indices	AVIS and others	Forests, British Columbia: comparison of spectral, spatial and radiometric characteristics	Teillet et al., 1997

Forest mapping	NOAA–AVHRR	Brazil: Amazon regeneration (resolution 1.1 km)	Lucas et al., 2000
Monitoring deforestation, forest degradation and regrowth	AirSAR (airborne radar)	Land cover type classification accuracy >90% Biomass classification: 8 classes differentiated at a high level of confidence	Hoekman & Quinones, 2000
TREES (Tropical Forest Mapping Project)	NOAA–AVHRR and other sensors	South East Asia, Africa and South America: continental-scale forest mapping	\<www.gvm.sai.jrc.it/Forest/asia/ rs.sources.htm\>; \<esapub.esrin.esa.it/eoq/eoq48/ mali48.htm\>

Landsat TM, Landsat Thematic Mapper; NOAA-AVHRR, National Oceanic and Atmospheric Adminstration – Advanced Very High Resolution Radiometer; Landsat MSS; Landsat Multispectral Scanner; JERS-1 SAR: Japanese Earth Resources Satellite – Synthetic Aperture Radar; SPOT (HRV): Système Pour l'Observation de la Terre (High Resolution Visible) imaging system; InSAR, Interferometric Synthetic Aperture Radar; ERASME, Electronically Steerable Antenna Copolar Scatterometer; IRS WiFS, Indian Remote Sensing Satellite – Wide Field Sensor; AVIS, Airborne Visible Imaging Spectrometer; EOS-MODIS, Earth Observing System – Moderate Resolution Imaging Spectroradiometer; LIDAR, Light Detection and Ranging; SLICER, Scanning Lidar Imager of Canopies by Echo Recovery; CASI: Compact Airborne Spectrographic Imager; ERS-2 SAR, European Remote Sensing satellites, Synthetic Aperture Radar; AirSAR, Airborne Synthetic Aperture Radar; LAI, leaf area index.

This chapter introduces the issues that need to be considered before fieldwork decisions are made. Crucially, these are always determined by the research question asked and the data required. A word of advice: optimal use of GIS and remote sensing techniques depends on understanding the underlying concepts coupled with plenty of hands-on practice.

GLOBAL POSITIONING SYSTEM

The GPS is a satellite-based navigation system using a network of 24 satellites operated by the US Department of Defense. In theory, the GPS operates under all weather conditions anywhere in the world, 24 hours per day. However, reality does not always match the rhetoric! Each satellite transmits microwave signals that carry unique identification information and the exact time of transmission. A GPS receiver uses the time delay before signals reach the earth to calculate the distance travelled. The combined positional data from three or more satellites gives the geographical position on the earth's surface as three-dimensional (XYZ) positional and altitudinal information.

SIGNAL RECEPTION AND ACCURACY

Signal reception is essential to the performance of any GPS device. Since the intentional degradation (selective availability) of the signal was removed in May 2000 (the predictable horizontal accuracy was previously 100 m), the positional accuracy of GPS for civilian users has improved significantly. Ideally, a receiver should have at least five, and up to eight, of the satellites in 'line of sight', preferably at high attitudes to the user (minimum elevation of 15°, not positioned towards the horizon). The receiver must be locked on to signals from a minimum of three satellites for two-dimensional positioning (XY), and four satellites for altitude (Z).

DATA COLLECTION

In the field, in addition to the GPS device, you will need notepad and pencil, spare batteries, compass, measuring tape, camera, plus hardcopy geo-referenced satellite images (if available), aerial photographs, and maps. If the procedure for fixing a location cannot easily be repeated, it is advisable to write down the most stable co-ordinates – just in case. The compass and measuring tape (or pacing) can be used to locate small-scale features relative to major features with a known location or located using the GPS.

When using a simple, hand-held device you will need to select the appropriate geodetic datum in the navigation set-up page (see the manual). Hold the receiver as high as possible, oriented towards a clearing or less dense part of the canopy. Unfortunately for primatologists, GPS signals are substantially degraded by scattering (e.g. by dense tree canopies: Peterson, 1990; Phillips *et al.*, 1998; Dominy & Duncan, 2001) and signals do not travel through solid objects (e.g. mountains). If you cannot get a signal, move until you can, then measure back to where you were. Researchers using a standard, low cost GPS device, and working in a forest clearing with good signal reception would, in theory, expect to get positional accuracy down to ±10 m. Accuracy can be increased with more costly, differential GPS (DGPS) devices and a reference receiver at a known position (base station), which can provide corrections and relative positioning data for the remote, or rover, receiver. Researchers working in Uganda using a 12 channel backpack GPS with satellite-based differential correction found that an antenna positioned 25–30 m above ground level (by climbing trees) provided acceptable positional fixes on individual forest trees. Very accurate locations (±1 m) were obtained once further differential corrections had been made to compensate for signal delay errors due to interference within the atmosphere (Dominy & Duncan, 2001). However, climbing trees is dangerous and not recommended. An antenna fixed to a telescopic pole is an alternative solution with more expensive systems, but height factors and stability may be a problem. A comparison of GPS receivers used under a forest canopy can be found at <www.fs.us/database/gps/mtdcgps2000/gps_comparison.htm>.

Readings may take some time to stabilise – if they do at all. There is no standard procedure: the British Columbia Ministry of Forests recommends that the minimum occupation time for a standard static point feature should be 30 seconds, with signals collected from four satellites (MOF website). Only you can decide what is appropriate for your purposes, equipment, study site and particular environmental conditions. Data from your GPS unit can be downloaded directly into a personal computer installed with the appropriate software, either in the field or later at base.

GPS ANIMAL TRACKING

Collars equipped with GPS transmitters or receivers can be used to satellite-track animals (Chapter 10).

GEOGRAPHICAL INFORMATION SYSTEMS

A GIS is a database with a sense of space, enabling you to map, explore, analyse and model data in different ways through the integration, interrogation and manipulation of multiple layers of information. At the most basic level, a GIS is an artificial representation of the real world, but not solely as a flat-earth two-dimensional map. Three-dimensional representation is a standard function in GIS so it is possible to configure ecologically significant features, such as canopy structures and vegetation patterns (Lucas et al., 2001). There are, of course, constraints in representational transformations. Essentially, spatial (areal) analysis and modelling within a GIS involves 'forcing' data into arbitrary spaces (entities) whose boundaries do not necessarily coincide with real world boundaries. In raster mode (as in remotely sensed imagery) the 'spaces' are grid cells; in vector mode the 'spaces' are polygons, and landscapes are represented as points, lines and areas. Researchers investigating the home range size of black howler monkeys (*Alouatta pigra*) found that there was no significant difference in the accuracy of the grid versus the polygon method, but that the latter may give the most accurate estimates (Ostro *et al.*, 1999).

Within a GIS, all things, including an animal's home range and all lacunae, are essentially features defined in terms of coherence or homogeneity, edges (boundaries), location and measurable attributes. The process of representing these features in two-dimensional space (entitation) can be problematic as it depends on recognising and defining boundaries – which may be fuzzy both through space and time. Furthermore, a primate is unlikely to perceive boundaries in the same way as an observer! Any definition will depend on an individual's perspective, chosen attributes or criteria, and will be dependent on scale.

PREPARATION

Field data collection must be 'fit for the purpose'. Data should be collected in a tabulated or graphical form that can later be transformed for use in the GIS. Before venturing into the field familiarise yourself with the GIS software you intend to use, run a test project, take a course, consult manuals, work through tutorials and, most importantly, talk to people who have used the system. Preparatory work might also include classifying habitat or vegetation using geo-referenced remotely sensed imagery. Copies of such thematic maps can be used in the field and later co-registered with multiple layers of field data. As spatial correspondence is essential, different Cartesian co-ordinate systems must be transformed to a common system within the GIS (see the manuals). You may also need to prepare other mapped information by scanning or digitising.

Mapping and GIS software packages are widely available. These range from the expensive (e.g. ArcInfo® and ERDAS Imagine®), to the less costly (e.g. ArcView® and Idrisi®), to free or free-trial packages. A comprehensive listing of free packages can be found on the Internet. Some of these are just for viewing maps/images and overlaying and are not fully functioning GIS but might be suitable for your needs. Cost will be a determining factor but, before buying, talk to people who have used the systems. GRASS® is a very popular and versatile raster (grid)-based system and is free. You will also need as much computer power and memory as you can afford (or access), a colour printer, a high resolution scanner and possibly a digitising tablet.

IN THE FIELD

Ideally, as all data for use in a GIS must be coded by geographical location, you will need a reliable base map or suitably geo-referenced image. Features of interest can be marked and a corresponding GPS fix obtained using a transparent grid overlay (registered to control points and/or known co-ordinates). The grid scale will be determined by the resolution required. If maps/images at a suitable scale are unavailable, then a sketch map based around known (located) control points will be indispensable. If GPS signal reception is blocked, then maps can be made 'by eye', using a compass and pacing or measuring distances from a located feature. In unmapped areas, establish a network of control points located using a GPS and use this as a base map. All features, attributes and sample sites should also be recorded in tabular format and cross-referenced to the GPS data: remember, everything must be located in absolute or relative geographical space.

Field data can be collected automatically using integrated GIS/GPS technology. Portable computers allow rapid capture of information and the ability to map features while in the field. An example of this is Cybertracker (<www.cybertracker.org>; <www.cybertrackerworld.com>), a hand-held device that allows you to record and construct a database of more than 300 observations per day. An optional, integral GPS automatically records the location of the observation, if signals are available. The data can be downloaded to a computer using a number of different formats, simple GIS functions can be performed and data can be exported to Excel® or ArcView® for further analysis.

BACK AT BASE

Data input, checking and editing in a GIS can be very time-consuming. Downloading GPS data to your personal computer requires specific software. Depending on how your field data are recorded, you are likely to have to edit them before importing them into a GIS. If necessary, data can be assigned to new values using an appropriate weighting method or classification scheme. Again, spatial correspondence of all data is fundamental and all the different layers of information must be co-registered in a common co-ordinate system.

INTERROGATION AND DATA ANALYSIS

A quick trawl through GIS applications on the Internet will give you an insight into the tremendous potential and creativity of a GIS for thematic mapping, three-dimensional modelling, interrogation and query–response mapping, derived parameter mapping, tabulation and statistical analysis. Furthermore, the functionality of GIS is still being improved. You are strongly advised to explore, experiment, take risks and think creatively about how you might exploit the visual and analytical opportunities of the system.

REMOTE SENSING

Remote sensing involves the gathering of electromagnetic energy or 'signals' associated with highly complex object–energy interactions (Wilkie & Finn,

1996; Curran, 2001). Optical and non-optical (e.g. radar) remote sensing devices, using visible, infrared and microwave energy, and carried both on space-borne (satellite) and airborne (aircraft) platforms, can provide environmental data unobtainable by other means (Freeman & Fox, 1994; Smith et al., 1997). Considerable manipulation of the digital data is needed to obtain information about environmental variables, and the effective application of remotely sensed data depends on their integration with other environmental information, most critically ground data and positional data acquired through GPS.

Remotely sensed imagery has a vital role to play in inaccessible and unmapped or poorly mapped areas. Tonal and textural differentiation, using visible and infrared imagery or aerial photographs (digital versions), is particularly useful and can be correlated with vegetation type and patterns in the field. Hardcopy versions of satellite and/or airborne imagery are useful for general differentiation of canopy features, preliminary mapping of distinctive clusters of trees and other vegetation, and detecting gaps and corridors. These can be updated in the field in real time: land cover changes can be noted and marked, routes can be sketched-in to give immediate visual feedback and, where possible, key locations fixed or tracked using the GPS. Transparent, grid overlays are extremely valuable for this type of observational work.

RADAR

Radar (RAdio Detection And Ranging) is the principal form of active remote sensing. It operates within the microwave part of the electromagnetic spectrum and is a powerful technique with many applications (Table 4.1). A major advantage of radar is that the atmosphere is transparent to very long wavelengths of microwave energy: in other words, clouds are not a problem. It is particularly useful for characterising forest canopies (Table 4.1).

SATELLITE IMAGERY

Satellite images can be purchased from commercial companies and distributors or from government agencies (a list of suppliers can be found on the Internet). Digital and hardcopy versions will be available. Useful data are also freely available on the Internet. The performance of multi-spectral passive sensors on board established earth observation platforms such as the Landsat Thematic Mapper (TM) and Système Pour l'Observation de la Terre (SPOT) is reduced significantly by cloud cover owing to the attenuation of electromagnetic energy. This represents a serious, but by no means insurmountable, impediment to effective passive remote sensing, particularly in the tropics.

The Tropical Rain Forest Information Centre (TRFIC) is an excellent source of data and metadata. The TRFIC is a NASA Earth Science Information

Partner (ESIP) that provides high resolution optical and radar data, digital deforestation maps and various databases via the Internet (<http://bsrsi.msu.edu/trfic>), and much of the data is easily accessed at no cost to the user. Under the TRFIC co-operative, users can also co-purchase high quality, low cloud cover Landsat TM scenes. The Vegetation Canopy Lidar (VCL) data will be of further interest. This mission (part of NASA's Earth System Science Pathfinder Program) aims to provide global maps showing the three-dimensional structure of forests using contiguous 25 m footprints every 2 km along the earth's surface (Drake *et al.*, 2002), and various vegetation measurements at high and medium resolutions (ESSP, 2001).

The acquisition of higher resolution data from airborne hyperspectral sensors will normally be very expensive, particularly if planned to coincide with your fieldwork. However, the potential use of this data for detailed vegetation studies is extremely exciting (Harding *et al.*, 2001).

AERIAL PHOTOGRAPHY

Aerial photographs, black and white, colour and sometimes infrared, are often available from national cartographic agencies or archives: stereo-pairs of photographs are especially useful.

CHOICE OF IMAGERY

Spatial resolution, spectral properties and repeat time will vary enormously depending on the system and sensor. Subscenes can be purchased for smaller areas; multi-spectral (e.g. visible red and infrared) data are usually purchased separately; high resolution images are more expensive; low resolution, synoptic (large area) images are often freely available from the Internet. You need to know what you need! A browse through the Internet is highly recommended, as the range of imagery now available is huge.

Spatial and temporal resolution

Spatial resolution depends on the sensing device and the platform's orbiting or flying height: the higher the resolution, the greater the amount of detectable detail. Sensors, and resolutions, appropriate for vegetation studies are indicated in Table 4.1. If you are interested in change phenomena, then temporal resolution is critical. Daily, weekly, monthly and yearly data are available but not always at the spatial resolution required. If individual tree species are relevant to your research, then you may choose airborne remote sensing systems, or possibly spaceborne photography. If forest fragmentation is important, high–medium resolution satellite data may be appropriate. If the overall ecological context or landscape is relevant, then coarse resolution data may be ideal. Satellite data is likely to be highly appropriate and cost effective for change detection; the costs of repeat usage of an airborne system would be prohibitive.

Spectral resolution

This is entirely dependent on the aspects of the environment in which you are interested: the visible red and near-infrared spectral bands are especially valuable for vegetation studies and are used to derive vegetation indices. Image processing and digital fusion techniques provide opportunities for making optimal use of data from multiple spectral bands acquired from different sensors at different spatial resolutions. Once data has been co-registered it can be displayed simultaneously. Such synergistic displays, which may include both optical and non-optical data, allow improved discrimination of features of interest.

DIGITAL TERRAIN (OR ELEVATION) MODELS (DTM/DEM)

These are excellent for landscape visualisation when used in conjunction with draped remotely sensed imagery, and for deriving attribute data about slope or shadowing. Global topographical data are freely available through the Internet but the trade-off is the coarse resolution (≥ 1 km). A three-dimensional 'virtual world' is extremely valuable for spatial analysis and may play a useful, as yet unexplored, role in addressing questions relating to primate perception and preferences.

IMAGE PREPARATION

Geometric rectification of remotely sensed imagery is essential prior to undertaking any image interpretation or using imagery in the field. These procedures are normally dealt with before the user receives the data but this should always be verified.

Preliminary analysis of a digital image is usually based on unsupervised or supervised classification techniques, functions available in most image-processing packages. The unsupervised method is based on the spectral characteristics of the scene; objects with similar spectral responses are categorised as the same object. Supervised classification depends on the use of *a priori* knowledge – either from additional resources (land cover maps, aerial photographs, archived descriptions and documentation) or data collected in the field. This additional information is used to 'train' the data such that unknown spectral data is allocated to known and defined clusters. As a rule of thumb, the more bands (different spectral images of the same scene) that are used, the more confident you can be about the relative accuracy of the classification. Many classification techniques exist, but results can never be 100% accurate. Basic mapping and classification can be achieved by eye using textural and tonal differences.

VERIFICATION IN THE FIELD

'The truth, the ground truth and nothing but the ground truth': field measurements are essential in all applications. Without ground information you can detect but not identify features. Habitat attributes must be observed and mapped, and salient features located as accurately as possible, ideally using a GPS device: a fundamental objective is to find features in the field that correspond to features in the image. The spatial resolution of the imagery will determine the accuracy required for ground data: high resolution imagery requires more accurate positional information. Acceptable errors are likely to be: ≤20 m for SPOT Multispectral Scanner data (MSS); ≤30 m for Landsat TM data (60–120 m for Band 6 thermal); and ≤80 m for Landsat MSS data.

Ideally, imagery should be calibrated by means of field spectroscopy or field radiometry. Milton (2001) provides excellent coverage of the principles of spectroscopy and details of instruments and methodology. Field spectroscopy of the canopy is unrealistic on anything but a very local scale and where a specific test site(s) has been set up using a tower or 'roaming' crane. Results from such test sites have demonstrated the value of spectral data for closed canopy ecological investigations (Blackburn & Milton, 1995, 1996) and the Smithsonian Tropical Research Institute's canopy crane in Panama provides excellent opportunities for very close contact with the upper canopy (Lowman, 2001). Open canopy areas can easily be investigated using a portable multiband radiometer (with GPS data) and the data used for later classification training of the imagery.

MAPS

If maps are available, correspondence between mapped features and those that can be detected in the imagery must be established using ground control points (GCPs) and a GPS device. Useful GCPs include permanent, or at least persistent, features such as: tracks/road junctions; track intersections with streams; small, tightly clustered settlements or individual buildings; and prominent rocky outcrops. Within a GIS, image features can be tied-in to such control points, providing an additional layer of essential information in the field.

CONCLUSIONS

One of the most important messages of this chapter is the need to dissolve the boundaries that isolate scientists within their different disciplines. Collaborative, integrative research involving primatologists, remote sensing scientists, ecologists, biogeographers and modelling (GIS) experts is needed to benefit from the opportunities afforded by current and emerging technologies. Reliable field data are essential for validating models, and remotely sensed data and fieldwork must be designed with these aims in mind. The spatial, spectral and temporal advantages of remotely sensed data can and should be employed synergistically with traditional, finer resolution ground data within a GIS environment, supported by accurate positional information. Computer-based visualisation techniques, or virtual worlds, have a major role to play in disseminating and communicating scientific information to the wider public, together with encouraging and supporting community participation in conservation strategies.

REFERENCES

Atkinson, P., Foody, G. M., Curran, P. J. & Boyd, D. (2000). Assessing the ground data requirements for regional-scale remote sensing of tropical forest biophysical properties. *Int. J. Remote Sen.* **21**, 2571–87.

Balzter, H. (2001). Forest mapping and monitoring with interferometric synthetic aperture radar (InSAR). *Progr. Phys. Geog.* **25**, 159–77.

Blackburn, G. A. & Milton, E. J. (1995). Ecology through spectrometry: the detection and classification of canopy gaps in deciduous woodlands. In *TERRA-2 Understanding the Terrestrial Environment: Remote Sensing Data Systems and Networks*, ed. P. M. Mather, pp. 175–87. Chichester: Wiley.

(1996). Filling the gaps: remote sensing meets woodland ecology. *Global Ecol. Biogeogr. Lett.* **5**, 175–91.

Bohlman, S. A., Adams, J. B., Smith, R. O. & Peterson, D. L. (1998). Seasonal foliage changes in the eastern Amazon Basin detected from Landsat Thematic Mapper satellite images. *Biotropica* **30**, 376–91.

Bongers, F. (2001). Methods to assess tropical rain forest canopy structure: an overview. *Plant Ecol.* **153**, 263–77.

Boyd, D. S., Foody, G. M. & Curran, P. J. (1999). The relationship between the biomass of Cameroonian tropical forests and radiation reflected in middle infrared wavelengths (3.0–5.0 μm). *Int. J. Remote Sens.* **20**, 1017–23.

Castel, T., Guerra, F., Caraglio, Y. & Houllier, F. (2002). Retrieval biomass of a large Venezuelan pine plantation using JERS-1 SAR data. Analysis of forest structure impact on radar signature. *Remote Sens. Environ.* **79**, 30–41.

Curran, P. J. (2001). Remote sensing: using the spatial domain. *Environ. Ecol. Stat.* **8**, 331–44.

Dale, M. R. T. (1999). *Spatial Patterns Analysis in Plant Ecology*. Cambridge: Cambridge University Press.

Dominy, N. J. & Duncan, B. (2001). GPS and GIS methods in an African rain forest: applications to tropical ecology and conservation. *Conserv. Ecol.* **5** [online] URL: <www.consecol.org/vol5/iss2/art6>.

Drake, J. B., Dubayah, R. O., Clark, B., Knox, R. G., Blair, J. B., Hofton, M. A., Chazdon, R. L., Weishampel, J. F. & Prince, S. D. (2002). Estimation of tropical forest structural characteristics using large-footprint LIDAR, *Remote Sens. Environ.* **79**, 305–19.

Du Puy, D.J. & Moat, J.F. (1998). Vegetation mapping and classification in Madagascar (using GIS): implications and recommendations for the conservation of biodiversity. In *The Ecology, Chronology and Taxonomy of the African and Madagascan Floras*, ed. D.F. Cutler, C.R. Huxley & J.M. Lock, pp. 97–117. Proceedings of the Frank White Memorial Symposium, Kew Bulletin Additional Series. Kew: Royal Botanic Gardens.

Dubayah, R.O. & Drake, J.B. (2000). Lidar remote sensing for forestry. *J. Forest.* **98**, 44–6.

ESSP (2001). VCL: The vegetation canopy Lidar mission. <http://essp.gsfc.nasa.gov/vcl/>.

Freeman, P.H. & Fox, R. (1994). *Satellite Mapping of Tropical Forest Cover and Deforestation: A Review with Recommendations for USAID*. Arlington, VA: Environment and Natural Resources Information Center, DATEX.

Gillespie, T.W. (2001). Remote sensing of animals. *Progr. Phys. Geog.* **25**, 355–62.

Harding, D.J., Lefsky, M.A., Parker, G.G. & Blair, J.B. (2001). Laser altimeter canopy height profiles: methods and validation for closed-canopy broadleaf forests. *Remote Sens. Environ.* **76**, 283–97.

Hoekman, D.H. & Quinones, M.J. (2000). Land cover type and biomass classification using AirSAR data for evaluation of monitoring scenarios in the Colombian Amazon. *IEEE Trans. Geosci. Remote* **38**, 685–96.

Joshi, P.K., Singh, S., Agarwal, S. & Roy, P.S. (2001). Forest cover assessment in western Himalayas, Himachal Pradesh using IRS 1C/1D WiFS data. *Curr. Sci. India* **80**, 941–7.

Lenton, S.M., Fa, J.E. & Del Val, J.P. (2000). A simple non-parametric GIS model for predicting species distribution: endemic birds in Bioko Island, West Africa. *Biodivers. and Conserv.* **9**, 869–85.

Lowman, M.D. (2001). Plants in the forest canopy: some reflections on current research and future direction. *Plant Ecol.* **153**, 39–50.

Lucas, P.W., Beta, T., Darvell, B.W., Dominy, N.J., Essackjee, H.C., Lee, P.K.D., Osorio, D., Yamashita, N. & Yuen, T.D.B. (2001). Field kit to characterize physical, chemical, and spatial aspects of potential primate foods. *Folia Primatol.* **72**, 11–25.

Lucas, R.M., Honzak, M., Curran, P.J., Foody, G.M., Milne, R., Brown, T. & Amaral, S. (2000). Mapping the regional extent of tropical forest regeneration stages in the Brazilian Legal Amazon using NOAA AVHRR data. *Int. J. Remote Sens.* **21**, 2855–81.

Martinez, J.M., Beaudoin, A., Durand, P., Toan, T. &. Stach, N. (2000). Airborne radar scatterometer for estimation of stand heights. *Can. J. Forest Res.* **30**, 1983–99.

Milton, E.J. (2001). *Methods in Field Spectroscopy*. <http://www.soton.ac.uk/~epfs/methods/spectroscopy.shtml>.

Ostro, L.E.T., Young, T.P., Siver, S.C. &. Koontz, F.W. (1999). A geographic information system method for estimating home range size. *J. Wildlife. Man.* **63**, 748–55.

Peterson, C. (1990). Into the woods with GPS. *GPS World* **1**, 31–6.

Phillips, K. A., Elvey, C. R. & Abercrombie, C. L. (1998). Applying GPS to the study of primate ecology: a useful tool? *Am. J. Primatol.* **46**, 167–72.

Plummer, S. E. (2000). Perspectives on combining ecological process models and remotely sensed data. *Ecol. Model.* **129**, 169–86.

Reybenayas, J. M. & Pope, K. O. (1995). Landscape ecology and diversity patterns in the seasonal tropics from Landsat TM imagery. *Ecol. Appl.* **5**, 386–94.

Riera, B., Philippe, M., Dosso, M., Bernard, R., Vidalmadjar, D. & Dechambre, M. (1994). Study of the ground vegetation system in the Guiana tropical rain-forest with scatterometer in a nadir looking mode. *Rev. Ecol. Terre Vie* **49**, 357–77.

Saatchi, S., Agosti, D., Alger, K., Delabie, J. & Musinsky, J. (2001). Examining fragmentation and loss of primary forest in the southern Bahian Atlantic forest of Brazil with radar imagery. *Conserv. Biol.* **15**, 867–75.

Sampson, P. H., Mohammed, G. H., Zarco-Tejada, P. J., Miller, J. R., Noland, T. L., Irving, D., Treitz, P. M., Colombo, S. J. & Freemantle, J. (2000). The bioindicators of forest condition project: a physiological, remote sensing approach. *Forestry Chron.* **76**, 941–52.

Sampson, P. H., Treitz, P. M. & Mohammed, G. H. (2001). Remote sensing of forest condition in tolerant hardwoods: an examination of spatial scale, structure and function. *Can. J. Remote Sens.* **27**, 232–46.

Smith, A. P., Horning, N. & Moore, D. (1997). Regional biodiversity planning and lemur conservation in western Madagascar. *Conserv. Biol.* **11**, 498–512.

Takeuchi, S., Suga, Y., Oguro, Y. & Konishi, T. (2000). Monitoring of new plantation development in tropical rain forests using JERS-1 SAR data. *Adv. Space Res.* **26**, 1151–4.

Teillet, P., Dtaenz, M. K. & Williams, D. J. (1997). Effects of spectral, spatial and radiometric characteristics on remote sensing vegetation indices of forested regions. *Remote Sens. Environ.* **61**, 139–49.

Thenkabail, P. S. (1999). Characterization of the alternative to slash-and-burn benchmark research area representing the Congolese rainforests of Africa using SPOT HRV data. *Int. J. Remote Sens.* **20**, 839–77.

Turner, D. P., Cohen, W. B., Kennedy, R. E., Fassnacht, K. S. & Briggs, J. M. (1999). Relationship between leaf area index and Landsat TM spectral vegetation indices across three temperate zone sites. *Remote Sens. Environ.* **70**, 52–68.

Velazquez, A., Romero, F. J., Rangel-Cordero, H. & Heil, G. W. (2001). Effects of landscape changes on mammalian assemblages at Izta-Popo volcanoes, Mexico. *Biodivers. Conserv.* **10**, 1059–75.

Wadsworth, R. & Treweek, J. (1999). *GIS for Ecology: An Introduction* Harlow: Longman Ltd.

Wilkie, D. S. & Finn, J. Y. (1996). *Remote Sensing Imagery for Natural Resources Monitoring: A Guide for First-Time Users*. New York: Columbia University Press.

Wulder, M. A. (1998). The prediction of leaf area index from forest polygons decomposed through the integration of remote sensing, GIS, UNIX and C. *Comput. Geosci.* **24**, 151–7.

Useful Internet sites

There is a wealth of exceptionally useful information covering all aspects of GPS, GIS and remote sensing on the Internet. It would be easy to list dozens of websites here but omission of others might suggest that they are of less value. If you start your search by accessing the following, they will lead you into current and regularly updated sources of information and key addresses:

<www.geog.nottingham.ac.uk/~mather/useful_links.html> [RS and GIS Links].
<www.geog.le.ac.uk/cti/gis.html> [GIS tutorial information and data sources].
<www.geo.ed.ac.uk/home/gishome.html> [GIS WWW Server].
<http://rst.gsfc.nasa.giv/> and other NASA sites [Excellent RS coverage].
<www.vtt.fi/tte/research/tte1/tte14/virtual/> [RS: Virtual Library].
<www.infoterra-global.com/satellite_data.html> [Useful source of information for satellite data].
<srmwww.gov.bc.ca/risc/pubs/teveg/opfield/assets/gps.pdf> [MOF website]
See also others mentioned in the text.

5 • Monitoring local weather and climate

JULIAN MAYES

Centre for Research in Ecology and the Environment, School of Life and Sport Sciences, University of Surrey Roehampton, London, UK

INTRODUCTION

Weather and climate have a profound influence on ecosystems. The climate of a region can act as a resource for an ecosystem as a whole, influencing the phenology of an area (Chapter 3), and as a determinant of food supply for primates, which can in turn affect reproduction, ranging and social interaction. While climate provides the background conditions, the real-time monitoring of local weather can provide a much more direct link to primate field studies at any given time. For example, daily temperature cycles can influence activity patterns (Chapter 17).

Local weather information can be straightforward to collect and the value of local observations at field study sites (as opposed to extrapolation of conditions from the nearest 'official' weather station, which may be far away) often offsets any deficiencies in observation technique. Weather conditions also vary within field study area as a result of the microclimate (e.g. Rosenberg *et al.*, 1983; Geiger *et al.*, 1995) and, as a result, local conditions can differ significantly from larger scale climatological surveys such as Buckle (1996) and McGregor & Nieuwolt (1998).

This chapter describes the types of local weather information that can be collected; shows how weather can vary over both space and time, enabling the interpretation of point observations in the context of wider conditions; and finally shows how local atmospheric conditions can easily be measured during fieldwork using different types of portable instrumentation.

Key terms for the main subjects covered in this chapter are:

Weather: the state of the atmosphere at a given place and time.
Climate: the aggregation of weather conditions over several decades (typically at least 30 years) comprising both averages and extremes of particular climatic parameters (temperature, rainfall, etc.).
Meteorology: the science of the atmosphere, concerned with explaining variations in weather in both horizontal and vertical planes (the latter extending from the soil to the upper atmosphere).

Field and Laboratory Methods in Primatology: A Practical Guide, ed. Joanna M. Setchell and Deborah J. Curtis. Published by Cambridge University Press. © Cambridge University Press 2003.

Microclimate: the variation of weather and climate over small areas – from tens of metres up to a few hundred.

Meteorological variation can be investigated at two distinct scales: local and regional. The local scale corresponds to variations in microclimate within a small study area, key influences being the type of land cover and the nature of the soil and ground surface. Most meteorological elements change abruptly between the surface and the first few metres of the atmosphere, within a zone easily accessed by human observers. While primates may often live above this level, the near-surface conditions may influence food supply potential, especially in forested environments in which microclimate is heavily modified by such factors as light attenuation. At the regional scale, variations in altitude exert a greater influence on local weather conditions. Measurement of weather variables therefore allows the findings from a specific study area to be placed within the context of a larger region or to compare different sites. The climatology of a site is expressed in terms of climatic averages for a given month, usually derived from daily measurements. It is important that these data are published in a consistent form throughout the world; standard terminology for the main climatic variables are defined as follows:

Mean daily maximum (minimum) temperature: the average of the highest (lowest) temperatures recorded over each 24 hour period within a specified period (e.g. one month).

Mean temperature: this is usually simply the average of the mean daily maximum and minimum temperature; with automatic weather stations it is more likely to be the mean of the hourly temperatures. It is often misinterpreted as referring to day temperature. Though frequently quoted in published data, the concept of mean temperature has less climatic relevance than mean daily maximum and minimum. For example, it is the minimum temperature that stimulates flowering of some species whose fruits are eaten by primates (Edwards & White, 2000).

Mean monthly precipitation: the sum of the daily precipitation totals over a month.

Sunshine: usually measured in terms of duration of bright sunshine in hours.

Solar radiation: typically expressed as a daily mean in watts per square metre. Because the attenuation of light and solar radiation is such an important facet of habitat modification in forest microclimates, measuring solar and net radiation can provide highly relevant data.

Humidity: the actual estimated water vapour content of the air, expressed as a proportion of the maximum amount that air can hold at that temperature. The latter is temperature dependent – as air cools, its capacity to hold water vapour diminishes (owing to changes in molecular bonding). Condensation of water droplets can occur when air approaches saturation (this can start at slightly below 100% relative humidity). If the relative humidity has been measured, it is possible to calculate the temperature at

Table 5.1. *Conversion factors between units of wind speed*

Unit	Conversion
1 m/s	= 2.237 miles per hour
	= 3.6 km/h
	= 1.94 knots (nautical miles per hour)
1 knot	= 1.15 miles per hour

which saturation will take place – the dewpoint temperature. This is the temperature below which dew, fog or low cloud may start to form. The high humidity of tropical rain forests means that dewpoint temperatures often reach 20 °C. Dewpoint temperature can be estimated from one of the following formulae:

when relative humidity >85%: $T_d = T_{air} - 0.133 \ (100 - RH)$
when relative humidity <85%: $T_d = T_{air} - 0.2 \ (95 - RH)$

where T_d is the dewpoint temperature, T_{air} is the air temperature, and RH is the percentage relative humidity.

Airflow: Wind direction is expressed as the direction from which the wind is blowing to the nearest 10° from north. Table 5.1 shows the relationship among the commonest units of wind speed. Scientific surveys frequently use metres per second but, for example, in the British Isles wind speed is officially recorded in knots. Take care when calculating average wind direction over several observations, particularly if winds blow from a northerly direction (N should be logged as 36, NNE as 02 etc.; it would be clearly erroneous to average 36 and 02 over successive sampling intervals and assume the mean direction was 190°).

Wind speed is usually measured over at least 15 seconds, in order to record a range of gusts and lulls. It is usual to note both the mean over this period and the maximum gust. The gust : lull ratio may also be quoted, especially in the turbulent wake downwind of obstacles to the airflow (such as a stand of trees or buildings) where the contrast between gusts and lulls typically increases. Similarly, a 'gust factor' expresses the 3 second maximum gust as a proportion of the mean speed (Linacre, 1992).

Cloud cover: Cloud cover is estimated by eye as an estimate of the number of eighths (oktas) of sky that are covered with cloud.

Climatic seasonality through the year is often depicted as a diagram (a climatogram) in which monthly temperature and precipitation are plotted on separate *y*-axes, with 10 degrees of temperature change against 20 mm rainfall change. Periods of drought ('dry season') are implied when rainfall falls below the day temperature (Brower *et al.*, 1989). Wet conditions occur where rainfall is >100 mm, regardless of the temperature (Edwards & White, 2000). However, this is only an approximate indication of water balance, since it takes no account of wind speed and solar radiation. Evaporation in millimetres per day from a wet vegetated surface can be estimated by the following formula (Linacre, 1992):

$$E = [0.015 + (4 \times 10^{-4}T) + 10^{-6}z] \times [380(T + 0.006z)/(84 - A)$$

$$- 40 + 4u(T - T_d)] \tag{5.1}$$

where T is the dry bulb air temperature; T_d is the dewpoint temperature; A is the latitude; z is the altitude in metres; u is the wind speed in metres per second.

METEOROLOGICAL VARIATIONS AT THE LOCAL SCALE

Microclimate variation occurs according to (among others) time of day, the effects of vegetation, altitude and topography. It can be investigated in terms of temperature, a variable that has an intricate relationship with the surroundings, depending on the nature of the local terrain and the role of land cover, radiation, airflow, and humidity.

SOME DEFINITIONS

Active surface

The active surface is the principal plane of meteorological activity in a system, where the majority of the solar energy is absorbed, reflected and emitted. It is where radiant energy is transformed into thermal energy (through absorption), where rainfall is intercepted and where drag on airflow is maximised. In the absence of vegetation, the active surface will be at the ground. Figure 5.1 shows the typical temperature profiles above and below the active surface of bare ground. It is the focal point of heat loss at night, since it is where radiation of long wave radiation from the earth's surface takes place. The concept of the 'active surface' identifies the energy transformations that create an area's microclimate. Most field surveys of temperature are carried out near the 'active surface', an area of sharp variations in temperature and other meteorological elements.

Albedo

The active surface warms as it absorbs a proportion of the solar radiation. While cloud cover determines the amount of direct solar radiation, the proportion of incoming radiation reflected by the surface also influences the degree of warming; this reflectance is expressed as the albedo of the surface.

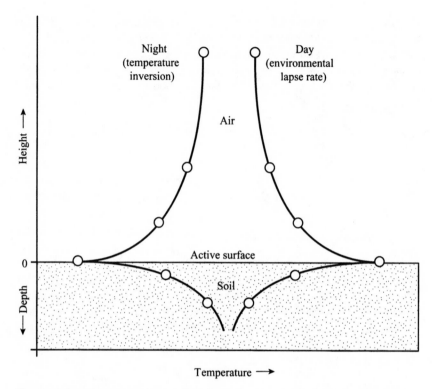

Fig. 5.1. Diurnal temperature profiles adjacent to an active surface (assuming light winds and clear skies).

Environmental lapse rate

The environmental lapse rate (ELR) is the drop in temperature that occurs as height increases. The average lapse rate in the global atmosphere is around 0.65 °C/100 m, with some variation by altitude and latitude (Linacre, 1992). It is greater when the surface is being heated by sunshine. At night, unless winds and cloud cover are high, temperature often rises with increasing height (a temperature inversion, or negative lapse rate), as shown in Fig. 5.1.

Net radiation

Heat energy is transferred through the atmosphere by radiation and is gained when a surface absorbs solar radiation. However, all surfaces emit radiation according to the surface temperature; the higher the temperature, the shorter the wavelength of radiation. Solar radiation is therefore termed short-wave radiation and that emitted by the surface of the earth is long-wave radiation. The balance between short-wave and long-wave radiation is termed net radiation, where positive values denote an excess of short-wave radiation (gain) over long-wave radiation (loss). The former promotes warming at the 'active surface' and the latter cooling. Only about 26% of the total short-wave radiation entering the atmosphere reaches the ground as direct solar radiation; the rest is diffused through cloud cover or scattered downwards (Robinson & Henderson-Sellers, 1999; Stoutjesdijk & Barkman, 1992).

DIURNAL VARIATIONS IN TEMPERATURE

Time of day is a fundamental factor in temperature surveys, it is the principal control on the balance of incoming and outgoing radiation, and hence on temperature at the active surface. In tropical environments diurnal energy exchanges are greater than seasonal changes (Oliver & Hidore, 2002). Net radiation remains negative for up to 30 minutes after sunrise because at low sun angles incoming solar radiation is insufficient to offset the continuing long-wave radiation. The result is that the active surface continues to cool. Net radiation reaches a peak on a cloud-free day towards midday in response to the elevation of the sun. Lapse rates in sunny conditions may rise well above average levels at this time; $1°C/100$ m is not unusual. Surface temperature will peak after a short delay; there is a further delay before the temperature peak is experienced further away from the active surface, such as at the height at which air temperature is officially measured, 1.25 m (Fig. 5.2).

VEGETATION EFFECTS

Vegetation raises the height of the active surface, typically to around two-thirds of the height of the vegetation or towards the tree canopy in the case of forest. This has a major effect on the vertical profile of temperature, wind speed and moisture above (and below) the ground. The principles of this modification apply to most vegetation types and environments.

Influence upon temperature and radiation balance
The obvious shading effect of vegetation arising from the vertical attenuation of solar radiation often results in a temperature inversion below the height of the active surface. As the active surface is approached, highest day temperatures tend to be recorded at around the height of the highest leaf density. The proportion of short-wave radiation reaching the ground through woodland is often less than 20% of that above the canopy. At the ground, the diurnal range in temperature is reduced and this dampening of the temperature extremes is carried down into the soil.

Night-time temperatures are elevated under a vegetation canopy, owing to reduced emission of long-wave radiation. This effect can be quantified by means of the sky view factor (SVF), where a completely unobstructed sky hemisphere is said to have a SVF of 1.0 and a site having a completely obscured sky has a SVF of 0.0. The lapse rates above the canopy can be reduced by evaporative cooling at the top of the vegetation (Oke, 1987).

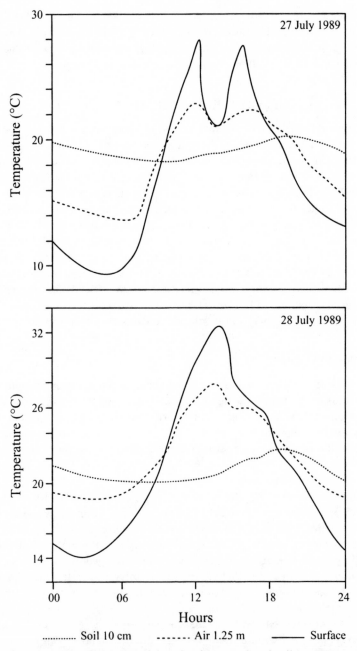

Fig. 5.2. Typical diurnal variations in air, ground and soil temperature measured by simple sensors on fine days during the UK summer (27 & 28 July 1989). Note the larger temperature range at the surface compared with 1.25 m and the sharp reduction of surface temperature during a cloudy period just after midday 27 July.

Albedo varies with both vegetation density and time of day. The albedo of an individual leaf is typically around 0.30, but a whole stand of crops or trees will generally have a value closer to 0.18–0.25 owing to trapping of solar radiation between leaves (Oke, 1987). This effect is greater at midday when the sun angle is higher. Type of vegetation also determines albedo and, in high latitude and high altitude environments, woodland can become markedly warmer than surroundings.

The thermal microclimate of individual leaves shows a sharp contrast to ambient air temperatures, since sunlit leaves take on the function of active surfaces. The upper surface of a sunlit leaf can be 5–10 °C warmer than the adjacent air (with undersides being a few degrees cooler). At night, leaves are an effective radiator of energy and can be up to 10 °C cooler than their surroundings with possible implications for arboreal primates.

Influences upon airflow

Vegetation has two effects upon airflow – the vertical profile of wind speed is modified in relation to vegetation density owing to frictional drag and the regularity of wind speed and direction is disturbed, with eddies forming in clearings.

The vertical profile of average wind speed shows a peak above canopy level, a sharp decrease below this – in the zone of maximum leaf density – and an increase closer to the ground if there is a substantial zone of trunks and stems clear of leaves. Values decrease again at the ground owing to frictional effects. In summary, both wind speed and direction in the stem zone can be independent of ambient conditions.

Frictional effects of whole stands of trees disturb the wider pattern of airflow, creating swirling eddies downwind of woodland. This can lead to a variety of wind directions in forest clearings and an increased gust : lull ratio. Both of these changes have important implications for the accurate measurement of wind speed and direction around forest edges and in clearings.

Influence upon moisture and relative humidity

Large forests are significant moisture stores; the leaf canopy intercepts incoming rainfall and the residual falls to the ground as stemflow (water flow down vegetation stems) and throughfall (flow directly to ground between leaves and stems). Interception rates are highest in short-lived showers when forest can be a temporary refuge from heavy rainfall. As rainfall duration increases, the infiltration ratio decreases.

REGIONAL VARIATIONS

Meteorological variations over tens of metres upwards are influenced by additional factors, fundamentally altitude and topography. Temperature variations at the surface are accompanied by density changes of the air close to the surface. In the absence of significant general winds, daytime heating results in upslope (anabatic) winds and nocturnal cooling produces a downslope counterpart (the katabatic wind). Height variations of 10 m or more within a study area may be sufficient to initiate such motion. This will be accompanied by the ponding of cold air in hollows or valley bottom areas, contributing to the nocturnal temperature inversion shown in Fig. 5.1. These changes are explored comprehensively in Geiger *et al.* (1995).

FIELD TECHNIQUES FOR MEASURING METEOROLOGICAL VARIABLES

This section introduces the key principles of meteorological measurement; these can be achieved either by traditional instruments that are read manually or by automated sensors (sometimes assembled as an automatic weather station) that work in conjunction with a data-logger to record observations at a predetermined frequency (Strangeways, 1997, 2000). Most of the key principles are common to both approaches; furthermore, portable manual instruments can be used as a back-up should automated systems fail (Crump, 1996).

KEY PRINCIPLES

Meteorological measurements in the field should be as comparable as possible with conventional meteorological observations at official observing sites. Traditional instruments are rarely portable or may be time-consuming and costly to set up in the field, but portable instrumentation can provide valuable data (Unwin, 1980). Whatever the type of instrumentation chosen, it is important to understand the principles upon which official observations are made. This allows the comparison of observations with official observing sites, and also informs the exact locations chosen for exposing the instruments (in the context of microclimatic variations considered above) and the time at which observations should be taken.

The basic assumptions of meteorological observation include:

* Air temperature is measured at a height of 1.25–1.5 m in shade (the temperature indicated by any thermometer exposed to direct sunlight is simply a function of the colour and construction of the thermometer itself).
* Precipitation (the generic term for rain, hail, sleet, snow and dew) is measured as a depth that accumulates in a straight-sided container.
* Wind speed is usually officially measured at a height of 10 m above the local ground surface. Anemometers measuring wind speed and direction erected at lower heights can provide readings having relative if not absolute validity.
* In the absence of data-logging, observations should be made at the same time each day (conventionally 0900 or 1800 hours local time).

Fig. 5.3. The interior of a Stevenson screen, showing dry-bulb and wet-bulb thermometers (vertical, in the centre), maximum and minimum thermometers (horizontal, in the centre), hygrograph (left) and thermograph (right).

TEMPERATURE MEASUREMENT

If time and finance permit, thermometers are exposed within a Stevenson screen – a ventilated wooden box, painted white to maximise reflection of direct radiation. These include dry- and wet-bulb thermometers (to measure air temperature and humidity) together with those for recording maximum and minimum temperature (Fig. 5.3). This standard for exposing thermometers is adapted in automatic weather stations with the use of the smaller and more portable Gill screen. It is important that thermometers exposed in screens are well ventilated in order to be representative of their surroundings. An additional refinement in poorly ventilated environments, such as forests, is an aspirated screen (fitted with a fan) if power is available.

Temperature measurements are often required from other locations. In unshaded terrain (where the active surface is at ground level) the ground surface is a zone of temperature extremes, where long-wave radiative losses are concentrated on a clear night. For measuring ground temperature, electrical sensors have the advantage of compactness and can be more easily shaded for daytime use than thermometers. Soil temperatures are most easily obtained by burying the small sensors of electrical thermometers to the required depth. To measure the microclimate of a forest, more information will be gained by traversing in and out of the woodland, recording temperatures in contrasting environments over a short period of time than by *in situ* measurements (an alternative is to establish separate observing sites).

There are three simple, low cost solutions to temperature measurement:

1. A Six's thermometer (mercury 'maximum–minimum' thermometer) is designed to record the maximum and minimum temperature since the instrument was last read and reset. These are intended mostly for amateur use and can give only a rough indication of ambient temperature.
2. A whirling psychrometer (whirling hygrometer). A psychrometer is a pair of thermometers exposed such that one records the air temperature (the dry-bulb thermometer) and the other records the temperature that results from evaporation at the thermometer bulb (the wet-bulb thermometer – the bulb is kept wet by a muslin wick connected to a small container of water, ideally distilled). The whirling action provides a means of shading and ventilating the thermometer bulbs (the latter increases the rate at which thermometers record ambient temperature after storage).
3. Simple battery powered electronic sensors: some of these can be read using a small remote unit operated by radio transmission at a distance of up to 30 m. These units may not be more accurate than other methods, but the digital displays reduce observer error.

Thermometers – accuracy, resolution and response time

Officially approved thermometers have an accuracy judged against reference instruments to within roughly ±0.3 °C across a wide range of temperatures. The resolution of observations refers to smallest change in a variable that can be measured or read; the precision rather than the accuracy of the reading. It is customary to read to one decimal place on the Celsius scale (an important requirement of digital display units). High resolution combined with low accuracy can result in spurious accuracy. However, consideration should also be given to whether it is essential for readings to be accurate in an absolute sense or merely to be internally consistent.

Response time is a critical factor in most field surveys. This is the time taken for the displayed temperature to come into equilibrium with the ambient temperature. Most electrical resistance thermometers used with sensors have a slightly faster response time than mercury thermometers. If one is undertaking a mobile survey, and carrying thermometers between observing sites, it is important that that the response time is known and allowed for before readings are made. The response time can be reduced if a whirling psychrometer is used, because the forced ventilation of the thermometers caused by the whirling action increases the cooling rate of the thermometers.

HUMIDITY MEASUREMENT

Humidity is measured by a set of wet-bulb and dry-bulb thermometers (a psychrometer), visible as the vertical thermometers in Fig. 5.3. The difference in temperature recorded by the two thermometers is related to the humidity of the air because if a thermometer bulb is wet, evaporation will increase as the surrounding air becomes drier. This has a cooling effect as latent heat is absorbed from the surroundings with the change of state from liquid to vapour. If air is saturated, no evaporation will take place and both thermometers will give the same reading. The wet bulb reading is not the same as the dewpoint temperature (this is lower) but the dewpoint can be calculated from hygrometric tables or slide rules supplied with a whirling psychrometer (or estimated using formula 5.1).

A surrogate for direct measurement of humidity involves the use of a hair hygrograph. Human hair expands in length as humidity increases. This movement can be shown by a moving arrow on a simple instrument (a hair hygrometer); when this is connected to a rotating drum and a pen arm, a humidity recorder is created (see Fig. 5.3), although data-logging now provides a more convenient method for monitoring humidity.

PRECIPITATION MEASUREMENT

Precipitation is measured in units of length (conventionally in millimetres), since it is expressed as the depth that would be measured in any straight-sided cylinder. It is conventional to measure precipitation every 24 hours. In general, the water collected (or melted) for a given diameter of funnel is measured in an appropriately calibrated measuring cylinder. The funnel has a narrow tube to minimise evaporation from the inner collecting vessel. Standard rain gauges in the British Isles have a 5 inch (12.7 cm) diameter funnel (those in the USA have an 8 inch (20 cm) diameter). A standard 5 inch gauge can hold at least 75 mm of precipitation, depending on the size of collecting vessel. The top of the funnel should be 30 cm above ground level. Gauges that are lower may record more due to in-splash and for this reason gauges should also be situated away from paved surfaces (ideally in short grass). Whilst it may seem convenient in field surveys to mount a gauge on a mast, any gauge higher than 30 cm will tend to record lower totals owing to overexposure in strong winds. This should be avoided if possible, even if readings are not intended for comparison with nearby official gauges, because of a lack of consistency between days of strong and light winds.

The resolution of all measurements should be to 0.1 mm (the measuring cylinder is narrower than the funnel in order to lengthen the vertical scale by a factor of more than 10). The top of the water column will form a meniscus around the edge of the cylinder and the level should be judged from the bottom of the meniscus.

Several types of recording rain gauge have been developed, of which the tipping bucket system is ideally suited to integration into an automatic weather station. A pulse is transmitted for each filling of a small cup corresponding to a known amount of precipitation (typically 0.2 mm). By recording the time of each tip, the intensity of rainfall can be deduced, though the precise duration of rainfall is easier to ascertain from traditional clock-driven tilting-siphon or natural-siphon gauges.

WIND SPEED AND DIRECTION

Wind speed is officially measured by an instrument called an anemometer on a mast at an effective height of 10 m. The effective height is the height above nearby obstructions such as trees or buildings. This is clearly not easily attained in field surveys, and automatic weather stations usually have much shorter masts. It is important to site the anemometer as far as possible from local features that might disturb wind speed and direction. Adequate estimates of wind speed can be obtained from hand-held anemometers, held at arm's length at right angles to the wind, although a 10% correction is sometimes applied to provide an estimate of the 10 m wind speed. However, estimation of wind speed in near calm conditions is more difficult using this method.

In traditional (non-electronic) anemometers, wind speed is displayed on a dial. Most instruments have a degree of damping to reduce the abruptness of fluctuations in wind speed and direction over short periods of time. It is usual to record both the mean wind speed and the maximum gust over a consistent length of time (for example 15–30 seconds). The maximum gust will often be 50% or more higher than the mean and the resulting gust:lull ratio is an important facet of the wind climatology of any region; it will tend to be higher in the vicinity of obstructions. Anemometer cups accelerate faster in gusts than they decelerate in lulls (Strangeways, 2000), leading to a possible risk of small over-estimation of mean speed.

Automated measurement

A major advantage of automated wind speed and direction instruments is the facility for automatic calculation and display of means, avoiding the need for observer estimation. This is a useful function of even inexpensive hand-held digital anemometers. Wind speed is measured by the closure of a switch contact generated by the revolution of the cups. The total number of contacts can be used to give a display of the mean wind speed and the time between pulses can indicate the maximum gust.

SUNSHINE AND RADIATION

Sunshine duration is measured by a Campbell–Stokes sunshine recorder. This is an expensive instrument ill-suited to field surveys. Automated sensors designed to estimate the duration of sunshine are available but the measurement of radiation is usually considered to be more useful in field observations. The development of automated weather observation has made it possible to measure the intensity of solar and net radiation in the field. These variables are more relevant for many purposes than sunshine duration, since they influence light levels. Solar and net radiation can be measured using pyranometers. These work on the principle that receipt of radiation will result in a temperature difference between adjacent black and white panels owing to contrasting surface albedo. Net radiation can be measured by exposing black discs to both solar and terrestrial radiation (i.e. facing up and down). Light-sensitive diodes can provide a cheaper alternative source of solar radiation or light data (Strangeways, 2000). If observations are taken manually at a series of different sites and/or times, the following should be noted: presence of sunlight on the area being observed, state of ground (dry, wet, nature of the surface) and SVF.

DATA-LOGGING AND AUTOMATED DATA ACQUISITION

Data-loggers can provide flexible round-the-clock monitoring of atmospheric conditions, and are invaluable at remote field sites or where regular fixed-point observations do not provide a sufficient monitoring of meteorological changes. The principles of data-logging are outlined in Strangeways (2000), where details of the technical characteristics of data-loggers can be found.

In general, the principles of correct observation using a data-logger are the same as for traditional instruments; automatic sensors can be viewed as miniature versions of the traditional instruments, designed to transmit analogue or pulse signals to a central data-logger. Analogue signals for variables such as temperature and humidity are first converted to digital format at the logger interface unit. Rain gauges and anemometers provide digital pulse signals from the sensors themselves.

USE OF MODERN DATA-LOGGERS

Data-loggers are housed in an insulated sealed box, varying in size up to the size of a brick, connected by cable to the appropriate sensors, and with an internal or external power supply. The latter may be replaceable alkaline cells, lithium batteries or solar panels (charging lead-acid batteries). It is important that the data-logger is not subjected to extremes of temperature. Operating ranges typically span $-20\,^{\circ}C$ to $+60\,^{\circ}C$, but battery lifetime is maximised when a data-logger can be sited in shelter (ideally indoors). In hot climates, data-loggers should be shaded from solar radiation and excess moisture. The length of cable connections from the sensors is an important constraint not only on data-logger location but on the siting of sensors, the general principles of which replicate the requirements of the corresponding traditional instruments.

Modern data-loggers are highly programmable. Two key variables are the scan and logging intervals. The scan interval is the frequency at which observations will be made, and battery lifetime is dependent on the chosen scanning interval. Collected data is downloaded to a laptop computer, or to a 'shuttle', removing the need to take a computer into the field.

CONCLUSIONS

Weather observation provides a rich source of local environmental data relevant to primate activity. Baseline local weather information is often unavailable from official sources in many locations in the field and meteorological conditions are highly point specific. This chapter has reviewed a variety of approaches to obtaining this local data. Though methods of instrumentation may vary, the principles of meaningful climate observation remain the same. Though individual field observations may be highly specific to particular times and locations, collectively observations made today also provide a potential source of data for future generations of primatologists who may be interested in investigating the degree to which global climatic change has altered local environments.

ACKNOWLEDGEMENT

I am grateful to Mrs Mary Mackenzie, who drafted the figures.

REFERENCES

Brower, J. E., Zar, J. H. & von Ende, C. N. (1989). *Field and Laboratory Methods for General Ecology*, 3rd edition. Dubuque, IA: William C. Brown Publishers.

Buckle, C. (1996). *Weather and Climate in Africa*. Harlow: Addison Wesley Longman.

Crump, M. L. (1996). Keys to a successful project: associated data and planning. In *Measuring and Monitoring Biological Diversity – Standard Methods for Mammals*, ed. D. E. Wilson, F. R. Cole, J. D. Nichols, R. Rudran & M. S. Foster, pp. 52–56. Washington, DC, and London: Smithsonian Institution Press.

Edwards, A. & White, L. (2000). Methods for recording the weather. In *Conservation Research in the African Rain Forests: A Technical Handbook*, ed. L. White & A. Edwards, pp. 85–92. New York: Wildlife Conservation Society.

Geiger, R., Aron, R. H. & Todhunter, P. (1995). *The Climate Near the Ground*, 2nd edition. Braunschweig: Vieweg Press.

Linacre, E. (1992). *Climate Data and Resources: A Reference and Guide*. London: Routledge.

McGregor, G. R & Nieuwolt, S. (1998). *Tropical Climatology*, 2nd edition. Chichester: Wiley.

Oke, T. R. (1987). *Boundary Layer Climates*. London: Routledge.

Oliver, J. E. & Hidore, J. J. (2002). *Climatology: An Atmospheric Science*, 2nd edition. Upper Saddle River, NJ: Prentice Hall.

Robinson, P. J. & Henderson-Sellers, A. (1999). *Contemporary Climatology*. Harlow: Pearson Education/Longman.

Rosenberg, N. J., Blad, B. L. & Verma, S. B. (1983). *Microclimate: The Biological Environment*. Chichester: John Wiley & Sons.

Stoutjesdijk, P. & Barkman, J. J. (1992). *Microclimate, Vegetation and Fauna*. Knivsta, Sweden: Opulus Press.

Strangeways, I. (1997). Ground and remotely sensed measurements. In *Applied Climatology: Principles and Practice*, ed. R. D. Thompson & A. H. Perry, pp. 13–21. London: Routledge.

(2000). *Measuring the Natural Environment*. Cambridge: Cambridge University Press.

Unwin, D. M. (1980). *Microclimate Measurement for Ecologists*. London: Academic Press.

List of suppliers and their Internet sites

RW Munro: 406 Roding Lane South, Woodford Green, Essex IG8 8EY, UK. Tel: +44 (0)208551 7000. <www.munro-group.co.uk>. (A wide range of traditional and automated meteorological monitoring equipment.)

The Casella Group Ltd: Regent House, Wolseley Road, Kempston, Bedford MK42 7JY, UK. Tel: +44 (0)1234 844 100. <www.casella.co.uk>. (Traditional and automated meteorological and environmental monitoring equipment.)

Philip Harris Scientific/Education: Novara House, Excelsior Road, Ashby Park, Ashby de la Zouch, Leicestershire LE65 1NG, UK. Tel: +44 (0)845 6040490. <www.philipharris.co.uk>. (A good source of low-cost portable instruments for field observations.)

The Weather Place at Ben Meadows Co.: <www.benmeadows.com/weather/default.htm>. (A range of automated meteorological observing sensors and equipment.)

6 • Survey and census methods: population distribution and density

CAROLINE ROSS[1] AND NIGEL REEVE[2]
[1]*Centre for Research in Evolutionary Anthropology,*
[2]*Centre for Research in Ecology and the Environment,*
School of Life and Sport Sciences,
University of Surrey Roehampton, London, UK

INTRODUCTION

A population study of a wild primate typically involves a considerable invest-ment of time and resources (i.e. money, equipment, labour) and it is vital to ensure that such effort is well targeted. When designing your study, a key is-sue is whether your study objectives genuinely demand an absolute estimate of the population density from either a census (a total count) or a survey (in which density is estimated from statistically valid samples), or whether less information will suffice. Relative estimates of density using data from methods such as 'catch-per-unit effort' from trapping or systematic searching do not provide absolute densities but, as long as the sampling methods and other conditions are standardised, can allow reliable comparisons between locations and monitoring of population change over time. Population indices are based on indirect indicators, such as the density of faeces or other charac-teristic signs, that can be correlated with population density. Such methods may be a more practical alternative to searching for secretive, hard-to-find animals.

In practice, no population survey or census is completely bias free and many studies may find that a reliable relative population estimate or index is more achievable than a reliable absolute estimate of the population size (Bibby *et al.*, 1992; Greenwood, 1996; Krebs, 1999). There is a trade-off between the depth of the data gathered and the number of replicate samples that can be obtained. The choice of method in the field will also be affected by the degree to which data from individuals are required (e.g. sex, age, reproduc-tive status, individual recognition) and whether capture and identification tagging is feasible (Chapters 7 and 10). Figure 6.1 should help you to choose the appropriate method for your own study.

To review this entire topic in one chapter would be impossible. Here we aim to provide a straightforward, non-mathematical guide to some of the key issues to consider in primate population studies and to direct the reader to detailed descriptions of the analysis methods and useful further reading.

Field and Laboratory Methods in Primatology: A Practical Guide, ed. Joanna M. Setchell and Deborah J. Curtis. Published by Cambridge University Press. © Cambridge University Press 2003.

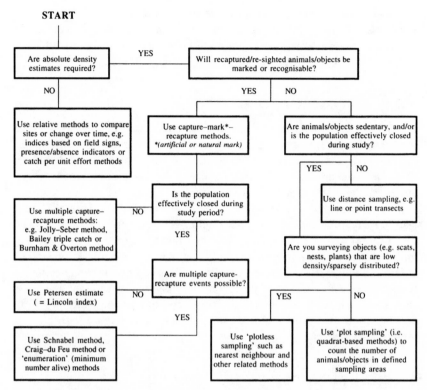

Fig. 6.1. Choice chart giving a general guide to measures of abundance most likely to be useful in primate-related studies of abundance. Note that the precise application of each method, and variants of those methods, will be governed by specific assumptions too detailed to be included in this chart.

BEFORE YOU START

- Decide what resources you can commit to the population study.
- Learn about the study site and target species (Chapter 2) and possibly also prepare for handling and trapping animals (Chapters 7 and 8).
- Think about the pros and cons of different census methods for your particular study. It is vital that you start thinking about statistical analysis at this stage. Your time, energy and money will be wasted if you cannot interpret your data.
- Practice using different techniques, perhaps in a local park where you could carry out a plot sample of birds' nests or spiders' webs or a line transect counting squirrels (or even people!). Play around with your data and do a variety of analyses.
- Use a pilot study to try out different methods of data collection, for example indirect versus direct observation methods, line transects in different

places, walking transects at different speeds, working alone (if it is safe) or with others, familiarising yourself with traps or other equipment. If possible, collect enough data for preliminary analysis – they may have a major effect on the type of study you eventually decide to carry out.

WHAT TO COUNT AND RECORD

The following should always be recorded: date and time, observer(s), weather conditions and other special circumstances (e.g. recent hunting). When animals are detected you may also record activity, height above ground, age/sex class, etc. Assessing age will require study of the age classes used by previous researchers and some practice (Chapter 9). Group size and composition may be counted and, if groups are stable, then repeated estimations should lead to increasingly accurate counts (Fig. 6.2). However these records may be inaccurate if some classes behave more conspicuously or avoid humans (e.g. mothers with infants) or because the group is widely dispersed and not all animals can be located. If accurate estimates of group numbers and compositions can be made, then the information may be used to estimate biomass – such estimates will be more accurate than those made using absolute numbers only, provided that good body weight data are available for animals of different age/sex classes.

SPECIES PRESENCE/ABSENCE RECORDS

Just recording presence or absence can help to establish the geographical range or habitat requirements of primate species (e.g. Singh *et al.*, 1999). Such data can also contribute to species lists and biodiversity evaluations as well as the identification of suitable sites for further ecological or behavioural studies (e.g. McGraw 1998; van Krunkelsven *et al.*, 2000). These surveys are comparatively simple to carry out, although the data yield is relatively low. Of course, proving absence conclusively may require a prolonged study or repeated investigations.

The presence of primates in the area may be clear from both direct and indirect evidence, such as:

- Direct observation of live animals.
- Indirect observation of characteristic signs (tracks, faeces, nests, carcasses, feeding signs, calls, etc.). See Wemmer *et al.* (1996) and Parnell (2000) for summaries of field signs used in mammal surveys.
- Observation of animals captured or killed (e.g. in markets or local villages). In this case care should be taken to ascertain where the animals were obtained.
- Reports from locals. The value of such reports will depend on the accuracy of identification of both the species of interest and the exact location of the sighting.

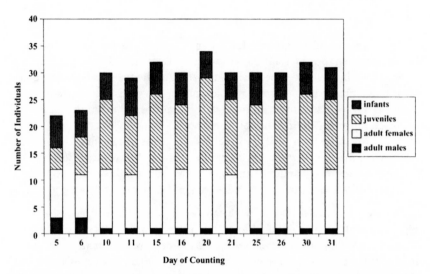

Fig. 6.2. Repeated counts of a single group of Hanuman langurs (*Presbytis entellus*) taken over a month in southern India. Early counts have clearly missed some animals whereas later counts show between 28 and 34 animals. On most days, one adult male is present. The 'extra' males present on days 5 and 6 were peripheral, chased repeatedly by the third male and once seen in association with another group. Juveniles were difficult to count as they were frequently playing and chasing on the edge of the group (e.g. on day 20, where juveniles were playing with those from another group the number recorded was probably too high).

For the sake of both internal consistency and comparability with other studies, survey techniques need to be standardised and all areas in the locality sampled equally.

ABSOLUTE ESTIMATES OF POPULATION DENSITY

For absolute density estimates, the choice of method depends on several factors, but two fundamental questions are:

- Are natality, mortality, emigration and immigration *during the study period* likely to be significant (open population) or can they be discounted as negligible (closed population)?
- Will the animals be identifiable by natural or artificial markers? (Chapters 2 and 10.)

TOTAL COUNTS

A total count of all animals in the study area (a census) is rarely feasible but a virtually complete count of all animals may be possible if the population is effectively closed, and in the following circumstances:

- Few animals will be missed, for example because of good visibility or a very high chance of trapping all animals. Detection of animals may be enhanced by well-developed field skills and knowledge of predictable daily or seasonal activity patterns – animals may be more visible at certain times of the day or year. Counts may target aggregations at known key locations, for example at sleeping sites or water holes.
- Animals can be identified from natural or artificial markings to prevent double counting. If stable social groups can be identified (by location or the presence of key animals) and accurately counted, then individual identification may not be needed to ensure a good estimate of numbers.

In long-term studies, census work can be combined with other research to give an overall picture of population demography and ecology. For example, research on mountain gorillas (*Gorilla gorilla beringei*) in the Virunga volcanoes has included both repeated censusing (of animals, nests and other signs) and the study of a number of focal groups. This has allowed almost all individuals to be counted, their distribution into social groups to be mapped and home range sizes to be estimated (Fossey, 1974; Fossey & Harcourt, 1977; Harcourt *et al.*, 1981, 1983; Sholley, 1991). However, even short studies can yield almost complete counts, provided that an area can be covered rapidly, for example using teams of workers (e.g. for mountain gorillas in Uganda, McNeilage *et al.*, 2001).

POPULATION ESTIMATES FROM SAMPLES

When total counts cannot be made for the whole study area, population density must be estimated from a survey of a statistically robust, representative sample of smaller areas within the main study area. Ideally these areas should be randomly located but stratified random sampling is often advisable. Stratification in this case involves the identification of the different habitat categories and the *separate* random sampling of these to ensure that each is sampled sufficiently well (Krebs, 1999). Two sampling methods are most usually used in studies of primates: plot sampling and distance sampling.

PLOT SAMPLING

Plot sampling is effectively the census (total count) of a representative set of replicate quadrats (standard sub-samples) of the study area. The population density estimate for the area is the mean of those obtained for the plots. Plots are typically either square (e.g. trapping grids) or rectangular (e.g. strip transects), but can be circular (e.g. point transects). Plot sampling is a good method to use when censusing inanimate objects (e.g. nests or plants), which are relatively easy to find and to mark, thus preventing double counting. For example, plot sampling in vegetation surveys is discussed in Chapter 3.

The major problems of plot sampling are related to difficulties in finding all the animals (or objects) within the area. If animals are hard to find, then extensive replication may be needed. In areas of good visibility, strip transects may be used and the observer may count animals within a defined distance from a central line – each transect can be a separate plot or several replicate transects may be included within a large plot. If several people are working together then a series of parallel transects can be traversed to carry out a sweep of the general area – although care must be taken that double counting does not occur.

Strip transects require knowledge of the area to be accurately sampled, i.e. the distance over which all animals can be seen from the line must be determined. This is derived from the distribution of the distances of the animals from the transect line (Fig. 6.3a). Here, the distribution suggests that all animals are visible up to a certain cut-off point, in this case 40 m on either side of the transect line. The population density in this example is simply the number of animals within a strip 80 m × 1000 m (8 ha). Because the strip width is determined by factors affecting visibility, there may be differences between transects in different habitats, times of day, seasons and weather conditions. If the results of transects located in different places or seasons are significantly different it may be necessary to truncate the strip width in some cases so that all widths are standardised to the width of the transect with lowest visibility.

Using a line transect and counting all animals may lead to the type of result illustrated in Fig. 6.3b, in which animals are increasingly difficult to see the further they are from the observer. If plot sampling were used to analyse these data, the area sampled would be an extremely thin strip transect (18 m wide in Fig. 6.3b) and very few of the observed animals would be included in the data. This is problematic because density estimates based on such a small, unrepresentative sample would be inaccurate. However, distance sampling seeks to overcome this problem by allowing a larger sample of animals to be included in the analyses.

DISTANCE SAMPLING

Distance sampling counts as many animals (or signs of an animal's presence) as possible from a point or line within an area. Whereas plot sampling assumes that all individuals or signs within the area are detected, distance sampling assumes that only a proportion will be seen. This proportion (which declines with distance from the observer) is then estimated and used to calculate an estimate of the total number of animals present.

There are two forms of distance sampling: line transect sampling, in which the observer walks (or drives) along a line and records the animals seen on either side of the line; and point sampling, in which the observer stands at one or more selected viewpoint and records all the animals seen from that location (effectively a transect of zero length). Point samples, often used in

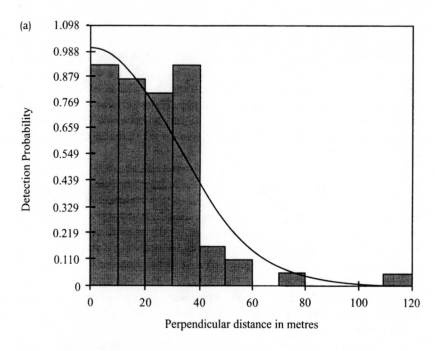

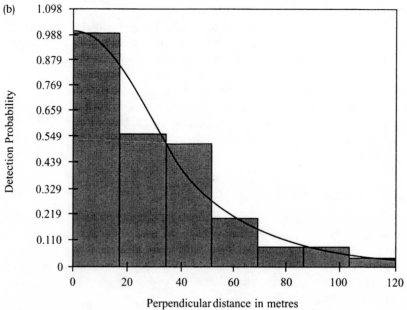

Fig. 6.3. (*cont.*)

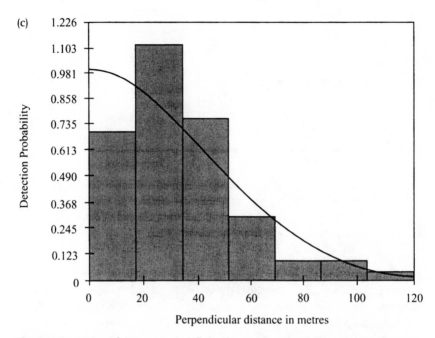

Fig. 6.3. Examples of data recorded from line transect sampling. In all cases a total of 62 observations has been used to generate the graphs shown. The computer program 'Distance' has been used to generate the detection functions (g), which is shown as a curve on each graph (when using the program the user can fit a number of models to the data and choose the best fit).

(a) Here, a high proportion of animals are seen up to 40 m away. If the data used are truncated at 40 m it could be used to give a total count of all animals within the area 40 m either side of the transect line. Alternatively a higher proportion of the data could be used if a model were used to estimate the proportion of animals not seen at greater distances. This can be done using the defection function (g) to estimate the number of animals nor detected.

(b) The detection function falls off more rapidly with distance than in (a).

(c) The data suggest that relatively few animals have been detected on the line (see text).

bird surveys, have certain design advantages in terms of being easier to locate randomly or systematically (especially in woodland or other dense habitats where the need for access may bias the location of transects) and detection is not compromised by the noise of walking, the distraction of obstacles or un- even ground (Bibby et al., 1992). However, primates are typically wide ranging and occur at relatively low densities, which would demand a large number of point samples placed over a wide area and travelling between dispersed points, whereas using line transects allows the observer to cover a wider area while simultaneously recording (Buckland et al., 2001). The use of point

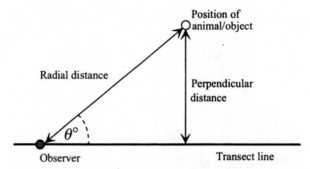

Fig. 6.4. Measuring the distance of sighted animals/objects from the transect line. Ideally the perpendicular distance (*x*) is measured directly, but it can also be calculated from the radial distance (*r*) and the bearing (θ, angle from the transect line).

sampling may sometimes be preferred because of difficult terrain or when movement is restricted for political reasons. Further information on point sampling can be found in Bibby *et al.* (1992), Buckland *et al.* (2001) and also see Bollinger *et al.* (1988) for a comparison with line sampling.

If you plan to rely heavily on distance sampling it is important to look at this subject in more detail, including the mathematical basis of population density estimations, not discussed in detail here (Anderson *et al.*, 1979, 1985; Bibby & Buckland, 1987; Buckland *et al.*, 1993, 2000, 2001). Comprehensive guides by Buckland *et al.* (1993, 2001) are supported by the computer software 'Distance' (Thomas *et al.*, 1998), which allows easy analyses of data obtained from distance sampling. This can be downloaded free from the URL at the end of this chapter, and is widely used (e.g. Blom *et al.*, 2001). Notation and abbreviations follow that used in Buckland *et al.* (2001) and the software.

The basic procedure for line transect sampling is as follows:

- Locate the transect line (L) randomly in the area.
- Traverse the line, recording the observed number (*n*) of objects on either side of the line.
- Each time an object is seen the observer records either (1) its perpendicular distance from the line (*x*) or (2) the sighting distance from the observer (*r*) and the sighting angle (θ) (Fig. 6.4). If (2) is chosen, both *r* and θ are then used to estimate the detection function.
- Calculate the 'detection function' (*g*) (see Fig. 6.3), using data for a number of objects detected at different distances, and use this to estimate the total number of animals in the area. Estimation of *g* is best done using 'Distance' (Buckland *et al.*, 1993, 2001). If you do not have access to a computer program, White & Edwards (2000) discuss alternative methods of estimating the detection function and hence population density.

For line distance sampling to be an effective means of estimating population density, a number of assumptions must be made:

- *Objects directly on the transect line have never been missed.* For arboreal species this assumption may be violated as animals above the transect line might be missed. Buckland *et al.* (2001) emphasise the importance of ensuring that this condition is met by paying particular attention to the area on and directly above the line. If objects on the transect line are missed then the distribution of animals against distance may reflect this, as shown in Fig. 6.3c.
- *Objects do not move before being detected.* If animals move before detection, the results will be biased towards far or near distance sightings.
- *Objects are not counted twice in a single transect walk.* Double counting is most likely to occur if the animals flee by moving ahead of the observer or because the transect is not straight and the same area is seen twice from different parts of the line. The second problem can be avoided by careful placing of transects and a mapping of areas visible from two parts of the same line. The first problem is less easily overcome but is avoidable if animals can be individually recognised. It is less important if objects are detected twice in two different walks of the same transect or in two walks of different transects, provided detection in the first walk is independent of detection in the second walk. In such cases each sighting should be recorded and treated as separate independent sightings for analysis.
- *Distances and angles are measured accurately.* Ideally use a tape measure or laser rangefinder and carry an alternative means of measurement. The angle of sighting can be measured with an angleboard or compass. Visual estimation of distances or pacing are less accurate. Avoid putting distances into bands (e.g. 0–5 m, 6–10 m, etc.) as this decreases the accuracy of the analyses. Check the accuracy of estimated distances frequently, to quantify measurement error, prevent 'drift' of estimations over time and to ensure intra-observer reliability. Normally, the simplest measure to take is the perpendicular distance of the object from the transect line, but if this is difficult to do the radial distance and the angle (θ) should be taken (see Fig. 6.4).
- *Sightings are independent events.* Analysis assumes that sightings are both randomly distributed and independent from each other. For social species, such as most primates, individuals are likely to be clustered and the group rather than each individual animal must be treated as the single object for analysis (e.g. Fashing & Cords, 2000) – with distances measured either to the group centre or to the nearest individual.
- *Sufficient sightings are made for an accurate estimate of the detection function.* To calculate an accurate estimate usually requires at least $n = 60$ to 80 individual sightings, although $n = 40$ may sometimes be sufficient (Buckland *et al.*, 2001). For group-living animals $n =$ the number of groups and the

data are used to calculate the mean group density. Population density can then be estimated using the mean number of animals per group – data that can be obtained either during the line transect sampling or from reliable group counts achieved in other situations.

Traversing the area and detecting animals

Most primate population surveys are carried out by walking through the study site and looking and listening for animals. The use of vehicles allows longer distances to be travelled but usually limits the survey to existing trails, roads or waterways. Although aerial surveys are commonly used to detect and survey large-bodied mammals, such as elephants or ungulates, they are less appropriate for smaller animals or dense habitats. In open-country species such as baboons, an aerial survey could permit rough population estimates or just indicate their presence.

Whatever the mode of transport, stop, look and listen at regular intervals. You may need to leave the transect line to count and detect animals or objects, although search effort should concentrate on areas near to the line (Buckland et al., 2001). In all cases the distance of the object from the line (not the observer) should be recorded and care should be taken to return to the line and not miss any part of the transect.

All nocturnal primates except for owl monkeys (*Aotus* spp.) and tarsiers (*Tarsius* spp.) have a reflective layer (the tapetum) at the back of the eye – creating 'eyeshine' in torchlight. Eyeshine has been used successfully to detect nocturnal primates in several studies, and it is particularly useful in species that freeze when threatened (Charles-Dominique, 1972, 1977; Singh et al., 1999). Thermal imaging, especially of nocturnal species, is not yet widely used to aid detection, but suitable portable devices (about the size of a video camcorder) are now available. Thermal imaging has been used successfully in both aerial and ground surveys of a number of mammal population studies (e.g. Garner et al., 1995; Havens & Sharp, 1998) and in difficult situations e.g. semi-fossorial ground squirrels (Hubbs et al., 2000) and forest deer (Gill et al., 1997).

Minimising observer bias and maximising the reliability of detection is particularly important. Key factors to standardise in the survey protocol include:

- Observer technique, vigilance, skill, motivation and other differences between individuals.
- Sampling effort, for example length of transects and time spent travelling them. Walking speed needs to be slow enough to detect enough animals while allowing a sufficiently large area to be surveyed. Most forest surveys of primates have found that a speed of 1–2 km/h is appropriate.

- Habitat conditions, for example seasonal variation. These can drastically alter the detection function.
- Weather conditions – differences in visibility and environmental noise will alter the detection function.

PLOTLESS SAMPLING METHODS

To estimate the density of immobile objects or plants, field signs or sedentary animals, there are several methods based on measuring the distance between randomly chosen objects and their nearest neighbour or between randomly chosen locations and the nearest object. These can allow the establishment not only of the density of objects but also the nature of their spatial distribution (e.g. random, clumped or regular). These techniques are also often referred to as 'distance' methods (e.g. Krebs, 1999; Southwood & Henderson, 2000), creating the potential for confusion with the distance sampling described above. A range of plotless sampling methods, such as the T-square method, are well explained by Krebs (1999).

CAPTURE–MARK–RECAPTURE METHODS

These methods are one of the commonest ways of providing absolute population density estimates in studies of mammals, but normally require that animals are caught (usually by systematic trapping), distinctively marked (Chapter 10), released and later recaptured on one or more regularly timed occasions. Trapping methods for primates are discussed in Chapter 7. Trapping is labour intensive, but capture creates opportunities for the additional collection of data (physical, physiological, genetic, parasitological, etc.), radio-tagging and identification marking. Capture–mark–recapture methods can be used without trapping (e.g. using plot or distance sampling) if the animals can be systematically detected and identified at a distance, and artificial marks may not be required if individuals can be distinguished by their natural features.

A real strength of capture–mark–recapture methods is the detailed information that can be obtained from longitudinal studies, allowing not only reliable absolute population estimates to be made but determination of demographic variables, for example birth and mortality rates, population age structure. However, one needs to be alert to the detailed assumptions of whichever technique is being used.

Choice of capture–mark–recapture analysis method
There are many methods, and variants of methods, to choose from and a vast primary literature covering their application to a variety of taxa. Begon (1979), Greenwood (1996), Krebs (1999) and Southwood & Henderson (2000)

are very useful overview texts that detail the assumptions and calculations for a wide range of methods. For a good review of techniques for mammals in general, see Nichols & Conroy (1996). A number of free software packages are also available (see Southwood & Henderson, 2000).

Table 6.1 summarises the basic characteristics of the principal, most commonly used methods (see also Fig. 6.1). Some of the key issues to bear in mind when planning your capture–mark–recapture study are:

- Is your study population open or closed (see above)?
- Can you assume that all animals in the population (irrespective of sex, age, social status, etc.) are equally trappable? This assumption is fundamental to most methods except the Burnham & Overton method (Greenwood, 1996).
- Is the probability of recapture affected by trapping and marking? Animals may learn to avoid capture (trap 'shy') or to exploit the traps for food and/or shelter (trap 'addicted').
- Marks must be permanent (at least during the study) and date specific. Marks can be used either simply to distinguish all animals caught at a particular time, or to identify certain sub-groups or individuals.
- Will multiple recapture events be possible? How many are appropriate and can be resourced? The simple Petersen estimate requires only one capture and subsequent recapture event, but more sophisticated methods will require several regularly timed recapture events.
- Trapping/catching effort must be enough initially to catch and recapture animals in sufficient numbers for good statistical power; as much as 50% of the population may need to be marked to obtain accurate estimates (for details, see Krebs, 1999).

INDIRECT OBSERVATION

If animals cannot be observed directly, calls may be used to identify species when visibility is poor and may also be used to estimate population density (Brockelman & Ali, 1987). If species identification is in doubt, then recordings of calls should always be made to allow later verification of the species (Chapter 15).

Observation of signs such as faeces, nests, tracks, etc. is useful for indicating presence, but the density of the objects (assuming a known probability of detection) may also be used to calculate population density if the following factors can be determined:

- Creation rate of the object per animal or group (e.g. number of nests built per week or number of defecations per day).
- Duration, i.e. how long does the object persist in the environment after its creation.
Thus:
population density = object density/(duration × creation rate)

Table 6.1. *Summary of some of the principal methods judged most likely to be useful in the estimation of primate population abundance from capture–mark–recapture data*

Method	Application, assumptions, key points
Closed populations	
Petersen estimate (Lincoln index)	Single capture–marking–single recapture events. Good for short studies, but restrictive assumptions and requires high levels of catchability
Schnabel method	Multiple capture–marking–recapture events. Successive Petersen estimates used to calculate a weighted mean population estimate. Graph plot of the proportion of marked animals per catch against total previously marked (each estimate is a point on the graph) is linear if assumptions are met
Craig–du Feu method	Multiple capture–marking–recapture events. Successive, cumulative population estimates are made until sufficient samples have been taken to stabilise the value with an acceptable standard error (Greenwood, 1996)
Enumeration (minimum no. alive methods)	Multiple capture–marking–recapture events. Based on a log of successive captures of individually marked animals. Those known to have been alive throughout any specified period are counted. Always an underestimate of the true population – a bias that worsens with low overall catchability (Krebs, 1999)
Open populations	
Jolly–Seber method	Multiple capture–marking–recapture events. Normally requires individually unique marks to determine when individuals were last captured. Calculates losses and recruitment to population. Equal catchability is a key assumption. Method of choice when assumptions are met and numbers captured are not too low
Bailey triple-catch	Three capture–marking–recapture events. Effectively a special case version of the Jolly–Seber method, allowing population losses and recruitment to be estimated quickly, but with less power (Begon, 1979)
Burnham & Overton method	At least four capture–marking–recapture events. Uses counts of number of animals caught exactly 1, 2, 3 and 4 times. Sensitive to change in capture probabilities over time and trapping effort must be constant. Allows individuals to differ in probability of capture (Greenwood, 1996)

All methods, unless otherwise stated, require that: (1) marks are not lost during the study, (2) marks are always correctly recognised, (3) all animals (whatever age, sex or status, marked or not) have an equal probability of capture, (4) the duration of sampling itself is brief in relation to the total study time. Multiple capture methods usually assume strictly regular spacing of sampling periods.

In the great apes, nest counts are used to estimate ape population numbers (Tutin & Fernandez, 1984; Tutin *et al.*, 1995), with the size of associated faeces being used to estimate the age class of the animals in each nest (Schaller, 1963; McNeilage *et al.*, 2001). Great apes usually build a new nest each night and only unweaned animals share a nest, hence the creation rate per weaned individual is relatively easy to calculate. Duration can be measured from the decay rate of marked nests, which depends on both the nest materials used and the climate (Tutin *et al.*, 1995). Detection rates are calculated as for the detection rates of animals (see Fig. 6.3). As with animal counts, groups of nests (found in gorillas and chimpanzees) should be counted as a single object and measurement made to the centre of the group. Discussion of the methods used to count nests and interpretation of the data can be found in Hashimoto (1995), Blom *et al.* (2001); and Furuichi *et al.* (2001); aerial counting of orang-utan (*Pongo* spp.) nests is discussed by Payne (1987).

GETTING A REPRESENTATIVE SAMPLE

Whatever sampling method is planned, you need to ensure that the locations of your sample areas are chosen to represent the habitat of the study area in an unbiased way, and that the sampling effort is sufficient for a robust population estimate. Pragmatic decisions about the size of the study area, sampling methods and the level of replication need to be balanced against, but not override, the need to avoid bias and ensure adequate statistical power of the study. If the area and/or species are previously unstudied then it is vital to conduct a pilot study before collecting large amounts of data. Further discussion on familiarisation with the study species can be found in Chapter 2.

The random location of a plot or transect can be achieved simply by using a map and random numbers (from tables or generated on a computer or calculator) to give grid references. Transects can be sited within the study area (or within an area of specific habitat if stratification is required) along a line joining two randomly generated grid references. However, plot or transect location may not be entirely random if the study aims dictate that it should cover a particular range of habitats (Chapter 3). For example, if you were interested in how population density of one or more species varied in relation to the proximity of human disturbance, transects might be placed perpendicularly to roads or fields (Fig. 6.5). Transects are often placed in regularly spaced parallel lines – but this does not seriously violate the assumption of randomness, provided that the first transect has been chosen at random, the distance between the lines does not replicate a feature of the landscape, and the lines are sufficiently separated to ensure independence.

Transects pose particular problems when one is attempting to avoid bias and properly represent the range of different vegetation types, altitudes or

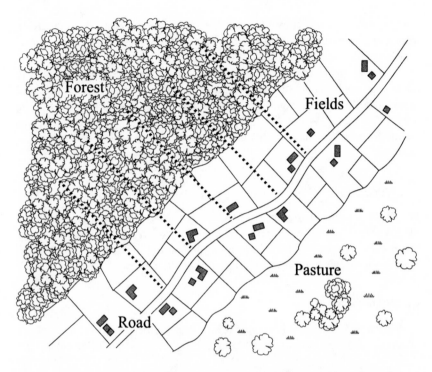

Fig. 6.5. A series of parallel transects (dotted lines) designed to investigate how species composition changes with variation in human disturbance.

other features within the study area. It is important that transects do not follow contour lines, water courses or run along established paths or tracks, as these are unlikely to be representative. In thick vegetation, trails may need to be cut in order to allow for easy movement, but disturbance should be minimised. If natural features prevent cutting of trails or movement along the planned line you must decide whether to truncate the transect, allow a detour around the barrier or abandon the line and start another. Whatever your decision, you should carefully consider the possible biases that might be introduced to your survey, for example avoiding marshy ground might result in loss of information on an important habitat. If no practical solution can be found, all you can do is to note that your data do not include marshy areas.

Transect length will depend on the population density and factors affecting detection rates – where detection rates are low the minimum n of 60 for calculating g will take longer to accumulate. In multi-species surveys, the species with the lowest detection rate will set the standard. A single transect through one particular habitat might be appropriate in some cases, but, if the study is investigating differences between a number of habitat types, either a long transect through different habitats or a number of shorter transects may be needed (note: $n = 60$ will be necessary for each habitat type).

Careful mapping of plots, transect lines and trap positions is important to allow later replication. This may be done using traditional means, artificial markers and/or photographs of the key features, or the Global Positioning System (Edwards & White, 2000; Chapter 4). The locations of observed animals should be noted on maps as this may allow analysis of habitat use or ranging patterns.

ACKNOWLEDGEMENTS

We thank Debbie Curtis and Jo Setchell for their invitation to contribute to this volume. Their comments and their thorough and patient editing of the manuscript were a great help to us. We would also like to thank Mary Mackenzie for her help in the preparation of the figures.

REFERENCES

Anderson, D. R., Laake, J. L., Crain, B. R. & Burnham, K. P. (1979). Guidelines for line transect sampling of biological populations. *J. Wildlife Manag.* **34**, 141–6.

Anderson, D. R., Burnham, K. P. & Crain, B. R. (1985). Estimating population size and density using line transect sampling. *Biom. J.* **7**, 723–31.

Begon, M. (1979). *Investigating Animal Abundance, Capture–Recapture for Biologists.* London: Edward Arnold.

Bibby, C. J. & Buckland, S. T. (1987). Bias of bird census results due to detectability varying with habitat. *Acta Ecol.* **8**, 103–12.

Bibby C. J., Burgess N. D. & Hill D. A. (1992). *Bird Census Techniques.* London: Academic Press.

Blom A., Almasi A., Heitkonig I. M. A., Kpanou J. B. & Prins H. H. T. (2001). A survey of the apes in the Dzanga-Ndoki National Park, Central African Republic: a comparison between the census and survey methods of estimating the gorilla (*Gorilla gorilla gorilla*) and chimpanzee (*Pan troglodytes*) nest group density. *Afr. J. Ecol.* **39**, 98–105.

Bollinger, E. K., Gavin, T. A. & McIntyre, D. C. (1988). Comparison of transects and circular-plots for estimating bobolink densities. *J. Wildlife Manag.* **52**, 777–86.

Brockelman W. Y. & Ali R. (1987). Methods of surveying and sampling forest primate populations. In *Primate Conservation in the Tropical Rain Forest* ed. C. W. Marsh & R. A. Mittermeier, pp. 23–62. New York: Alan R. Liss.

Buckland, S. T., Anderson, D. R., Burnham, K. P. & Laake, J. L. (1993). *Distance Sampling: Estimating Abundance of Biological Populations.* London: Chapman & Hall.

Buckland, S.T., Goudie, I.B.J. & Borchers, D.L. (2000). Wildlife population assessment: past developments and future directions. *Biometrics* **56**, 1–12.

Buckland, S.T., Anderson, D.R., Burnham, K.P., Laake, J.L., Borchers, D.L. & Thomas, L (2001). *Introduction to Distance Sampling: Estimating Abundance of Biological Populations.* Oxford: Oxford University Press.

Charles-Dominique, P. (1972). Écologie et vie sociale de *Galago demidovii* (Fischer 1808, Prosimii). *Z. Tierpsychol* Suppl. **9**, 7–41.

(1977). *Ecology and Behaviour of Nocturnal Primates.* London: Duckworth.

Edwards, A. & White, L. (2000). Maps, compasses, GPS units and principles of navigation. In *Conservation Research in African Rain Forests: A Technical Handbook*, ed. L. White & A. Edwards, pp. 31–51. New York: Wildlife Conservation Society.

Fashing, P.J. & Cords M. (2000). Diurnal primate densities and biomass in the Kakamega Forest: an evaluation of census methods and a comparison with other forests. *Am. J. Primatol.* **50**, 139–52.

Fossey, D. (1974). Observation on the home range of one group of mountain gorillas. *Gorilla gorilla beringei. Anim. Behav.* **22**, 568–81.

Fossey, D. & Harcourt, A.H. (1977). Feeding ecology of free-ranging mountain gorillas (*Gorilla gorilla beringei*). In *Feeding Ecology*, ed. T.H. Clutton-Brock, pp. 415–47. London: Academic Press.

Furuichi, T., Hashimoto, C. & Tashiro, Y. (2001). Extended application of a marked-nest census method to examine seasonal changes in habitat use by chimpanzees. *Int. J. Primatol.* **22**, 913–28.

Garner, D.L., Underwood, H.B. & Porter, W.F. (1995). Use of modern infrared thermography for wildlife population surveys. *Environ. Manag.* **19**, 233–8.

Gill, R.M.A., Thomas, M.L. & Stocker, D. (1997). The use of portable thermal imaging for estimating deer population density in forest habitats. *J. Appl. Ecol.* **34**, 1273–86.

Greenwood, J.J.D. (1996). Basic techniques. In *Ecological Census Techniques: A Handbook*, ed. W.J. Sutherland, pp. 11–110. Cambridge: Cambridge University Press.

Harcourt, A.H., Fossey, D. & Sabater Pi, J. (1981). Demography of *Gorilla gorilla. J. Zool. Lond.* **195**, 215–33.

Harcourt, A.H., Kineman, J. & Campbell G. (1983). Conservation of the Virunga gorilla population. *Afr. J. Ecol.* **21**, 139–42.

Hashimoto, C. (1995). Population census of the chimpanzees in the Kalinzu Forest, Uganda: comparison between methods with nest counts. *Primates* **36**, 477–88.

Havens, K.J. & E.J. Sharp. (1998). Using thermal imagery in the aerial survey of animals. *Wildlife Soc. Bull.* **26**, 17–23.

Hubbs, A.H., Karels, T. & Boonstra, R. (2000). Indices of population size for burrowing mammals. *J. Wildlife Manag.* **64**, 296–301.

Krebs, C.J. (1999). *Ecological Methodology*, 2nd edition. Menlo Park, CA: Addison Wesley Longman.

McGraw, W.S. (1998). Three monkeys nearing extinction in the forest reserves of eastern Côte d'Ivoire. *Oryx* **32**, 233–6.

McNeilage, A., Plumtre, A.J., Brock–Doyle, A. & Vedder, A. (2001). Bwindi Impenetrable National Park, Uganda: Gorilla census, 1997. *Oryx* **35**, 39–47.

Nichols, J.D. & M.J. Conroy (1996). Techniques for estimating abundance and species richness. In *Measuring and Monitoring Biological Diversity: Standard Methods for Mammals*, ed. D.E. Wilson, F.R. Cole, J.D. Nichols, R. Rudran &. M.S. Foster, pp. 177–234. Washington, DC: Smithsonian Institution Press.

Parnell, R.J. (2000). Information from animal tracks and trails. In *Conservation Research in African Rain Forests: A Technical Handbook*, ed. L. White & A. Edwards, pp. 157–89. New York: Wildlife Conservation Society.

Payne J. (1987). Surveying orang-utan populations by counting nests from a helicopter: a pilot survey in Sabah. *Primate Cons.* **8**, 92–103.

Schaller, G.B. (1963). *The Mountain Gorilla: Ecology and Behavior*. Chicago: University of Chicago Press.

Sholley, C.R. (1991). Conserving gorillas in the midst of guerillas. In *American Association of Zoological Parks and Aquariums, Annual Conference Proceedings*, pp. 30–7.

Singh, M., Lindburg, D.G., Udhayan, A., Kumar, M.A. & Kumara, H.N. (1999). Status survey of slender loris *Loris tardigradus lydekkerianus* in Dindigul, Tamil Nadu, India. *Oryx* **33**, 31–7.

Southwood, T.R.E. & Henderson, P.A. (2000). *Ecological Methods*, 3rd edition. Oxford: Blackwell Science.

Thomas, L., Laake, J.L., Derry, J.F., Buckland, S.T., Borchers D.L., Anderson, D.R., Burnham, K.P., Strindberg, S., Hedley, S.L., Marques, F.F.C., Pollard, J.H. & Fewster, R.M. (1998). *Distance 3.5. Research Unit for Wildlife Population Assessment*. St Andrews. University of St Andrews.

Tutin, C.E.G. & Fernandez, M. (1984). Nationwide census of gorilla (*Gorilla g. gorilla*) and chimpanzee (*Pan t. troglodytes*) populations in Gabon. *Am. J. Primatol.* **6**, 313–36.

Tutin, C.E.G., Parnell, R.J., White, L.J.T. & Fernandez, M. (1995). Nest building by lowland gorillas in Lope Reserve, Gabon: environmental influences and implications for censusing. *Int. J. Primatol.* **16**, 53–76.

van Krunkelsven, E., Inogwabini, B.-I. & Draulans, D. (2000). A survey of bonobos and other large mammals in Salonga National Park, Democratic Republic of Congo. *Oryx* **34**, 180–8.

Wemmer, C., Kunz T.H., Lundie–Jenkins G. & McShea W.J. (1996). Mammalian signs. In *Measuring and Monitoring Biological Diversity: Standard Methods for Mammals*, ed. D.E. Wilson, F.R. Cole, J.D. Nichols, R. Rudran &. M.S. Foster, pp. 157–76. Washington, DC: Smithsonian Institution Press.

White, L. & Edwards, A. (2000). Methods for assessing the status of animal populations. In *Conservation Research in African Rain Forests: A Technical Handbook*, ed. L. White & A. Edwards, pp. 225–73. New York: Wildlife Conservation Society.

Useful Internet site

Distance sampling: <www.ruwpa.st–and.ac.uk/distance/>.

7 • Trapping primates

CLIFFORD J. JOLLY[1], JANE E. PHILLIPS-CONROY[2] AND
ALEXANDRA E. MÜLLER[3]

[1]*Anthropology, New York University, New York, USA*
[2]*Anatomy and Neurobiology, Washington University School of Medicine, St Louis,
Missouri, USA*
[3]*Anthropological Institute and Museum, University of Zürich, Zürich, Switzerland*

INTRODUCTION

There are many reasons to capture your study animals, including marking or
radio-collaring individuals (Chapter 10), taking morphological measurements
(Chapter 9) and biological sampling (Chapters 1 and 8). For small nocturnal
primates, trapping may be the only way to gather data for density estimates
(Chapter 6). Furthermore, it is essential for the determination of spatial distribution and social interactions of individuals, as the most effective method
uses direct observation of radio-tagged animals (Sterling *et al.*, 2000). Histor-
ically, studies in which wild, larger-bodied non-human primates (hereafter
called primates in this chapter) are habituated for long-term observation have
rarely included capture, perhaps because researchers have been understand-
ably wary of its effects on subsequent behaviour and habituation. However,
our survey (Jolly & Phillips-Conroy, 1993; C.J. Jolly and J.E. Phillips-Conroy,
unpublished data) of more than 120 studies that combined observation with
capture, and which involved about 65 primate species, showed that a careful
capture–release programme using trapping will not cause a previously habituated population to change its behaviour towards human observers, and will
not be associated with excess mortality or serious injury. Changes in ranging habits caused by baiting and trapping will be temporary at worst, and
basic social organisation and structure will not be affected. The survey also
showed that trapping has been used most often to catch diurnal–terrestrial
and nocturnal–arboreal species. Diurnal–arboreal primates (apart from callitrichines) have generally been captured by darting (Chapter 8), a bias that
seems unjustified. While in skilled hands either trapping or darting can produce good results, trapping is much more productive, far less hazardous to
the subjects, and less likely to cause them to become frightened of an observer. It is true that some species are notoriously difficult or even impossible
to trap – either because they will not approach a strange object such as a trap
(e.g. owl monkeys, *Aotus* spp., and titis, *Callicebus* spp.), or because they are
not attracted to foods offered as bait (most folivores and gummivores) – but

Field and Laboratory Methods in Primatology: A Practical Guide, ed. Joanna M. Setchell and
Deborah J. Curtis. Published by Cambridge University Press. © Cambridge University
Press 2003.

trapping should certainly be attempted for arboreal and semi-arboreal, frugivorous–omnivorous primates such as lemurids, platyrrhines such as *Cebus* and catarrhines such as forest cercopithecines and some colobines.

Here, we outline trapping methods, drawing both on the results of our survey and on our experience of catching cercopithecines (C. J. J. and J. P.-C.) and nocturnal lemurs (A. E. M.). Although details vary greatly with the biology of the species and the study's objectives, all live-trapping projects involve similar considerations: when to capture; which animals to capture; trap design; number of traps; trap placement; baiting, trapping and processing the animals; and a protocol of holding, recovery and release.

WHEN TO CAPTURE

Since trapping is unlikely to prejudice subsequent observation, it is best to trap early in the study programme, after the animals have been habituated to human contact (Chapter 2). This gives the study the immediate advantage of marked and radio-collared individuals, and the support of a comprehensive data set that includes individuals that emigrate or die during the study. Data for animals that immigrate or are born later must of course be added by individual trapping or by non-invasive sampling of 'bio-detritus' (Chapters 19 and 20). While it can be carried out at any time of year, trapping – especially the first trapping of a group unfamiliar with bait – is generally most productive when natural food is less abundant (Charles-Dominique, 1977).

WHICH ANIMALS TO CAPTURE

There are obvious economies of scale to be derived by sampling the population comprehensively, and no sex or age category needs to be excluded. Some capture programmes have avoided pregnant females and young infants, because these animals are felt to be particularly at risk from trapping and handling. This restriction is unnecessary, however, as long you are careful to avoid separating dependent infants from their caregivers – usually their mothers, but (especially among some platyrrhines) other group members also. We (C. J. J. and J. P.-C.) have successfully captured, processed (and observed after recovery) pregnant female baboons (*Papio* spp.) that gave birth the next day, and at least one infant with its umbilical cord still attached. Contrary to expectation, the most vulnerable animals among cercopithecines seem to be the adult males, which are most likely to sustain minor trauma and also most susceptible to overheating under ketamine sedation (Chapter 8).

Even if observation is to focus on a single social group, it is advantageous to trap neighbouring groups as well. Not only will this document the wider demographic and genetic context, it will also allow animals that are likely to immigrate into the primary study group to become accustomed to traps and bait. Unless previously familiarised in this way, immigrants tend to be excluded from bait and traps by more confident, acclimatised, residents.

All studies report that the vast majority of individuals simply resume their previous social roles after trapping and release. There is, however, a possibility that, unless the group is captured in its entirety at one time, the temporary absence of key individuals may cause changes in the status of other group members. In particular, consortships may be disrupted long enough for mating to be affected, and in species such as gelada (*Theropithecus gelada*) and hamadryas baboons (*Papio hamadryas*), 'harems' may disperse or be taken over by rival males. Such events can be considered analogous to natural injuries or predation events that similarly affect individual life histories, but do not alter the overall structure of society. By timing the trapping season as early as possible, one ensures that any capture-induced changes in status occur before individual identities have become a crucial component of the observational data.

TRAP DESIGN

Trapping primates depends upon conditioning them to recognise traps as desirable and 'friendly' feeding sites, which means that traps can be entirely visible and do not have to be disguised. Two major kinds of trap are used. Gang traps are designed to catch several animals at a time, while individual traps are meant to catch one animal per closure, though in practice multiple captures often occur. In theory, gang traps have the advantage of a high capture rate, and of catching whole social units together, thus minimising the chance of social disruption. Though several investigators (J. Bert, personal communication, 1970; T. Shotake, personal communication, 1999) have reported using gang traps successfully to catch macaques (*Macaca* spp.) and baboons, our experience of them has not been positive. Large traps are difficult to transport and set up in the field, and removing animals from the trap is much more difficult, and correspondingly more stressful to both the operator and the animals. There is also a greater risk that smaller individuals will be bitten when trapped *en masse*. An alternative is to use a multi-chambered trap to capture entire social groups (e.g. Garber *et al.*, 1993; Savage *et al.*, 1993).

Most traps used to catch monkeys consist of a simple, custom-built, individual cage with a hinged or (more usually) vertically sliding door (Brett *et al.*, 1982; de Ruiter, 1992). An alternative design catches the animal under a shallow tray-like box that is propped on edge when set (Brett *et al.*, 1982). Commercially available Tomahawk traps, wrapped with 5 mm^2 wire mesh to prevent bait stealing, have been used for callitrichines (Dawson, 1977; Albernaz & Magnusson, 1999; J. M. Dietz, unpublished data).

Custom-built cage traps, probably the design of choice for most species, will obviously vary greatly in size. For convenience, and for the animal's safety, they should be as small and light as possible. For example, a cage that is a cube with sides of about 90 cm, and a door measuring 50 cm × 50 cm, can catch and hold a 30 kg baboon. The height should be sufficient to allow the animal to sit comfortably, but need not be much greater than that. Traps are best constructed with a rigid frame but flexible mesh walls, using materials as light in weight as is compatible with necessary strength. If securely tied down to prevent tipping, a trap of chain link, or even good-quality chicken wire, on a light metal frame, will hold any primate up to baboon size. Fine mesh netting should screen the bait to prevent animals reaching it from outside.

Most traps are designed so that they can be closed either automatically (tripped by the animal itself) or manually (released from a distance by an operator). An automatic trigger can be a treadle, operated by the animal's weight, or a simple catch that the animal releases by pulling on a bait. Manual closure usually involves the operator pulling a string attached to the catch, from a distance (commonly approximately 25–100 m) that must be determined empirically in the field. The operator can be concealed in a blind (hide) (Garber et al., 1993), but this restricts vision and movement, and we have not found it necessary for catching baboons or vervet monkeys (Chlorocebus aethiops).

For nocturnal species, the classic ('Chardonneret') trap is a 30 cm × 35 cm × 50 cm box made of wire mesh stapled to a wood frame, with a large, automatically triggered entrance on top (Charles-Dominique & Bearder, 1979; Müller, 1999). This has proved effective for most galagines, pottos (Perodicticus potto) and small cheirogaleids, and has also caught pygmy marmosets (Cebuella pygmaea; Zingg, 2001) and brown lemurs (Eulemur fulvus; A. Müller, unpublished data). Appropriately sized commercial traps such as Sherman and Tomahawk Live Traps are also commonly used to catch cheirogaleids (e.g. Wright & Martin, 1995; Kappeler, 1997; Atsalis, 1999; Fietz, 1999; Fietz & Ganzhorn, 1999; Ehresmann, 2000). They have the advantage of being portable, convenient and comparatively light, but importation can cause bureaucratic problems and expense.

Whatever the design, it is important that the traps have no sharp edges inside and the use of plastic-covered wire can prevent some injuries. If the animal's tail is likely to lie across the threshold, a device should be incorporated to prevent guillotining when the door shuts. Trapped animals often try to enlarge any small break in the netting, so it is very important to fix the netting securely to the frame (binding it with wire is usually more reliable than welding), and to mend breaks promptly. Escape efforts are usually concentrated on the door, which must be rigid enough not to spring out of its runners when impacted on vigorously from inside. Many primates become adept at raising the door manually. This can be prevented with a simple catch that locks when the door drops.

NUMBER OF TRAPS

The number of traps employed will obviously vary with the demography of the target species and anticipated capture rates. For group-living species, more traps will produce a larger per-day yield and may make it easier to catch the shyest individuals. Processing the captives then becomes the limiting step, especially if they are to be released immediately. About 20 traps per group will probably be a practical maximum. For nocturnal prosimians, some studies have used up to 200 portable, commercial traps, when animals are processed in the camp and released the following night (e.g. Fietz & Ganzhorn, 1999; Schwab, 2000). Other studies, in which up to 80% of the traps were successful and animals were processed at the capture site, have used 20–40 traps (Müller, 1999; A. Müller, unpublished data).

SELECTION OF THE TRAP SITE

The prime consideration here is that the animals should visit the trap site frequently, so that they can be acclimatised to taking bait – a site near a favoured sleeping-place is ideal. It should be close enough that it is usually traversed as the animals enter and leave their sleeping-place, but not so close that a visit at dusk or dawn by the trappers is likely to scare the animals. If a cluster of manually closed traps is to be used, the site should be chosen so that all can be scanned and operated from a single position.

Arboreal species must be trapped in the trees, at a height where the animals are accustomed to feed. Traps are positioned so as to afford easy access, for example fixed between two or more small trees, in a fork of a branch, or wired securely to a horizontal branch or vine. For group-living species, it is often advantageous to construct an elevated baiting platform (1 m × 2 m), and place traps on it and in the surrounding trees (Savage *et al.*, 1993; Dietz *et al.*, 1994; J.M. Dietz, unpublished data). When trapping nocturnal prosimians that forage alone, a large number of traps can be dispersed in a grid (e.g. Fietz, 1999; Fietz & Ganzhorn, 1999), or fewer traps can be grouped at a feeding site, so as to minimise trapping animals other than the target species (Müller, 1999).

HABITUATION AND PRE-BAITING

This is probably the most crucial phase in determining the success of a trapping campaign; obviously, if no animals take the bait none will be caught. With a new species or a naïve population, one should plan on a month or more of patient and careful preliminary observation, non-interactive habituation, and evaluation of potential baits. It may take some time for naïve animals to try an unfamiliar food, and it is a common mistake to abandon a potential bait after insufficient trials.

Sweet fruits such as bananas are a highly attractive bait for many species, including most nocturnal prosimians and callitrichines (Savage *et al.*, 1993; Dietz *et al.*, 1994; Wright & Martin, 1995; Müller, 1999; J.M. Dietz, unpublished data), but they are also perishable and attract insects such as ants. For species (especially macaques, baboons and guenons, *Cercopithecus* spp.) that find dried grains attractive, loose maize (corn), rice or wheat kernels are very convenient to handle and store, and also hold the animals' attention to the trap site, since they cannot easily be carried off for consumption elsewhere. Some studies have used a 'caller' – a captive of the target species – to attract wild groups to the baiting area, but this raises issues of logistics and animal welfare (e.g. Savage et al., 1993) and it has been shown repeatedly to be unnecessary with careful pre-baiting (Garber et al., 1993; Savage et al., 1993; J. M. Dietz, unpublished data).

Once a viable bait has been identified, pre-baiting commonly takes at least one to two weeks per group. If traps are to be concentrated at a feeding site or platform, bait is usually offered at first at this locality without traps present. When animals visit the site regularly, unarmed (fixed open) traps are introduced and bait is placed around, on and in them. Multiple traps on a grid are locked open and individually pre-baited. If possible – especially for group-living, diurnal species – the later phases of pre-baiting should be monitored by an observer, to ensure that all target animals are entering the traps to feed. Individuals that do not enter traps during pre-baiting are unlikely to be caught when traps are set.

CAPTURE, TRANQUILLISATION AND RELEASE

In most reported studies of nocturnal prosimians, portable, automatic traps have been set at dusk and checked at dawn. Captives are carried to the field laboratory in the traps, processed, and released at dusk at the site of their capture. Alternatively, and preferably, traps are monitored regularly during the night and captives processed immediately, at the capture site (Müller, 1999). As most animals are trapped soon after nightfall, this minimises holding time, does not prevent lactating females from attending to their 'parked' infants, and allows the animals the rest of the night for foraging. Small nocturnal prosimians can be taken out of the trap easily and tranquillisation is only necessary (if at all) when the animals are radio-collared. With larger species, one person can push the animal to the side of the trap with a cloth or glove while a second person gives a subcutaneous injection (O. Schülke, personal communication, 2001). After processing, the animals are usually held in their individual traps (covered with a cloth) or in dark cotton bags until release.

When callitrichids are to be caught, traps are set before dawn. Traps can be closed manually, by pulling a string (e.g. Garber et al., 1993; Savage et al., 1993) or automatically (Albernaz & Magnusson, 1999; J. M. Dietz, unpublished data). In the latter case, traps are checked in the late morning. At dusk, traps are closed and trapped animals brought to the field laboratory for processing. After spending the night in a cloth-covered trap, cloth bag or cage, they are released at their capture site the next morning (e.g. Savage et al., 1993; J. M. Dietz, unpublished data).

Much the same protocol can be used for cercopithecine monkeys. Animals are captured in the morning as they leave their sleeping-place. Traps are closed manually, and catching continues until all traps are filled and closed, or the group has left, or no 'new' animals are entering the traps. Animals in excess of the processing team's daily capacity should not be caught, because if released without being tranquillised they will be difficult to retrap (it is unclear why tranquillisation should make monkeys easier to retrap – perhaps drugs such as ketamine and phencyclidine hydrochloride reduce or eliminate the adverse memories of capture). Captives are tranquillised with a pole- or hand-syringe (Jolly, 1998) at intervals determined by the pace of processing, examined at a field laboratory close to the capture site, then placed in a recovery cage and released individually as soon as the effects of sedation wear off. Small, nursing infants are not tranquillised, and usually cling to their mother while she is sedated and processed. As a precaution against an infant jumping off and running away, it should be secured to its mother's wrist by a cord tied around its waist or ankle. The cord is removed when the mother and infant are placed in the recovery cage.

If complete ascertainment is important, and if the target species (like many terrestrial and semi-terrestrial cercopithecines) lives in groups consisting of more animals than can be trapped and examined in a single day, then immediate release has some drawbacks. Daily capture rates rapidly dwindle as untrapped animals are outnumbered by those that have already been released (and which then tend to 'hog' the traps). A residue of timid or subdominant individuals will probably never be caught. Moreover, because traps must be closed manually, the group has to be thoroughly habituated to human presence before trapping begins.

For monkey species that live in large groups, and are known to tolerate captivity, a hold–release strategy may be preferable (Brett *et al.*, 1982; Melnick *et al.*, 1984). In this case, traps are set to close automatically. After the group has moved away from the trapping area, all captives are tranquillised, transported to a processing site, examined and put into holding cages where they will live until trapping of their group is completed. A major advantage is that no operator need be present during capture, allowing minimally habituated groups to be trapped. On the other hand, extended holding involves expenditure on cages and food, and possibly increased stress and health risks to the captives. Another potential problem is that as a group is progressively deprived of its most powerful members – inevitably trapped first – it may be displaced from the trap site by neighbours, resulting in 'mixed bags' drawn from more than one social group.

If a hold–release strategy is adopted, dedicated holding cages must be built. They should be as small as is consistent with comfort, and should be placed so that group members are able to see each other and communicate. To avoid injury, co-housing arrangements should be compatible with the species' social behaviour. Often, adult males are best housed singly, while juveniles and compatible females should be co-housed. The holding area should, of course, be sanitary and provide protection against weather, snakes and predators, and should be off-limits to all non-essential human visitors.

Whether released singly or *en masse*, animals must be released in their home range, usually at their capture site. In a mass release, the youngest animals should be liberated first to reduce the chance of adults leaving them behind (J. M. Dietz, unpublished data). With wide-ranging species such as baboons the group may move a considerable distance before captives are released. We have not, however, found it necessary – or practicable – to keep animals under sedation and carry them back to their group before release. Once recovered from sedation, they are much more efficient than we could be at finding their troop mates.

DATA GATHERING

Because trapping is costly, it makes sense not only to mark or radio-collar animals, but also to collect as comprehensive a data set as is compatible with the animal's well-being (e.g. Chapters 8, 9, 10, 18, and 20). Even materials of no immediate relevance to the study's objectives often prove to be useful to later investigations, some using techniques undreamed of at the time of collection. The complexity of the data to be collected, the number of animals to be caught each day and the size of individual animals will determine the size and composition of the field team. In the simplest cases, trapping and data collection can be carried out by one researcher and an assistant. When sampling larger and more numerous animals, it is preferable to organise separate trapping/sedating and data-gathering teams. At the examination table, tasks such as applying an eartag or other marker, making dental moulds and casts, blood sampling, weighing and measuring are individually assigned. As far as is practical in the field, the usual precautions against zoonotic infection (gloves, masks, goggles) should be observed. One team member should be assigned to monitor the sedated animal's condition (Chapter 8), and another should remain 'clean' in order to record data and to check off procedures. An experienced team can gather an animal's full data set in about 30 minutes, which imposes a practical limit of about 15–20 animals per day. A hold–release protocol affords a second opportunity to collect data when animals are tranquillised again for transportation to the release site.

Retrapping animals in subsequent surveys is rarely problematic in the case of cercopithecines, callitrichines (Savage *et al.*, 1993; J. M. Dietz, unpublished data) or nocturnal prosimians. Indeed, conditioning may even be reinforced by the experience of repeated trapping, tranquillisation and handling, as animals tend to become attached to traps and trap-sites as favoured feeding places. A more likely problem than trap aversion is that animals become 'trap-happy', or learn to take bait without entering or closing the trap. There is, however, much inter-individual variation in reaction to traps; some individuals are much harder to trap/retrap and do not enter the same trap twice.

CATCHING SMALL, NON-TRAPPABLE PRIMATES

Some species (typically those with a natural diet composed largely of small animals, exudates, or leaves) apparently cannot be lured into baited traps, and many are also too small to be darted easily, but can be caught in other ways. Slender lorises (*Loris tardigradus*), for instance, seem to be relatively easy to catch by hand if the animal is low in the tree (Petter & Hladik, 1970). Mist-netting captures tarsiers (*Tarsius* spp.) safely, but recapture is difficult (Gursky, 2000). Slow lorises (*Nycticebus coucang*) can be caught in commercial traps but success is low, and requires many traps (approximately 200) and creative tactics (Wiens, 2002). Aye-ayes (*Daubentonia madagascariensis*) can be captured from their nests during the day (Petter & Peyrieras, 1970; Ancrenaz *et al.*, 1994). Schülke succeeded in capturing fork-marked (*Phaner* spp.) and sportive lemurs (*Lepilemur* spp.) by leading them into a Tomahawk trap via a tunnel of netting stapled around the entrance to their nest hole in a hollow tree (O. Schülke, personal communication, 2001). Inventiveness is the name of the game, and it would also be worth while experimenting with lures other than conventional baits – natural or artificial scents, for example, live insect bait, recorded calls or images of conspecifics – to attract animals into traps.

CLOSING THOUGHTS

A trapping programme in a populated area always attracts the attention of local people, and it is important to convey the study's objectives clearly to them – in particular, to stress that tranquillised animals are sleeping, rather than dead or injured, and that animals will be released unharmed, will not be eaten and will not be killed (even at the request of local farmers). Often, people who have never seen a non-human primate close up are amazed by its human-like anatomy and physiognomy, and this can be turned to advantage in stimulating empathy. Such demonstrations should, of course, be confined to a very select audience; it is disastrous if a trapping operation becomes a public spectacle. Between trapping seasons, traps should be securely stored to prevent illicit uses.

Although trapping primates is generally safe, it can be stressful for both the captive and the captor. Primatologists accustomed to observing their subjects with minimum contact and disturbance are seldom psychologically prepared for the kind of interactions required by trapping. A trapped animal (and some-times its group mates, if present) will scream blood-curdlingly when the trap is approached, and will certainly bite in self-defence if given the chance. Since it is the fear of accidentally injuring or even killing a study animal that makes capturing stressful for the novice trapper, it is sensible wherever possible to call upon experienced trappers for help when beginning a project. Trapping primates can be very easy, avoiding associated problems and knowing when not to trap is the real challenge.

ACKNOWLEDGEMENTS

C.J.J. and J.P.-C. thank the many colleagues who contributed unpublished data to the survey of trapping, and the numerous colleagues, students and volunteers who have assisted in trapping seasons. A.E.M. thanks Simon Bearder, Jim Dietz, Joanna Fietz, Leanne Nash, Oliver Schülke and Frank Wiens for their information on trapping techniques, Urs Thalmann for his assistance in trapping dwarf lemurs during her own field study and Beno Schoch for his help with preparing the traps. The authors' fieldwork has been supported by the NSF, the NIH, the Harry Frank Guggenheim Foundation, Washington University, New York University, and Earthwatch (C.J.J. and J.P.-C.) and the A.H. Schultz Foundation, Family Vontobel Foundation, Goethe Foundation, G. & A. Claraz Donation and the Swiss Academy of Natural Sciences (A.E.M.).

REFERENCES

Albernaz, A.L. & Magnusson, W.E. (1999). Home-range size of the bare-ear marmoset (*Callithrix argentata*) at Alter do Chão, Central Amazonia, Brazil. *Int. J. Primatol.* **20**, 665–77.

Ancrenaz, M., Lackman-Ancrenaz, I. & Mundy, N. (1994). Field observations of aye-aye (*Daubentonia madagascariensis*) in Madagascar. *Folia Primatol.* **62**, 22–36.

Atsalis, S. (1999). Seasonal fluctuations in body fat and activity levels in a rainforest species of mouse lemur, *Microcebus rufus*. *Int. J. Primatol.* **20**, 883–910.

Brett, F.L., Turner, T.R., Jolly, C.J. & Cauble, R.G. (1982). Trapping baboons and vervet monkeys from wild, free-ranging populations. *J. Wildlife Manag.* **46**, 164–74.

Charles-Dominique, P. (1977). *Ecology and Behaviour of Nocturnal Primates*. London: Duckworth.

Charles-Dominique, P. & Bearder, S.K. (1979). Field studies of lorisid behavior: Methodological aspects. In *The Study of Prosimian Behavior*, ed. G.A. Doyle & R.D. Martin, pp. 567–629. New York: Academic Press.

Dawson, G.A. (1977). Composition and stability of social groups of the tamarins (*Saguinus oedipus geoffroyi*) in Panama: ecological and behavioral implications. In *The Biology and Conservation of the Callitrichidae*, ed. D.G. Kleiman, pp. 23–8. Washington, DC: Smithsonian Institution Press.

de Ruiter, J. (1992). Capturing wild long-tailed macaques (*Macaca fascicularis*). *Folia Primatol.* **59**, 89–104.

Dietz, J. M., Baker, A. J. & Miglioretti, D. (1994). Seasonal variation in reproduction, juvenile growth, and adult body mass in golden lion tamarins
(*Leontopithecus rosalia*). *Am. J. Primatol.* **34**, 115–32.

Ehresmann, P. (2000). Ökologische Differenzierung von zwei sympatrischen
Mausmaki-Arten (*Microcebus murinus* und *M. ravelobensis*) im Trockenwald
Nordwest-Madagaskars. Ph.D. thesis, University of Hanover.

Fietz, J. (1999). Mating system of *Microcebus murinus*. *Am. J. Primatol.* **48**, 127–33.

Fietz, J. & Ganzhorn, J. U. (1999). Feeding ecology of the hibernating primate
Cheirogaleus medius: how does it get so fat? *Oecologica* **121**, 157–64.

Garber, P. A., Encarnación, F., Moya, L. & Pruetz, J. D. (1993). Demographic and
reproductive patterns in moustached tamarin monkeys (*Saguinus mystax*):
implications for reconstructing platyrrhine mating systems. *Am. J. Primatol.*
29, 235–54.

Gursky, S. (2000). Sociality in the spectral tarsier, *Tarsius spectrum*. *Am. J.
Primatol.* **51**, 89–101.

Jolly, C. J. (1998). A simple and inexpensive pole syringe for tranquilizing primates. *Lab. Primate News.* **37**(2), 1–2.

Jolly, C. J. & Phillips-Conroy, J. E. (1993). The use of capture in field primatology.
Am. J. Phys. Anthropol., Suppl. **16**, 158 [Abstract].

Kappeler, P. M. (1997). Intrasexual selection in *Mirza coquereli*: evidence for
scramble competition in a solitary primate. *Behav. Ecol. Sociobiol.* **45**, 115–27.

Melnick, D. J., Pearl, M. C. & Richard, A. F. (1984). Male migration and inbreeding avoidance in wild rhesus monkeys. *Am. J. Primatol.* **7**, 229–43.

Müller, A. E. (1999). Aspects of social life in the fat-tailed dwarf lemur
(*Cheirogaleus medius*): inferences from body weights and trapping data. *Am.
J. Primatol.* **49**, 265–80.

Petter, J. J. & Hladik, C. M. (1970). Observations sur le domaine vital et la den-
sité de population de *Loris tardigradus* dans les forêts de Ceylan. *Mammalia*
34, 394–409.

Petter, J. J. & Peyrieras, A. (1970). Nouvelle contribution à l'étude d'un lémurien
malgache, le aye-aye (*Daubentonia madagascariensis* E. Geoffroy). *Mammalia*
34, 167–93.

Savage, A., Giraldo, H., Blumer, E. S., Soto, L. H., Burger, W. & Snowdon,
C. S. (1993). Field techniques for monitoring cotton-top tamarins (*Saguinus
oedipus oedipus*) in Columbia. *Am. J. Primatol.* **31**, 189–96.

Schwab, D. (2000). A preliminary study of spatial distribution and mating
system of pygmy mouse lemurs (*Microcebus* cf. *myoxinus*). *Am. J. Primatol.* **51**,
41–60.

Sterling, E. J., Nguyen, N. & Fashing, P. J. (2000). Spatial patterning in nocturnal
prosimians: a review of methods and relevance to studies of sociality. *Am.
J. Primatol.* **51**, 3–19.

Wiens, F. (2002). Behavior and ecology of wild slow lorises (*Nycticebus coucang*):
social organization, infant care system, and diet. Ph.D. thesis, University of
Bayreuth.

Wright, P.C. & Martin, L.B. (1995). Predation, pollination and torpor in two nocturnal prosimians: *Cheirogaleus major* and *Microcebus rufus* in the rain forest of Madagascar. In *Creatures of the Dark: The Nocturnal Prosimians*, ed. L. Alterman, G.A. Doyle & M.K. Izard, pp. 45–60. New York: Plenum Press.

Zingg, J. (2001). Nahrungsökologie und Sozialverhalten von Zwergsei-denäffchen (*Cebuella pygmaea*) in Ecuador. Diploma thesis, University of Zurich.

Useful Internet sites

Sherman traps: <www.shermantraps.com>.
Tomahawk Live Traps: <www.tomahawklivetraps.com>.

8 • Handling, anaesthesia, health evaluation and biological sampling

MARC ANCRENAZ[1], JOANNA M. SETCHELL[2] AND DEBORAH J. CURTIS[2]

[1]*Kinabatangan Orang-utan Conservation Project, Sabah, Malaysia*
[2]*Centre for Research in Evolutionary Anthropology, School of Life and Sport Sciences, University of Surrey Roehampton, London, UK*

INTRODUCTION

Infectious agents affect population dynamics, ecology, behaviour and individual reproductive success, and disease and health issues are widely recognised as important factors in wildlife conservation (Daszak *et al.*, 2000; Deem *et al.*, 2001). Non-human primates (NHPs) deserve special attention for general health issues because of their close relatedness to humans and potential disease exchange between NHPs and humans (zoonosis) and vice versa (anthropozoonosis) (Chapter 1). However, there are very few published data on the health status and diseases of free-living NHPs, the interrelationships between diseases of wild NHPs and other biological parameters, and more generally on the baseline biological parameters of NHPs under natural conditions. This paucity of information hampers our understanding of natural epidemiological situations and especially the understanding of new emergent diseases (Wolfe *et al.*, 1998), the impacts of habitat modification, the success of introduction, reintroduction or translocation programs (Tutin *et al.*, 2001), interactions between domestic and wild animals, and interactions between humans and NHPs (Chapter 1).

Primatologists capture wild NHPs for a number of reasons other than medical screening, including sampling for endocrinology, genetics (Chapter 20) and physiology (Chapter 18), or for marking and radiotelemetry (Chapter 10). This provides a valuable opportunity to collect biological samples, and since the capture and handling of free-ranging NHPs is always accompanied by the risk of injury or mortality, it is ethically important to maximise the amount of information gathered during these procedures (Karesh *et al.*, 1998). Biological samples should therefore be collected opportunistically every time wild NHPs are handled.

This chapter gives basic methods for handling NHPs, clinical examination, and the collection of biological samples. For certain procedures, it is necessary to collaborate with a wildlife veterinarian. However, primatologists can gain useful basic veterinary knowledge by attending training courses at zoos or at

Field and Laboratory Methods in Primatology: A Practical Guide, ed. Joanna M. Setchell and Deborah J. Curtis. Published by Cambridge University Press. © Cambridge University Press 2003.

other captive facilities prior to departure for the field. Researchers also need to contact the relevant authorities in the habitat country to obtain permission for the capture, anaesthesia and sampling of free-ranging animals, as well as for exportation and importation of biological samples.

HANDLING NHPs

As a general rule, NHPs are sensitive to stress, and (except for younger animals or smaller species) are very strong and able to inflict severe injuries on human handlers. Handling is a compromise between human and animal safety in a situation where stress levels must be kept as low as possible. Physical handling without tranquillisation is possible for smaller individuals, but chemical tranquillisation is necessary for larger individuals, or for major procedures.

PHYSICAL HANDLING

NHPs weighing less than 3 kg can easily be handled and manipulated by one person. However, most of these species are agile and will not hesitate to bite if they feel threatened. Always handle primates with protective leather gloves. The smallest animals can be gripped around the neck with the thumb and forefinger, or by a fold of skin on the flank (especially prosimians). Larger prosimians can be held by the scruff of the neck with one hand, and by the feet with the other. Simians should be grasped just under the arms, with the fingers encircling the upper chest. This position is convenient for intramuscular or intraperitoneal injection. However, for more effective immobilisation, the arms are held together behind the animal's back in the same fashion as that used with larger primates. A second person is always required to carry out procedures, and restraint should be kept to as short a time as possible to minimise stress. Most prosimians have an explosive flight reaction when disturbed in their nest or their captive box and considerable care should be taken when capturing individuals manually (Bearder & Pitts, 1987).

Animals weighing 3–10 kg can be caught in a heavy net (or blanket or piece of fabric) and held by pinning the arms back. The two arms are held together with one hand behind the primate's back whilst the two feet are held together with the other hand. Never hold the animal by only one arm, as this could result in fracture or dislocation of the humerus.

For obvious safety reasons, species weighing more than 10 kg need to be chemically restrained.

CHEMICAL HANDLING

Chemical immobilisation of wild primates is a difficult, risky and hazardous procedure. Besides the risks related directly to the capture itself, several authors have stressed the problem of behavioural disruption of a group following an anaesthetic event, resulting for example in a change in the social status of the darted individual (Hamadryas baboon, *Papio hamadryas*; M. Ancrenaz, personal observations), or an altered response to the human observer (Karesh *et al.*, 1998).

Methods for trapping free-ranging NHPs are described in Chapter 7. Procedures for darting small arboreal primates are described by Glander *et al.* (1991), Jones & Bush (1988) and Karesh *et al.* (1998); guidelines for chemical capture of large terrestrial NHPs can be found in Sapolsky & Share (1998) and Sleeman *et al.* (2000); practical tips for chemical restraint of mammals as well as a list of the necessary equipment for field anaesthesia may be found in Osofsky & Hirsch (2000); finally, practical information on chemical immobilisation is also available from various websites (see 'Useful Internet sites' at the end of this chapter).

Choice of drugs

The choice of substance and dosage is always a compromise between safety and efficiency, and depends on the delivery system, skill and preparation of the human team, ecological conditions (temperature), and the biological features of the primate itself. There are wide differences in physiological characteristics (body temperature, heat deperdition, heart rate, etc.) between a 60 g galago (*Galago demidovii*) and a 350 kg gorilla (*Gorilla gorilla*). The equivalent dosage of anaesthetic will have a variable effect, depending on the metabolic size of the individual: all aspects of anaesthesia are accelerated in smaller individuals and slowed in larger individuals. Thus dosage rates should be increased for smaller animals and reduced for larger ones (Sedgwick, 1985). Several other biological features (age, body condition, pregnancy, lactation, etc.) must also be considered.

The first priority when choosing an anaesthetic is to maximise the safety margin of the drug (the difference between an efficacious and a lethal dose), since the exact weight of the animal to be caught is rarely known before darting takes place in the wild. Two main types of anaesthetic drugs are used for NHPs: dissociative agents (also called cyclohexamines), and alpha-2 agonist agents.

The two main dissociative agents used to anaesthetise wild NHPs are ketamine hydrochloride and tiletamine hydrochloride. They are called dissociative, since (in contrast to classical anaesthetics) despite a good stage of analgesia, thermoregulation, blood pressure, respiratory and spinal reflexes are well maintained (Sapolsky & Share, 1998), and the persistence of these reflexes can sometimes be confused with insufficient anaesthesia (Schobert, 1987). Anaesthetised individuals have increased skeletal muscle tone, with slight tachycardia and hypertension (Dripps *et al.*, 1977). Dissociative agents are easily available, relatively cheap, easy to use, and have a short pre-anaesthesia time and a very wide safety margin. The main side effects and treatments are listed in Table 8.1. To decrease these side effects, dissociative agents are typically used in association with a tranquilliser: xylazine for ketamine (1 unit of xylazine per 10 units of ketamine) and zolazepam for tiletamine (1 : 1 ratio). However, these tranquillisers have a strong depressive effect on the cardiorespiratory system.

Alpha-2 agonist drugs, developed only recently, have been used successfully to anaesthetise wild primates (Jalanka & Roeken, 1990; Vidal *et al.*, 1998). Medetomidine has a wide safety margin, induces good myorelaxation and can be efficiently reversed with atipamezole (an alpha-2 antagonist agent) (Table 8.1). However, some cardiorespiratory side effects can occur with alpha-2 agonists (Muir *et al.*, 1995). Medetomidine can be used in association with ketamine to achieve better and longer anaesthesia.

Administration
Theoretically, the easiest way to anaesthetise an NHP is orally. However, baiting an NHP with an anaesthetic drug is almost impossible because of the bad taste of most drugs. Medetomidine is the exception and is well accepted by chimpanzees (*Pan troglodytes*) when it is mixed with chocolate or concentrated milk at a mean dosage of 70–90 µg/kg body weight (Vidal *et al.*, 1998). However, tranquillisation takes at least 15–30 minutes, and the wrong animal may take the food.

The most common approach to capturing wild NHPs is remote injection using blowguns or capture rifles. Description of these systems, their advantages and disadvantages can be found in Sapolsky & Share (1998) and Karesh *et al.* (1998). Remote delivery systems should be used only by experienced people owing to the high risks of injury to the animal from a badly located dart, and hazards related to the pre-anaesthetic stage (e.g. falling from a tree or into water, aggression from a conspecific or a predator). NHPs also often remove the dart following impact, before the full dosage has been injected, and only a mild tranquillisation stage is achieved. In this case, the animal must be monitored until completely recovered.

Examination
The anaesthetised primate should be brought to a safe place for a general veterinary examination. To prevent false regurgitation, empty the oral cavity of food, place the animal in a lateral position (Fig. 8.1a), and monitor respiration closely. During the general body examination, pay special attention to the dart injection site and to any injury resulting from the capture. Check the palpebral, corneal and pain reflexes regularly to detect early signs of recovery.

Recovery
During the post-anaesthetic period, place the primate in a quiet area, either in a cage or in a cloth bag, depending on its size. Animals should only

Table 8.1. *Main anaesthetic agents used for chemical immobilisation of free-ranging non-human primates*

		Dissociative agents		Alpha-2 agonist:
		Ketamine hydrochloride	Tiletamine hydrochloride	Medetomidine
Dosage	i.m.	5 mg/kg (larger species) 20 mg/kg (smaller species)	2–4 mg/kg (larger species) 12–15 mg/kg (smaller species)	25–50 µg/kg
	i.v.	2–10 mg/kg	2–8 mg/kg	5–20 µg/kg
	p.o.	10–15 mg/kg	6–17 mg/kg	75 µg/kg
Tranquillisation	i.m	Xylazine: ratio (X:K) 1:10 to 1:15	Zolazepam: ratio (T:Z) 1:1	
Recumbency time	i.m.	±10 min	±5 min	Instant
	i.v.	Instant	Instant	Instant
	p.o.	Variable	±5 min	>15 min
Reversal		No	No	5 units of atipamezole per unit of medetomidine
Disadvantages		Insoluble: large NHPs require large volumes Sensitivity to heat and daylight Possible excitatory pre-anaesthetic phase Possible psychotic effects during recovery period	Short duration of anaesthesia Long recovery period	Cost Possible depressive effects
Side effects and remedies		Extra-pyramidal effects: inject diazepam (dosage 0.3 mg/kg, i.v. or 1 mg/kg, i.m.) Excess salivation: atropine 0.02–0.06 mg/kg, i.m. Hyperthermia: move to a cool place Hypothermia: cover the body and inject dexamethasone	Zolazepam can have a depressive action on the respiratory system: reverse with flumazenil: (F:Z) 1:20 to 1:50	

i.m., intramuscular; i.v., intravenous; p.o., *per os.*

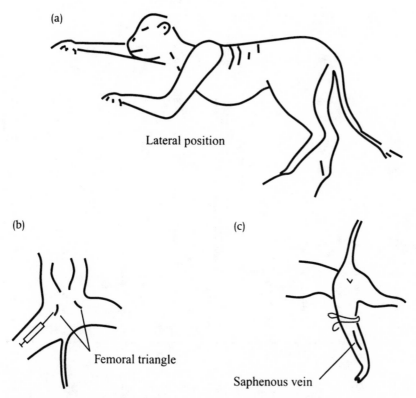

(a)

Lateral position

(b)

(c)

Femoral triangle

Saphenous vein

Fig. 8.1. Illustrations of (a) the lateral position, (b) the femoral triangle, and (c) the position of the saphenous vein.

be released when fully awake, to decrease the risks of aggression from conspecifics (Sapolsky & Share, 1998). If everything goes wrong, the only way to euthanase an NHP humanely is by intravenous injection of lethal drugs.

CLINICAL EXAMINATION

Record on a data sheet all observations made during the physical examination and all biological samples collected (Fig. 8.2). Take close-up photographs or a video of the primate's body, face, hands, feet, etc., and of any special findings during the clinical examination (Chapter 16).

EXTERNAL EXAMINATION

Includes general appearance, sex, fur colour, and presence of scars, injuries and disabilities.

A. GENERAL INFORMATION

Species: Weather Conditions:
Human Team: Date and Time:

Location (GPS position, name of the area, etc.) of Capture:

Method of Capture:
Physical (device):
Chemical (drug, dosage, administration route, etc.):

Miscellaneous Information:

B. PHYSICAL EXAMINATION

1. *External Aspect:*
   –sex:
   –approximate age:
   –fur:
   –presence of injuries, scars:
2. *General Body Condition:*
   –fat:
   –level of hydration:
   –weight
3. *Morphometrics (mm):**
4. *Oral Cavity:*
   –colour of the mucosa:
   –dentition: Maxillary set: Molars Premolars Canines Incisors
              Mandibular set: Molars Premolars Canines Incisors
   –eruption stage, wear, abnormalities:
5. *Cardiorespiratory System:*
   –auscultation:
   –percussion:
6. *Locomotor System:*
   –reflexes:
   –manipulation of bones and articulations:

7. *Digestive System:* results of transabdominal palpation:

8. *Reproductive System:* results of rectal palpation:

9. *Miscellaneous:* eyes, ears, remarks:

10. *Clinical Parameters*

Time after anaesthesia:

	T0	T0+10 min	T0+20 min	T0+30 min	T0+40 min	T0+50 min
Body temperature:						
Heart rate:						
Respiratory rate:						

C. CHECKLIST OF BIOLOGICAL SAMPLES

Type of sample	Number of samples	Label
Blood		
Faeces		
Swabs		
Other		

Photographs:

Ectoparasites:

Other Findings/Samples:

General Comments:

Fig. 8.2. Example of a data sheet for the clinical examination of a NHP in the field. *See Chapter 9 for details of morphometrics. GPS, Global Positioning System.

GENERAL BODY CONDITION

Includes body mass (Chapter 9), abundance of subcutaneous fat deposits, and hydration level. To determine the hydration level, pinch the skin with two fingers and release. If the fold disappears instantaneously, hydration is normal. However, if the fold holds a few seconds after release, the animal is dehydrated and parenteral infusion or oral hydration is required.

MORPHOMETRICS

See Chapter 9.

ORAL CAVITY AND DENTITION

The colour of the oral mucosa is a good indicator of the level of blood oxygenation. Record the dental formula (number of teeth), the stage of eruption and wear of the teeth (Chapter 9).

BODY TEMPERATURE

Rectal temperature is easily recorded on anaesthetised primates using a digital thermometer. The mean body temperature of NHPs ranges from 37 °C to 40 °C. There are wide inter-specific fluctuations, for example the body temperature of mouse lemurs (*Microcebus* spp.) and dwarf lemurs (*Cheirogaleus* spp.) can fall as low as 12 °C during winter periods of inactivity (aestivation) (Charles-Dominique, 1977). Moreover, NHP body temperature also fluctuates across the day: it will be minimal around 3 a.m. and maximal around 3 p.m. for diurnal species and the opposite for nocturnal species (Chapter 17). Temperature is higher in neonates, but lower in younger and juvenile individuals, as compared with adults. Stress and muscular activity (prior to capture) also induce transient hyperthermia. It is important to record the body temperature of anaesthetised primates at intervals in order to detect early signs of hypo- or hyperthermia.

EXAMINATION OF THE DIFFERENT SYSTEMS

Cardiorespiratory system

Examine the heart and lungs using a stethoscope. Record heart and respiratory rates (heart beats per minute and breaths per minute, respectively) at intervals throughout anaesthesia. As a general rule these decrease with increased body size. These parameters are of limited clinical interest owing to the wide fluctuations resulting from the capture itself: stress and physical effort prior to the capture induce tachycardia and hyperventilation, and anaesthetic drugs can have a depressive effect. For example, in rhesus

Table 8.2. *Methods for taking biological samples, analysis, storage, and associated costs*

Sample	Analysis	Equipment required	Storage[a]	Time	In situ analysis	Costs[b] Sampling	Costs[b] Storage	Costs[b] Analysis
Whole blood	Haematology	Red blood cell count, white blood cell count	EDTA[c] or heparinised tube, microscope, slides, coloration kit	Fridge	Max. 4 days	Possible	$	$$	$$
		Haematocrit	Capillary tubes, microcentrifuge	Fridge	Max. 4 days	Easy	$	–	$
		Erythrocyte sedimentation rate	Wintrobe tubes	Fridge	Max. 4 days	Easy	$	–	$
		Thin blood smear: differential white blood cell count	Microscope, slides	Room temp.	Max. 4 days	Easy	$	–	$
	Pathogens	Thin blood smear: bacteria	Slide, coloration kit, microscope	Room temp.	Months	Easy	$	$	$
		Thick blood smear: haemoparasites	Slide, coloration kit, microscope	Room temp.	Years	Easy	$	$	$
		Virus or bacteria isolation	Aseptic equipment, aseptic transport medium	Wet ice	Few days	No	$$	$$	$$
				Frozen	Months	No	$$	$$$	$$
	Toxicology	Pesticides	Dry collection tube	Frozen	Months	No	$	$$$	$$$
	Immunology	Antibodies	Whatman-type filter paper	Cool place	Months	Possible	$	$	$$
	Genetics	DNA	Storage medium tube	Room temp.	Months	No	$	$$	$$$
Plasma	Chemistry	Vitamins, minerals and metals	Plasma collection tube, centrifuge	Frozen	Months	No	$	$$$	$$$
Serum	Chemistry	Enzymes, electrolytes	Dry collection tube	Fridge	Few days	Possible	$	–	$$
	Endocrinology	Hormones	Dry collection tube	Frozen	Months	No	$	$$$	$$
	Immunology	Globulins, antibodies	Dry collection tube	Fridge	Few days	No	$	$$$	$$$
				Frozen	Months				

Sample	Discipline	Analysis	Collection material	Temperature[a]	Storage time	Home analysis	Cost	Cost	Cost[b]
Faeces	Toxicology	Pesticides	Dry collection tube	Fridge Frozen	Few days Months	No	$	$	$$$
	Parasitology	Intestinal parasites	Sedimentation kit, slide, microscope, 10% formalin or 70% alcohol	Room temp.	Months	Easy	$	$	$
		Protozoa	Slide, coloration kit, microscope	Room temp.	Months to years	Easy	$	$	$
		Pathogen culture and isolation	Special transport medium	Fridge	Few days	No	$	$$	$$
	Others	Genetics, toxicology, immunology, endocrinology, etc.	Depends on methods	Depends on methods	Few days to years	No	$$	$$	$$
Tissues	Pathogens	Virus or bacteria isolation	Swab and aseptic equipment	Room temp. (aseptic transport medium) or frozen	Years	No	$$	$$	$$
		Bacterial identification	Swab, slide	Room temp.	Years	Possible	$	$	$
	Cytology	Reproductive cycle	Swab, slide	Room temp.	Years	Possible	$	$	$
		Presence of pathogens							
	Histology	Tissue	Plastic containers with Bouin solution or 10% formalin	Room temp. (do not freeze)	Years	No	$	$	$$
Ectoparasites	Parasitology	Ectodermic agents	Plastic container with 70–95% ethanol	Room temp.	Years	Possible	$	$	$

[a] Fridge, +4 °C; frozen, −20 °C or lower; room temp., room temperature.
[b] Figures are indicative of the relative costs; from very low cost ($) to expensive ($$$).
[c] EDTA, ethylenediaminetetra-acetic acid, disodium salt.

monkeys (*Macaca mulatta*), the heart rate ranges between 150 and 333 beats/minute in restrained individuals, but only between 98 and 108 beats/minute in radio-monitored free-ranging individuals (Whitney & Wickings, 1987). In practice, the most important feature to consider is the depth and the regularity of the respiratory movements during anaesthesia.

Locomotor system
Investigate muscular reflexes and the integrity of bones and articulations using manipulation and palpation.

Digestive system
Use a transabdominal palpation to examine the size and shape of internal organs (liver, spleen, kidneys, presence of abnormal mass).

Reproductive system
Use a rectal palpation (animal lying in a lateral position, Fig. 8.1a) to explore the genital tract and determine stage of pregnancy in females.

Other
Examine the ears and eyes, and apply an ophthalmic ointment to protect the cornea from desiccation in anaesthetised animals.

BIOLOGICAL SAMPLING

Most biological samples require specific storage conditions (refrigeration or freezing) and rapid processing following collection. However, relatively inexpensive field-compatible portable fridges, freezers and liquid-nitrogen containers are available, and several biological parameters can be screened easily *in situ* with diagnostic field kits, meaning that samples do not necessarily have to be shipped to a distant laboratory.

Table 8.2 summarises the biological samples that can be taken from an anaesthetised NHP, the information that they can yield, and appropriate storage methods. All samples should be carefully labelled (using a solvent-proof pen) with the date, location, species, individual identification, and the type and number of the sample taken.

BLOOD SAMPLING

Whole blood
For live individuals, whole blood is collected after venipuncture. Collection is made from the femoral vein for the smallest species, or when large quantities of blood are required for other species. The femoral vein is located in the femoral triangle, immediately adjacent to the femoral artery (Fig. 8.1b). When the primate is lying on its back, its legs in slight extension, the pulsation of

the artery can be felt with the index finger, establishing the location of the vein. Prior to venipuncture, the sampling site should be washed and disinfected with iodine solution and/or alcohol. The smallest species (prosimians, marmosets and tamarins) are sampled with a 0.4 mm gauge needle and a 1 ml syringe, squirrel monkeys (*Saimiri* spp.) are sampled with a 0.9 mm needle and a 1 or 2.5 ml syringe or a small vacuum tube, and larger species with a 1.2–2 mm needle and a 2.5–10 ml vacuum tube. Apply strong pressure to the vein for at least one minute following venipuncture to prevent haematoma formation, especially when the femoral artery has been inadvertently punctured. Other possible sites are the jugular vein (especially for the smallest species) or the saphenous vein (guenons or larger species). The saphenous vein lies just under the skin on the caudal surface of the gastrocnemius muscle (Fig. 8.1c), and is easy to locate after compression of the upper thigh. However, this vein is rather small and collapses easily when large quantities of blood are collected. This is the site of choice for intravenous injections for all species of NHPs. In recently dead primates, blood can be obtained easily from the heart.

As a general rule, to prevent risk of hypotension and associated heart failures, blood sampling should never exceed 1 ml per 100 g body weight per month.

Haematological parameters
To prevent clotting, whole blood is placed in a collection tube containing sodium heparin or EDTA (ethylenediaminetetra-acetic acid, disodium salt) immediately following venipuncture. Red and white blood cell counts (RBCC and WBCC, respectively) are determined with a Coulter Counter or a Malassez' Cell slide after staining. Differential WBCC are made by examining 100 leucocytes in smears stained using the Wright–Giemsa method. These results provide valuable information on the medical status of the individual (presence of anaemia, infection, parasitism, neoplasm and other haematological abnormalities). Whole blood also allows the determination of other parameters (e.g. mean corpuscle volume, mean corpuscle haemoglobin) that explore haematological function more precisely. Erythrocyte sedimentation rate (ESR) can be determined using disposable Wintrobe tubes; sedimentation rates are recorded at 15 minute intervals for a period of 1–2 hours. Haematocrit or packed cell volume (PCV) gives information on the general hydration level of the organism, and is determined by centrifuging blood-filled capillary tubes in a portable microcentrifuge. These analyses are carried out from a few hours to a few days (two to five days) following venipuncture (if the samples are stored in a fridge at 4–8 °C). To prevent haemolysis of blood cells, tubes must not be frozen or shaken.

The physiological range for biological parameters in captive NHPs can be found in the literature, or the ISIS/MedArks system (available from all zoos and most captive facilities holding NHPs).

Bacteria and virus isolation

The isolation of bacteria and viruses from whole blood requires strictly aseptic conditions and sterile sampling equipment to prevent any bacterial contamination. Blood is collected in a sterile container with transport medium (sterile buffered glycerine or other buffered preparations used in cell cultures, with added bovine serum albumin (BSA) and antibiotics in the case of virus isolation). Samples are placed on wet ice (but not frozen) and sent directly to the laboratory since most viruses and bacteria remain viable for two to four days if kept cold. If a delay is expected, freeze the blood in a mixture of CO_2 and alcohol, or in liquid nitrogen, and store at -70 °C or below.

Blood smears

Blood smears are easy to make, and yield information concerning haematology, blood parasites and bacterial infection. Peripheral or capillary blood is collected from the tip of a finger or an ear. Disinfect the sampling site, slightly puncture the skin with a small needle, and collect a drop of blood. To make a thin blood smear place a single drop of whole or peripheral blood near one end of a horizontal microscope slide. Bring the end of a second slide up to the drop at 45° until the drop disperses along its edge. Then push the second slide quickly and evenly towards the opposite end of the first slide. To make a thick smear, place a drop of blood on a slide, and spread in a small circle with the tip of a needle or the corner of a second slide for at least 30 seconds. After a few minutes at ambient temperature the smears are dry and can then be transported to a laboratory or analysed in the field after staining or fixation in 90% methanol. Take a minimum of three to five blood smears per individual.

Blood drops

Small amounts of blood can be collected onto specially designed filter paper to test for antibodies against specific pathogens, or for hormonal analysis. When the blood spots are dry, place the filter papers in individual plastic bags or in normal envelopes with a desiccant and store in a dry place.

Plasma

Plasma is obtained by centrifugation of unclotted whole blood, and separation of the yellow/tan liquid material (plasma) from the clotted component (cell membranes and other blood composites). This is divided into aliquots in small plastic vials (0.5–2.5 ml), and stored frozen (-20 °C or below).

Serum

To collect serum, place whole blood in special serum separator tubes or dry sterile blood collection tubes immediately after venipuncture. Leave the tubes undisturbed for at least one hour at ambient temperature to encourage clot formation and then centrifuge at 2000 *g* for 15–20 minutes. When a centrifuge is not available, serum can still be obtained by letting the clot or blood cells settle for few hours, then aspirating the liquid phase (serum). In order to maximise the quantity of serum, allow the blood to clot with the collection tube inverted (rubber stopper down). After a few hours, turn the tube stopper up and carefully remove the stopper with the blood clot attached, leaving the serum in the tube (Munson, 2000). Divide the serum into several aliquots in small vials and refrigerate or freeze.

FAECAL SAMPLES

Faecal samples are a 'goldmine' for biologists owing to the range of information they can provide concerning intestinal parasites, feeding ecology (Chapters 12, 13 and 14), seed dispersal (Chapter 11), DNA (Chapter 20), and endocrinology (Chapter 19) (Table 8.2). Faeces can be collected directly from the rectum when a primate is handled, or non-invasively during fieldwork. They are stored easily in the field and most analyses can be performed long after sample collection. Faeces should never be handled without gloves because of the risk of zoonosis transmission.

When collecting faeces, note the colour, odour, size and consistency at the time of collection, along with major macroscopic elements, including the presence or absence of worms. Worms (if any) should be preserved in a 10% formalin solution for identification.

Direct microscopic observation of intestinal parasites is possible by simply immersing a small portion of fresh faeces in two drops of distilled water on a microscopic slide. However, this technique lacks sensitivity and sedimentation or flotation methods are preferred. For sedimentation (Brumpt method), 5–10 g of fresh or formalin (10%) fixed faeces are mixed with distilled water and sieved (1 mm wire mesh, the approximate size of a tea sieve). The filtrate is left undisturbed for a maximum of two hours (fresh sample), or up to 12 hours (fixed sample). Because of their weight, parasites and ova sink and sediment on the bottom of the vial. Dispose of the upper part of the mixture, homogenise the bottom part, and place two drops on a microscopic slide for examination. A centrifuge can be used for better results (1500 *g* for 3 minutes). For the flotation method, the water is replaced with a dense solution. This can be water saturated with Sheather's sugar (density (d) 1.18 g/cm^3), sodium (d = 1.19 g/cm^3), magnesium sulphate (d = 1.28 g/cm^3), or zinc sulphate (d = 1.18 g/cm^3), all of which are easy to prepare in the field. Better results are achieved with an iodomercurate potassium solution (150 g of mercurate biiodure, 111 g of potassium iodure, 399 g of water), but this solution is highly hazardous. Dilute 5–10 g of faeces in the dense solution and pour it into a laboratory tube until it is completely full. Place a small microscope slide in contact with the solution at the top of the tube for about 30 minutes, and then remove it for microscopic examination.

Proper examination of a microscope slide is achieved with the regular screening of the entire field along parallel stripes. However, this gives only a qualitative analysis of parasite load. Quantitative analysis requires a 'MacMaster slide'.

The presence of parasites in faeces does not necessarily mean that an animal is sick. Under natural epidemiological conditions, primate populations can be infested with parasites without adverse effects. Indeed, the pathogenic effects of many parasite species depend mainly on their abundance in the host (strongyds, ascarids) and only quantitative coprology will assess the associated disease risks. In addition, parasites and their ova are not permanently excreted in faeces, so a negative coprology does not necessarily mean that the individual is free from internal parasites. To assess the true parasitological status of an individual, samples should be collected for the same individual at regular intervals.

SWABS

A swab can be moistened with any visible lesion or physiological secretion. Throat swabs are obtained by carefully swabbing the posterior pharyngeal tonsillar area. Rectal swabs are obtained by carefully inserting a swab in the rectum, rotating it with care to avoid puncturing the bowel wall, and removing the stained swab. Swabs are then placed in 0.5–1 ml of transport medium. With liquid medium, swabs are rotated vigorously in the fluid, and the fluid expelled from the swab by pressing it against the inside of the tube. When no transport medium is available, the swab can be pressed against a microscope slide to make a smear. Smears can also be made directly from pus or infected tissues (abscesses, skin ulcerations, etc.). Clean the tissue with a scalpel blade, dry the surface with paper towel and gently touch it in several locations with clean slides. Air-dry the smears and fix them in 95% alcohol or 100% acetone. Vaginal swabs are of special interest to determine the reproductive status of a female. A dry swab can also be moistened with physiological secretions from different types of glands (e.g. scent-marking glands) before being placed in a dry collection tube. These swabs can be stored in the field for long periods.

ECTOPARASITES

Check carefully for the presence of ectoparasites (fleas, lice, or ticks). If present, these should be collected and stored in 70–95% ethanol for identification. Where a tick is already fixed inside the skin, apply ether or alcohol before removal to prevent the rupture of the parasite's sucking organ and head inside the primate's cutaneous tissue. For intradermic parasites (mites, acarids, dermatophytes), scrape lesions with a scapel blade, and place the scrapings in lactophenol. For leishmaniosis and cutaneous microfilaria diagnosis, collect a thin piece of skin with sharp scissors. Only the superficial layers of skin are collected and this does not induce bleeding. For leishmaniosis diagnosis, press the sample against a slide to make a smear, and stain with May Grünwald Giemsa. For microfilaria, place the sample in physiological serum for 15 minutes, and then examine under a microscope.

SAMPLING DEAD ANIMALS

Dead NHPs may be found in the field, brought to the scientist for a number of reasons, or may result from a failed capture procedure. Post-mortem examinations of wild NHPs yield valuable data about cause of death and the health of the individual prior to death. Very little information is available on the natural pathology of wild NHPs and field studies offer one of the only opportunities for complete pathological examinations. Necropsy methods have been described elsewhere (Rabinowitz, 1993; Munson, 2000). The person performing the necropsy should wear a mask, gloves and protective clothing. This is especially important when a carcass is found in the field and no detailed information is available on the case history. Do not touch or necropsy any carcass found with symptoms of severe hemorrhagic diseases. In such cases, reports must be sent to the relevant authorities as soon as possible.

REFERENCES

Bearder, S. & Pitts, R. S. (1987). Prosimians and tree shrews. In *The UFAW Handbook on the Care and Management of Laboratory Animals*, 6th edition, ed. T. Poole, pp. 551–67. Harlow, Essex: Longman Scientific and Technical.

Charles-Dominique, P. (1977). *Ecology and Behaviour of Nocturnal Prosimians*. London: Duckworth.

Daszak, P., Cunningham, A. A. & Hyatt, A. D. (2000). Emerging infectious diseases of wildlife: threats to biodiversity and human health. *Science* **287**, 443–9.

Deem, S. L., Karesh, W. B. & Weisman, W. (2001). Putting theory into practice: wildlife health in review. *Conserv. Biol.* **15**, 1224–33.

Dripps, R. D., Eckenhoff, J. E. & Vandam, L. D. (1977). Dissociative anesthesia. In *Introduction to Anesthesia. The Principles of Safe Practice*, ed. W. D. Saunders, pp. 187–8. Philadelphia: W. B. Saunders.

Glander, K. E., Fedigan, L. M., Fedigan, L. & Chapman, C. (1991). Field methods for capture and measurements of three monkey species in Costa Rica. *Folia Primatol.* **57**, 70–82.

Jalanka, H. H. & Roeken, B. O. (1990). The use of medetomidine, medetomidine-ketamine combinations, and atipamezole in nondomestic animals: a review. *J. Zoo. Wildlife Med.* **21**, 259–82.

Jones, W. T. & Bush, B. B. (1988). Darting and marking techniques for an arboreal forest monkey, *Cercopithecus ascanius*. *Am. J. Primatol.* **14**, 83–9.

Karesh, W. B., Wallace, R. B., Painter, R. E., Rumiz, D., Braselton, W. E., Dierenfeld, E. S. & Puche, H. (1998). Immobilization and health assessment of free-ranging black spider monkeys (*Ateles paniscus chamek*). *Am. J. Primatol.* **44**, 107–23.

Muir, W. W., Hubbell, J. A. E., Skarda, R. T. & Berdnaski, R. M. (1995). *Handbook of Veterinary Anesthesia.* St Louis, MO: Mosby.

Munson, L. (2000). Necropsy procedures for wild animals. In *Conservation Research in the African Rain Forests: A Technical Handbook*, ed. L. White & A. Edwards, pp. 203–24. New York: Wildlife Conservation Society.

Osofsky, S. A. & Hirsch, K. J. (2000). Chemical restraint of endangered mammals for conservation purposes: a practical primer. *Oryx* **34**, 27–33.

Rabinowitz, A. (1993). *Wildlife Field Research and Conservation Training Manual.* New York: Wildlife Conservation Society.

Sapolsky, R. M. & Share, L. J. (1998). Darting terrestrial primates in the wild: a primer. *Am. J. Primatol.* **44**, 155–67.

Schobert, E. (1987). Telazol use in wild and exotic animals. *Vet. Med.* **82**, 1080–8.

Sedgwick, C. J. (1985). Scaling and anesthesia for primates. In *Primates: The Road to Self-Sustaining Population*, ed. K. Benirschke, pp. 815–22. Berlin: Springer-Verlag.

Sleeman, J. M., Cameron, K., Mudakikwa, A. B., Nizeyi, J.-B., Anderson, S., Cooper, J. E., Richardson, H. M., MacFie, E. J., Hastings, B. & Foster, J. W. (2000). Field anesthesia of free-living mountain gorillas (*Gorilla gorilla beringei*) from the Virunga Volcano Region, Central Africa. *J. Zoo. Wildlife Med.* **31**, 9–14.

Tutin, C. E. G., Ancrenaz, M., Paredes, J., Vacher-Vallas, M., Vidal, C., Goossens, B., Bruford, M. & Jamart, A. (2001). Conservation biology framework for the release of wild-born orphaned chimpanzees into the Conkouati Reserve, Congo. *Conserv. Biol.* **15**, 1247–57.

Vidal, C., Paredes, J. & Ancrenaz, M. (1998). Anesthesia en chimpancés: primer estudio comparativo de varias técnicas en trabajos de campo. *Consulta* **6**, 1670–3.

Whitney, R. A. & Wickings, E. J. (1987). Macaques and other old world simians. In *The UFAW Handbook on the Care and Management of Laboratory Animals*, 6th, edition, ed T. Poole, pp. 599–627. Harlow, Essex: Longman Scientific and Technical.

Wolfe, N.D., Escalante, A.A., Karesh, W.B., Kilbourn, A., Spielman, A. & Lal, A.A. (1998). Wild primate populations in emerging diseases: the missing link? *Emerg. Infect. Dis.* **4**, 149–58.

Useful Internet sites

Chemical immobilisation: <www.wildlifeservices.co.za/captureandcare> and <www.wildnetafrica.com>.

9 • Morphology, morphometrics and taxonomy

COLIN GROVES AND JOANNE HARDING
School of Archaeology and Anthropology, Australian National University, Canberra, Australia

INTRODUCTION

Fieldworkers are often unaware of the value of making accurate records of their study animals' morphology. External morphology includes measurements, observations of glandular activity and detailed description of the pelage; internal morphology may include skull and postcranial measurements, recording of suture closure, epiphyseal fusion, dental eruption and wear, and observations on the gut. All these observations will yield information on taxonomy, age–sex class (Table 9.1), reproductive status, individual variation, growth, development, sexual dimorphism and so on. It may also be interesting to look at individuals' behaviour in the light of their external differences: does facial coloration correlate with behaviour in mandrills (*Mandrillus sphinx*), does flange development correlate with behaviour in orangutans (*Pongo* spp.), and so on?

Measurements may be made either on dead specimens, or on living animals when they are captured (Chapter 7) and, preferably, anaesthetised (Chapter 8). Valuable descriptive information, other than measurements, can be gathered from simple observations, or from photographs (Chapter 16). On occasion, measurements can even be made in this way. For example, Markham & Groves (1990) cite a personal communication from Herman Rijksen of how he weighed four wild orang-utans (worth repeating here because it is such a wonderful example of lateral thinking) 'by "measuring" the bending arc of particular branches when supporting the full weight of the animal and hoisting up buckets of sand to match the same arc'.

The accurate identification of study animals has always been vital for their proper understanding, especially today when remnant populations of once widespread species are in need of strict conservation, scientific names are needed for them, and the question of their uniqueness arises. Is this the last remaining population of its species or subspecies? If the patterns of ecology and behaviour of the study population are different from those of others reputed to be of the same species, are we sure that the populations really

Field and Laboratory Methods in Primatology: A Practical Guide, ed. Joanna M. Setchell and Deborah J. Curtis. Published by Cambridge University Press. © Cambridge University Press 2003.

Table 9.1. *Age classes. Definable age classes will of course vary greatly among taxa; not only between major categories such as families, but even between quite closely related species. The terms in this table are often used*

Age class	Definition
Neonate	Infant showing signs of having been very recently born (a few days)
Infant	Unweaned; not independently moving (carried about, in nest, etc.)
Juvenile	Immature but independently moving
Immature	Any individual not evidently sexually mature
Sub-adult	Individual that is apparently sexually mature, but not physically mature
Adult	Both sexually and physically mature
Old adult	Adult showing apparent signs of age degeneration (like the first author)

are the same, or could there be some undetected taxonomic differences after all? It is increasingly important for fieldworkers to be aware of taxonomic niceties, and of morphological and morphometric matters in general, for a full appreciation of the biology of any study population.

TAXONOMIC DATA: SPECIES AND SUBSPECIES

Measurements and other observations will, in the first line, be relevant to taxonomy. This may matter crucially as far as interpreting behaviour is concerned: different species, for example, may differ from each other in behaviour and unless taxonomic differences are appreciated such behavioural differences might unwittingly be ascribed to different ecological settings or, worse, to chance. So a brief introduction to taxonomy is in order.

There is no such thing as 'the official' taxonomy; a taxonomic scheme is a hypothesis, a way of interpreting data, just like an ecological or a behavioural model. Most biologists who are not specialists in taxonomy probably think of species as populations, or groups of populations, that are reproductively isolated from others (the so-called Biological Species Concept); but they misunderstand what this means. It does not mean that species cannot interbreed with each other, nor that if they do interbreed their hybrids are sterile. Species of *Macaca* or of *Cercopithecus* that are widely sympatric may on occasion interbreed, and not merely F_1 hybrids but even back-crosses are to be seen mingling with pure-bred troops of one or other parental species. So species are not invariably reproductively isolated!

Then recall that there are innumerable examples of populations that are clearly closely related, but look different from each other, which are totally allopatric – separated from each other by rivers, or by unsuitable habitat, or on islands. How can one possibly judge whether or not they are reproductively isolated?

If this 'reproductive isolation' criterion is taken too seriously, it is obvious that the vast majority of organisms could never be classified as to species by any objective measure – even if we place asexual organisms on one side. This is a major reason why specialist taxonomists increasingly recommend the Phylogenetic Species Concept: species are what can be diagnosed. The meaning of diagnosability is simply that every single individual, within its age–sex class, can be allocated to its species. Any character, as long as it can be plausibly assumed to be genetic, is a candidate for the diagnosis: size, shape, colour, colour pattern, vocalisation, or a particular nuclear DNA sequence. The implication is that a species has one or more fixed genetic differences from other species; it is genetically unique, not (or not necessarily) reproductively isolated.

At this point, we must specify that we exclude mitochondrial DNA (mtDNA) sequences. These are inherited uniparentally, not biparentally, which is different from every other DNA sequence (Y-chromosome DNA of course excepted) and so also from all observable (phenotypic) features. The time to coalescence of the nuclear DNA sequences of two different populations is a function of their effective population sizes (N_e), while the N_e of mitochondrial DNA is effectively one-quarter that of a nuclear sequence, with a consequently much shorter time to coalescence (Nichols, 2001). Moreover, mtDNA is inherited, to all intents and purposes, only through the female line; it tells us nothing about what the males have been doing, nor (because there is no recombination) has it anything to say about population integration. MtDNA, consequently, offers us only incomplete information about populations; as species and subspecies are population concepts, it cannot, strictly speaking, be used in a taxonomic sense.

Subspecies are geographical populations of a species that differ as a whole, but not absolutely. Note that they are geographical: by definition, they cannot be sympatric. As is the case for species, mtDNA should not by itself be used to characterise subspecies.

PRESERVING SPECIMENS

If fieldworkers have any doubt as to the taxonomic identity of their study animals, they should photograph them and try to compare the photographs with the type specimen. This is the specimen, lodged in a museum somewhere, on which the species was first described and named: it is the ultimate decider of a species/subspecies. A voucher specimen is even better than a photograph: we are absolutely not recommending that a fieldworker should shoot a specimen for comparison, but all too often there are poachers around, and if a specimen is confiscated (note: *not* purchased) from a poacher the best disposal of it is in some museum where it can be studied, and from there it can be borrowed for comparison with other specimens (especially the type specimen), or at least photographed. Birds of prey also take many non-human primates and, usefully, they bring up the bones in pellets underneath their nest perches, or pieces of prey may simply fall to the forest floor during dismemberment. Primatologist fieldworkers should assiduously collect such specimens, and present them to museums where they will form permanent vouchers.

BONES

Bones are easy to preserve. If they are found in an already macerated state (i.e. separated from the soft tissues), they should be clearly labelled and placed in a box or phial, padded with cotton wool or some other soft material – if all else fails, then scrunched-up paper will do. If a specimen is found dead, or a live animal has to be humanely dispatched for some reason, the skin should be separated from the bone in some way.

If the skin is not required, soak the specimen in warm water until the soft tissues float off; preferably it should not be boiled, as, although many times quicker, this method destroys *all* non-bony tissue, so that sutures spring apart and teeth fall out. Bio-active washing powders, which contain digestive enzymes, will clean off soft tissue from bones very well, but should be used carefully as ultimately they have the same effect as boiling. Finally, in dry environments an ant-heap serves very well as a bone-cleaning laboratory, but the bones will be shifted about by the ants and should be held down in some way (e.g. wrapped in wirenetting). Be warned that in too dry a climate bone will flake and ultimately falls apart, and bony splinters can penetrate your skin and set up infection.

SKINS

If the skin is required, then the body is turned on its back, an incision made up the ventral surface from groin to sternal notch, and transverse incisions between the two arms, to at least halfway down towards the elbows, and between the two legs to at least halfway down towards the knees. Insert the fingers carefully between skin and underlying fatty tissue, and separate them, peeling back the skin. The less fat that remains on the underside of the skin, the easier it will be to preserve, and the less unpleasant it will be to the touch. The phalanges can be preserved as part of the skeleton, or can stay inside the skin; the latter makes skinning a whole lot easier, but will work only with small primates, as there is too much soft tissue in the digits of larger species. The ear cartilages come away with the skin, as do those of the nose in the case of strepsirrhines. It is often convenient to leave most of the tail vertebrae in the skin, at least until the specimen can be got to a museum.

Skinning should be performed, and the skin preserved, as soon as possible, because tropical heat and humidity cause the hair to come loose from the skin. The important thing is to keep it dry and make it unappealing to insects and mould. For best results, turn the skin inside out, and hold the hair surfaces apart with loose material such as scrunched-up tissue paper. Rub in salt, and leave it for about 24 hours, pouring away any brine that is formed. Dry it either in the open (but not in full sunlight) or near a fire. Make sure it is dry and as fat-free as possible. Now rub in some preservative: in former days, an arsenic preparation was used, but borax is fine. Turn the skin back right-way out and (optionally) stuff it loosely with cotton-wool or tow (thick, dry, somewhat shredded rope). Store it in a dry place, preferably with naphthalene (mothballs).

WET SPECIMENS

Wet specimens are sometimes useful, if special anatomical study is required. Specimens should be stored in fresh formalin (formaldehyde solution, stock 37–40% diluted 10 parts water to 1 formaldehyde). You will need twice the volume of fixative as specimen in your container. Inject the thorax with preservative and slit open the abdomen to allow the fixative to penetrate to internal organs. The concentration is important: too weak and the specimen rots, too strong and dehydration makes specimens hard and brittle. Do not keep the specimen in formalin too long (or at all, if DNA samples are required): it destroys the pelage colour and makes the bones brittle, and as for the human collector, it brings tears to the eyes; remember that it is **POISONOUS**! After a maximum of 30 days, rinse your specimen in water and store it in 70% ethanol. Change the ethanol once after 30 days to eliminate the formalin. Other alcohols may be used temporarily if ethanol is not available.

GENETIC TYPE SPECIMENS

There has been some discussion about the possibility of a 'genetic type specimen'. The *International Code of Zoological Nomenclature* (1999), which came into force on January 1, 2000, specifies that a type specimen must be 'an animal, or any part of an animal...' (Article 72.5), which leaves it a little unclear as to whether a DNA sample would count. We think not. Certainly simply photographs or a detailed morphological description and measurements of a living animal would not be eligible.

If you think you may have found a new species, but cannot collect a specimen, what do you do? This is a quandary that no one has yet answered satisfactorily. When Thalmann & Geissmann (2000) met with an undescribed species of *Avahi* at Bemaraha they took good, crisp photographs of it, but could not collect a specimen, both because the taxon is endangered (so killing one would have been, in effect, illegal as well as immoral) and because they quite simply could not bring themselves to kill one because of their moral scruples (which, incidentally, we share). Faced with the same quandary in the case of their new species of *Hapalemur*, Meier *et al.* (1987) captured animals alive and lodged them in a zoo; but this was not an option for *Avahi*, as they do not survive at all well in captivity. The only option would seem to be to keep the population and its predators under close observation, and keep an eye out for new and exciting bones under raptors' roosts, or individuals that are obviously suffering some terminal condition and are candidates for euthanasia for humane reasons.

LABELLING

Accurate and detailed labelling of specimens is extremely important. Standard printed labels can be obtained from any Natural History Museum; these have spaces for geographical and habitat data, circumstances and date of collection, sex, and body measurements. Even if these printed labels are not available, it is essential to record the following data:

Date of collection
Collector
Method of collection (pick-up, euthanasia, confiscation etc.)
Locality: name (if there is one); distance from nearest settlement, river, forest
    block or other location easily found on a map or on the ground by another
    visitor; geographical coordinates; altitude
Habitat: forest type, other environmental details
Measurements: at the very least, head and body, tail, hindfoot, ear, weight
    (see below).
Other: details of phenotype, especially eye colour, pigmentation of exposed
    skin (ears, nose, palms and soles); whether pregnant.

The data should be recorded in the field and copied into a notebook or computer as soon as possible. A separate label should be made out for skin, skull and any other sizeable parts that are kept (pelvis, skinned body, etc.). The specimen should be allocated a collector's number, and this should be written on *every* item as far as possible. Keep all the parts in the same specimen box, avoiding damage by wrapping fragile items (especially the skull) in cotton wool or other soft material; small bones, such as hand and foot bones and vertebrae, should be put in special bags (one for each extremity, one for each spinal region).

SKULLS

Being the most complicated part of the skeleton, the skull (including the dentition) offers the most opportunities for diagnosis not only of taxonomic identity but also of sex, age and health status. Museum specimens commonly consist of the skin and the skull; preservation of the postcranial skeleton is unusual, of the soft tissues even more so.

Many traditional hunting and gathering cultures collect trophy skulls as a sign of their hunting success. The identification of these skulls to species level can provide important information about species present in an area and can be used to confirm or contradict information acquired through interviewing local people.

Skulls are measured with calipers; these are a kind of metal ruler with two downward-pointing arms: a fixed arm and a sliding one. Some calipers have a digital read-out dial, which goes down to 0.1 mm or less; some have a vernier scale. Measurements should always be recorded in millimetres, or fractions thereof; be only as accurate as the material warrants: a 300 mm gorilla (*Gorilla* spp.) skull is not worth measuring down to 0.1 mm! Calipers are small and light enough to take on any field trip, but in an emergency calipers used to measure machinery can be used, and are more easily available in local towns.

When it is not possible to remove specimens from the field good photographs are imperative. Black-and-white slide film is preferable. Photographs of crania should be taken from the side (*norma lateralis*), front (*n. facialis*) and underneath (*n. basalis*), while mandibles should be photographed from the side, front and above (*n. dorsalis*), so that the teeth can be clearly seen. Photographs are best taken with a plain backdrop and it is important that they are taken square on to avoid distortion. Place the calipers or a matchbox in the photograph as a scale.

The skull consists of two moveable parts: cranium and mandible. The cranium is composed of many bones, joined together along sutures, which eventually fuse with age. Surrounding the brain is the neurocranium; hafted in front of it is the facial skeleton or splanchnocranium. These and some other standard anatomical terms are illustrated in Fig. 9.1.

In smaller strepsirrhines and in *Tarsius*, fusion of the two rami of the mandible is incomplete at the symphysis and the two halves can move separately. In larger strepsirrhines, and in all Simiiformes, the symphysis is firmly fused and the mandible is a single bone.

MEASURING THE SKULL

There is, unfortunately, no standard set of measurements for skulls. Figure 9.2 illustrates some that are used in craniometric studies, and will enable a fieldworker to make initial comparisons, before turning the skull(s) over to a public collection. Numbers in parentheses in the headings below refer to Fig. 9.2.

Greatest length, total length, skull length (1–1)
From prosthion (the most anterior point on the bone between the upper central incisors) to opisthocranion (the most posterior point on the skull).

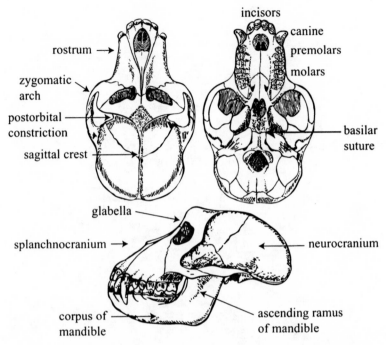

Fig. 9.1. A skull (*Macaca assamensis*), to show important anatomical terms.

Hold the fixed point of the calipers on prosthion, and pull the sliding arm back until its point rests as far back on cranium as possible.

Condylobasal length (2–2)

From prosthion to a tangent across the posterior borders of the occipital condyles. Lay the calipers sideways down; hold the fixed arm across prosthion, and pull the sliding arm back until it rests against the posterior margins of both condyles.

Basal length (2'–2')

An alternative to condylobasal length; the two measurements can substitute for each other when different parts of the (often fragile) basicranium are missing. Place the fixed point on prosthion, and move the sliding arm back till the point is on basion, the anterior edge of the foramen magnum.

Bizygomatic breadth/width, zygomatic breadth/width, greatest breadth (3–3)

The greatest width of the skull across the zygomatic arches. Slide the calipers back and forth along the outer sides of the zygomatic arches (perpendicular to the skull length, of course) until you have the greatest reading.

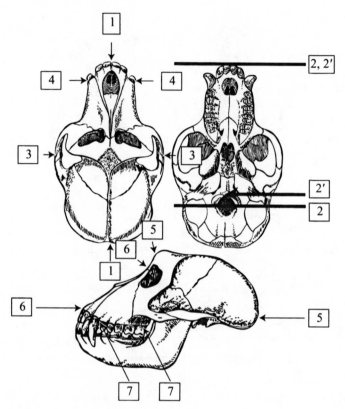

Fig. 9.2. Recommended cranial measurements (for explanation, see the text).

Bicanine breadth/width (4–4)
Greatest width of the muzzle, across the canine alveoli.

Cranial length, braincase length (5–5)
From either glabella or nasion to opisthocranion. Glabella is the midline swelling between the brow ridges; nasion is the point where nasal bones and frontal bone(s) meet or, if you like, the point where internasal and frontonasal sutures meet. These two measurements may or may not be virtually the same.

Facial length, facial height, upper facial height (6–6)
From prosthion to nasion.

Maxillary toothrow length (7–7)
From the anterior end of the anterior upper premolar (P^3 in catarrhines and indriids, P^2 in most other primates) to the posterior end of the posterior upper

molar (M^3 in nearly all primates, M^2 in callitrichines). Measure along the alveolar margin; this way, the measurement can be taken even if all the teeth have dropped out.

Others

Some authors take mandibular toothrow length; some include the canine in their toothrow lengths; some measure molar rows. Some authors take a cranial height measurement (usually basion to bregma); some favour a cranial width on the braincase itself, and/or on the external or internal margins of the orbits. However, by taking the above seven or eight measurements, you cannot go far wrong and, of course, if the specimen is preserved in a museum or other public institution, it is always available thereafter for more detailed studies.

Cranial capacity

Cranial capacity is a close approach to actual brain size. For this you will need a measuring cylinder or cup, a funnel, and some filler: birdseed is ideal, as it packs well – sand is alright, but may slip out through cracks such as incompletely fused sutures. Block up the optic foramina, preferably with plasticene; at a pinch, simple mud will probably do this effectively. Inspect the neurocranium to see whether there are any other holes. Hold the cranium upside down so that the foramen magnum is horizontal, and gradually pour in the filler. Every so often, shake the cranium, to make sure that the filler settles and packs in. When the filler has reached the rim of the foramen magnum, and cannot pack down any further, pour it out through the funnel (to make sure none is spilled) into the measuring cylinder, shaking this every so often to pack it down. Read off the volume.

AGEING AND SEXING THE SKULL

The rate at which cranial sutures fuse varies both inter- and intra-specifically. The only suture whose fusion is within narrow age limits is the basilar suture (strictly, not a suture at all, but a synchondrosis: the spheno-occipital synchondrosis), between the basioccipital and basisphenoid bones, some way in front of the foramen magnum, the hole through which the spinal cord passes. This fuses at maturity; afterwards, little or no further cranial growth takes place. It is therefore important to note whether it is widely open, in process of fusion, or fully fused.

The skull is impossible to sex without some prior knowledge of the way in which sexual dimorphism is expressed in that species. In most (not all) Simiiformes, males are larger than females and, because the facial skeleton grows with positive allometry, the jaws are relatively longer.

CRANIAL PATHOLOGY

A more or less symmetrically developed pitting or channelling on the roofs of the orbits is called cribra orbitalia, and is due to anaemia. Anaemia (lowered haemoglobin levels) is a symptom of a whole range of conditions: dietary insufficiency, chronic disease, abnormal haemoglobins (commonly indicating a genetically determined response to malaria, in the human case at least). More severe anaemia results in thickening of the cranial vault and widescale pitting, especially on the parietals (porotic hyperostosis).

Thin, light, somewhat pitted skull bones are due to osteoporosis, indicating some type of dietary malabsorption.

A cancerous lesion (osteosarcoma) on the skull takes the form of bony swelling, which has often burst to form a wide, saucer-like crater.

Healed fractures are detectable because part of the skull, such as the zygomatic arches or orbital margins, is markedly asymmetrical, and one side is irregularly shaped where the other is normal.

In species with strong muscular development, such as gorillas, damage to the muscle results in underdevelopment of the supporting bone on that side, making the entire skull asymmetrical.

INFORMATION FROM TEETH

Teeth consist of crowns (the part visible above the jawline) and roots (in alveoli, or sockets, in the jaw). A tooth is formed mainly of dentine, which is yellowish; in primates, the crowns are coated with enamel, a hard, white, shiny mineralised substance; the roots are coated with cement, a grayish, rough-cast coating.

The teeth are deciduous in infants; they are shed, usually around the time of weaning, and replaced by permanent teeth. Record the eruption and wear status of the dentition in any skulls obtained: which teeth are present, which are in process of eruption, whether they are worn. Third molars erupt on the threshold of physical maturity and usually begin to show signs of wear shortly before the basilar suture begins to fuse. The sequence of eruption of both deciduous and permanent teeth varies interspecifically and may be slightly variable intraspecifically.

Tooth wear is initially in the form of faint scratch lines on the enamel, then as shiny facets; later, the enamel becomes worn through, revealing the softer, yellow-brown dentine beneath. Cusps on the cheekteeth (premolars and molars) wear at different rates; typically, the lingual (tongue-side) cusps on the upper (maxillary) cheekteeth are more worn than those on the buccal (cheekside), and it is the reverse for the lower (mandibular) cheekteeth. Cusp shape and disposition is, within reason, a good indicator of diet. High, pointed cusps suggest insectivory; low, rounded cusps suggest frugivory. Taxa with a strongly folivorous diet commonly have high crowned cheekteeth, with the cusps to some degree joined by ridges (lophs); this condition is most strongly developed in the Old World Monkeys (Cercopithecoidea), where the joining of the molar cusps in transverse pairs, accompanied by constriction of the crown between the lophs, is called bilophodonty. Different dietary items leave characteristic scratches on the crown, which can be analysed under a microscope (especially a scanning electron microscope).

Incisor teeth may occlude end-to-end (topbite), or the uppers in front of the lowers (overbite) or, rarely (but commonly in langurs of the genus *Presbytis*), the lowers in front of the uppers (underbite). If some incisors are missing, or only the cranium or the mandible is present, the bite is indicated by the face of the remaining incisors that has wear marks.

In most Strepsirrhini, the lower incisors and canines are strongly procumbent (forming a tooth comb or scraper) and typically do not occlude at all with the uppers. In the true marmosets (genus *Callithrix*) the lower incisors and canines are also undifferentiated, compressed and slightly procumbent, though they do not form a dental comb in the same way; a closer approach to a tooth comb is seen in the sakis and uakaris (Pitheciinae), in which the incisors (but not the canines) of both jaws are procumbent and compressed.

Teeth must be measured with a dial or vernier caliper to tenths of a millimetre. The usual measurements are mesiodistal length, buccolingual (labiolingual) breadth, and crown height, but others are possible; for example, in the case of lower molars with marked differentiation between trigonid and talonid (lemurs, for example), or a bilophodont construction (Old World monkeys), both mesial and distal breadths are often taken. Take the length and breadth measurements at the level of the alveolar line, i.e. just above the cemento-enamel junction. With age, teeth (especially cheekteeth) decrease in length owing to interproximal wear (wearing against the teeth in front and behind); as the crown is worn away, the breadth may change too, first increasing (the crown usually flares somewhat below the occlusal surface), then decreasing. But these changes are minor, except with extreme wear. The only major change is the reduction in crown height; crown height measurement is of value mainly for documenting the degree of wear.

Teeth can also be measured in live, anaesthetised animals (see e.g. Phillips-Conroy & Jolly (1986), who also made dental casts of their study animals).

SEXING THE TEETH

The canine teeth are reliable indicators of sex in all Catarrhini except gibbons (Hylobatidae) and humans (*Homo*). The male in all other catarrhines has longer canines than does the female, and the anterior lower premolar is modified to act as a hone for it; this premolar is narrow, typically unicuspid, and with the enamel dipping down below the usual alveolar margin. This sexual dimorphism is gross in Cercopithecinae and many Colobinae, but in a few colobines the sexual difference is marginal. In many, but not all, Platyrrhini there is likewise sexual dimorphism in the canines. There is no such sex difference in tarsiers or strepsirrhines, and these are impossible to sex using the skull.

DENTAL PATHOLOGY

Look for irregular, shallow discoloured pits on the walls of the teeth (usually low down, near the neck of the tooth where it emerges from the alveolus), sometimes on the occlusal surface in the fissures between the cusps. These indicate caries.

Dental abscesses affect the surrounding bone and take the form of large crater-like holes in the alveolar bone overlying the infected tooth roots.

Empty alveoli (tooth sockets) generally mean that the tooth has fallen out after death (postmortem), but an alveolus that has been partly or completely filled in by bone indicates antemortem tooth loss. Teeth may be lost because of infection, or as a result of trauma.

Transverse lines across teeth, most obvious on incisors and canines, are enamel hypoplasia (inadequate formation of enamel). This condition is to be distinguished from simple traumas on the teeth because the lines are symmetrical on the jaws, meaning that the lesion occurs at the same level on both antimeres (both lateral upper incisors, or whichever tooth is affected). They result from some acute stress, such as disease or poor nutrition. Because teeth form at regular rates, the measurement from the lesion to the cemento-enamel junction indicates the age of the animal when the hypoplasia-inducing stress occurred. Weaning, which often involves transferring to a poorer diet, is a likely cause of hypoplasia.

Look for broken teeth. Unlike bones, teeth do not regrow, so a breakage remains visible and, if the pulp cavity has been exposed, potentially a cause of infection. The jagged edges wear and become rounded. A badly decayed (carious) tooth may break, but the usual cause is either fighting (especially canines), or trying to bite too hard an object.

MEASURING THE BODY

POSTCRANIAL BONES

The five bones commonly measured (lengths only, or shaft widths as well) are humerus, radius, femur, tibia and clavicle. Simply take the greatest length, from end to end, parallel with the shaft.

Humerus + radius expressed as a percentage of femur + tibia is called the Intermembral Index. A very low index (below 70) tends to characterise vertical clingers; a very high one (above 100) indicates a brachiator, clamberer or knuckle-walker; the quadrupeds have indices in between, and the lower the index the more of a leaper the animal (as a rule of thumb).

Radius expressed as a percentage of humerus is the Brachial Index; tibia as a percentage of femur is the Crural Index.

GROWTH OF THE BODY

During growth, the articular ends, and some other protuberances, of most postcranial bones remain separated from the shafts by cartilaginous plates. These are known as epiphyses, and they fuse to the shafts when growth is complete. As well as limb bones (including metapodials and phalanges as well as long bones), the clavicle, scapula and pelvis grow by means of epiphyses, and can therefore be used to indicate age.

FRACTURES

Note the presence of healed fractures in the skeleton. Jagged edges indicate that a bone has broken after death, or very shortly beforehand. When a breakage occurs during life, and the animal not only survives but lives for some time afterward, there are signs of healing: the broken ends grow over and join together again, but always leave some irregularity; in the case of a long bone or rib, the two halves of the bone override to some extent, instead of rejoining end to end. Illustrations of a variety of healed fractures and other pathologies are given by Schultz (1944) for white-handed gibbons (*Hylobates lar*); this classic study shows how unexpectedly common breakages and healing are in at least one species, and the grotesque injuries and deformities with which some primates can live.

EXTERNAL MEASUREMENTS

There are four standard linear measurements that are taken in all mammals:

Head and body length (H+B, or HB). Tip of snout to anus (when head is stretched out as far as possible in line with back).
Tail (T). From anus to tail tip, excluding hair.
Hind foot (Hf). From heel to tip of longest toe, excluding nail. Record which toe is longest.
Ear (E). Behind the conch, from the base of the ear to the tip.

These are taken straight, with a caliper if possible, not with a tape. They are best done in the flesh, as the skin stretches somewhat in skinning.

Additional measurements often taken are head length (an approximation to greatest skull length), trunk height (manubrium to pubis), hip breadth, hand length and breadth, foot breadth, and lengths of individual limb segments (thigh, lower leg, upper arm, forearm). Though not standard, they may be useful as approximations to bone lengths (see McArdle, 1981).

Weight (body mass) is nowadays, for better or worse, the most commonly employed standard size measurement. The animal should preferably be weighed at once; or, if the gut has been removed, the guts should be weighed separately, losing as little of their contents as possible.

From the weight and body length, the body mass index can be calculated as (kilograms weight)/(metre body length)^2. Skin-folds and limb circumferences can also be useful measures of body condition.

GENITAL ANATOMY

Many primates, especially strepsirrhines, show marked changes in testicular size according to breeding condition. Length and width of testes can be measured with calipers, and testicular volume (TV) calculated from this as a regular ellipsoid:

$$TV = [\pi \times (width)^2 \times length]/6$$

A scrotal skin fold is often subtracted from the length and width of the testes.

Females may undergo periodic opening and closing of the vulva, or exhibit sexual swellings, and some fairly standard scores exist to describe these (see e.g. Scott, 1984). Nipples become elongated with suckling, and can indicate whether a female has borne young; pregnancy should be noted.

GUT MORPHOLOGY

If a dead animal is to be preserved whole, the alimentary tract (guts) should always be removed first and preserved separately, being a prime site for bacterial activity. Gut contents can be useful for dietary studies (Chapter 11). Gut size and morphology is subject to significant variations across the Primates as a whole, and even between quite closely related taxa. In most primates the stomach is a simple sac, somewhat elongated transversely; the oesophagus enters it from above, and the duodenum (the first part of the small intestine) leaves it at one side; the two surfaces between oesophagus and duodenum are called the lesser and greater curvatures (self-explanatory terms). In the subfamily Colobinae (Catarrhini, Cercopithecidae), however, the stomach is sacculated, and divided into three or four compartments: saccus (the blind, sac-like anterior compartment), tubus (the middle section) and pars pylorica (the posterior compartment); the fourth compartment, the praesaccus, if it occurs at all, is located at the junction of the saccus and the oesophagus. The saccus (and praesaccus) and anterior tubus are fermentation chambers, where cellulose and hemicellulose are fermented by symbiotic bacteria into short-chain fatty acids and waste gases. A groove, the reticular groove, whose edges may be contracted to make it into a tube, runs along the dorsal side of the tubus, and conducts fluid ingesta that need no fermentation.

The other main gut variable is the caecum, which may be extremely long, and full of cellulolytic bacteria (e.g. *Lepilemur*), or very reduced but with a vermiform appendix, containing lymphatic tissue (all the Hominoidea). The colon may also be a significant site for fermentation, and varies from thick and sacculated (in Simiiformes) to slender but folded into an M-shaped loop, the ansa coli (all Lemuriformes except the Cheirogaleidae).

Traditionally, the lengths of each of the components of the alimentary tract are measured: oesophagus, stomach, small intestine, caecum and colon (= large intestine), rectum. The small intestine is complexly coiled, and the colon has ascending, transverse and descending portions (plus a median loop, the ansa coli, in ·strepsirrhines), and the mesenteries holding them in place must be cut through before they can be straightened out for measuring.

Chivers & Hladik (1980) attempted a more sophisticated method of measurement. First of all they flattened each segment and measured its breadth at points, from which an average breadth was calculated, and a surface area (breadth × length); except for the stomach, where they measured the length of the greater curvature as a measure of its circumference. From these areas, they calculated volumes using the formula for the volume of a sphere in the case of the stomach, and that of a cylinder in the case of the other components. This is probably the best that can be achieved in the field, but is still in the end a rather crude approximation to the complexities of villous formation (small intestine) and sacculation (large intestine).

PELAGE

The pelage in most primates is patterned; this may consist only of the ventral surface (plus, usually, the inner aspects of the limbs) being lighter in colour than the dorsal, or the entire body may be divided into different chromogenetic fields. The boundaries between any such fields should be recorded, along with their approximate colour. Colour photographs vary deceptively in different light or under different exposures, and some workers include a standard colour strip in their photographs.

Some or all of the dorsal hairs are agouti patterned: that is, with alternating light (phaeomelanic) and dark (eumelanic) bands. The number of band-pairs should be recorded, along with the approximate tone of each band. Non-agouti hairs are called saturated.

Hair tracts on the cheeks, crown, nape, throat and other places are usually in the form of whorls. These should be recorded, and so should the often quite complicated arrangements of tracts on the top and sides of the head.

In some species, the sexes are different colours and differently patterned (most strikingly in some lemurs, genus *Eulemur*; sakis, genus *Pithecia*; and gibbons, Hylobatidae). Age changes in pelage are usual in catarrhines, especially from neonate coat to juvenile/adult coat.

GLANDS

An important feature of many primates, but one liable to be overlooked by casual observers, is skin glands. Their presence and location is not always obvious, but even if there is not a clearly demarcated glabrous area there is generally indication in the form of irregularity or sparse hair-covering of the skin, and sometimes the secretion itself is visible. Lemuriformes and Lorisformes have glands on the perineum, scrotum, labia, wrist, inner elbow, shoulder, jaw angle or throat. Probably all platyrrhines have a gland on the sternum, and so do some catarrhines including mandrills, gibbons and orang-utans. The activity of glands can be described for individual animals, or scored, and samples of glandular exudates collected (Chapter 8).

REMOVAL OF SPECIMENS FROM THE FIELD

It is often impractical owing to a large number of samples, or unethical, to remove trophy skulls or skins from the field. It is therefore imperative that as much information as possible about the specimens can be collected while in the field. If there is a facility that can keep them, specimens should be kept in their country of origin so that other researchers can benefit from them. If all measurements and photographs can be taken before leaving the research country there should be no need to remove the specimens from their country of origin. If specimens do need to be exported from the research country there are a number of procedures to follow (e.g. CITES (Convention on the International Trade in Endangered Species) permits, health/agriculture department authorisations, and other permits may be required for import and export).

REFERENCES

Chivers, D.J. & Hladik, C.M. (1980). Morphology of the gastrointestinal tract in primates: comparisons with other mammals in relation to diet. *J. Morphol.* **166**, 337–86.

International Commission on Zoological Nomenclature (1999). *International Code of Zoological Nomenclature*, 4th edition. London.

Markham, R. & Groves, C.P. (1990). Weights of wild orang utans. *Am. J. Phys. Anthropol.* **81**, 1–3.

McArdle, J.E. (1981). Functional morphology of the hip and thigh of the Lorisifermes. *Contrib. Primatol.* **17**, 1–132.

Meier, B., Albignac, R., Peyriéras, A., Rumpler, Y. & Wright, P. (1987). A new species of *Hapalemur* (Primates) from South East Madagascar. *Folia Primatol.* **48**, 211–15.

Nichols, R. (2001). Gene trees and species trees are not the same. *Trends Ecol. Evol.* **16**, 358–64.

Phillips-Conroy, J. E. & Jolly, C. J. (1986). Changes in the structure of a baboon hybrid zone in the Awash National Park, Ethiopia. *Am. J. Phys. Anthropol.* **71**, 337–50.

Schultz, A. H. (1944). Age changes and variability in gibbons. A morphological study on a population sample of a man-like ape. *Am. J. Phys. Anthropol.* **2**, 1–129.

Scott, L. (1984). Reproductive behavior of adolescent female baboons (*Papio anubis*) in Kenya. In *Female Primates: Studies by Women Primatologists*, ed. M. F. Small, pp. 77–100. New York: Alan R Liss, Inc.

Thalmann, U. & Geissmann, T. (2000). Distribution and geographic variation in the Western woolly lemur (*Avahi occidentalis*) with description of a new species (*A.unicolor*). *Int. J. Primatol.* **21**, 915–41.

10 • Marking and radio-tracking primates

PAUL E. HONESS[1,2] AND DAVID W. MACDONALD[2]
[1]*Department of Veterinary Services, University of Oxford, Oxford, UK*
[2]*Wildlife Conservation Research Unit, Department of Zoology, University of Oxford, Oxford, UK*

INTRODUCTION

Although the past two decades have seen a revolution in many aspects of field biology due to advances in radio-tracking and telemetry, including the development of satellite and Global Positioning System (GPS) technology, these techniques have been applied less to primates than to other orders of mammals or to vertebrates in general (Casperd, 1992). A comprehensive account of animal tagging, or even of radio-tagging, would warrant a book in itself, but here we aim to draw the attention of primatologists to the advances in this family of techniques. We review a number of important studies of primates that have used them, and introduce the practicalities involved. The reader considering the use of these techniques should advance no further without exploring the wider, and vast, literature associated with radio-tracking other mammals. General reviews of radio-tracking are presented in Amlaner & Macdonald (1980), Kenward (2001), and Millspaugh & Marzluff (2001) (see 'List of suppliers and useful Internet sites' at the end of this chapter).

The term 'radio-tracking' is correctly applied only to the use of radio-transmitters and receivers to record location information. Traditionally, most field biologists have used VHF (very high frequency) or UHF (ultra high frequency) radio-tracking transmitters, but recently there has been a rapid growth in satellite tracking systems and associated technology (Chapter 4). 'Biotelemetry' is, strictly, the remote measurement of biological, particularly physiological, data (e.g. heart rate, blood pressure, body temperature, brain wave activity) (Amlaner, 1978; Chapter 18).

An almost invariable prerequisite to marking animals, by whatever means, is capture and restraint (Chapters 7 and 8). It is important to ensure that capture, handling and, ultimately, wearing the tag are minimally stressful (Halloren *et al.*, 1989), and appropriate veterinary guidance is essential (Karesh *et al.*, 1998; Chapter 8). For both ethical and scientific reasons, the impact of the research on the animal, and of the number of subjects involved, must be minimised as a priority, while still achieving results with adequate statistical power.

Field and Laboratory Methods in Primatology: A Practical Guide, ed. Joanna M. Setchell and Deborah J. Curtis. Published by Cambridge University Press. © Cambridge University Press 2003.

MARKING

Individual characteristics may allow the accurate identification and recognition of non-human primates (hereafter primates, in this chapter) (Ingram, 1978; Chapter 2), but this is not generally applicable and artificial marking is frequently required. Marking provides a method for the unambiguous recognition of individuals. Circumstances under which this may be useful include those where the animals under study:

- do not have readily recognisable naturally occurring individual characteristics (e.g. scars, missing digits, etc.);
- live in large social groups;
- live in social/foraging groups of changing composition such as fission/fusion societies (e.g. black-handed spider monkeys, *Ateles geoffroyi*; Campbell & Sussman, 1994);
- are nocturnal and/or live in habitat that affords poor visibility, but where individuals are frequently encountered;
- are the subjects of long-term studies, where different researchers may be involved;
- are the subjects of capture–mark–recapture studies.

The following methods of marking have been applied to primates. With all methods, it is important to consider the need to identify individuals from whichever side they are viewed (Rasmussen, 1991).

FUR SHAVING

Charles-Dominique & Bearder (1979) identified mohol galagos (*Galago moholi*) by shaving patterns of rings on limbs. This technique suffers from limited longevity owing to hair regrowth, as well as risks associated with exposing skin to sunburn, insect bites and abrasion (Rasmussen, 1991).

FUR DYEING

Application of picric acid to white hair turns it yellow. Halloren *et al.* (1989) recommend using this dye for marking captive callitrichid infants soon after birth. They report that at 12 months of age there was no difference in the body weight of those that had and those that had not been marked, nor was there any difference in survivorship or age of emigration.

Permanent black dye (Nyanzol-D, J.Belmar Inc.) lasts up to one year in adults but less than four months in young monkeys (species not specified; Rasmussen, 1991). Megna (2001) recommends the following preparation: 'Mix 12 cc water, 12 cc over-the-counter hydrogen peroxide, 48 cc rubbing alcohol and 1 heaped teaspoon of the dye. Stir well, allow to sit a while, stir again and apply with a sponge-type paint/craft brush.' This dye is reported to work well on skin and fur, and has been used on adults and juveniles, but not infants (Drea & Wallen, 1999).

Savage *et al.* (1993) used a commercially available hair dye, Redken DecoColors (Redken Laboratories Inc.) in different combinations of red, yellow, orange, purple and brown to mark the naturally white parts of the pelage of cotton-top tamarins (*Saguinus oedipus oedipus*). This form of marking was used in addition to tattooing and radio-tagging and no difference was found in the time spent carrying infants between marked males with transmitters and marked males and females without transmitters.

FREEZE BRANDING

Here the fur is first clipped away and then the skin is sprayed with freon, through a template of a unique code, for 12 seconds (Jones & Bush, 1988a). This technique 'bleaches' the hair follicles, which subsequently produce white hair even after moulting and so it is really only valuable for marking dark-haired species. Freeze branding when done properly will not cause blistering or permanent damage to the skin, but anaesthesia or sedation should be considered for restraint and to allow accurate branding. Large brand marks may be necessary to ensure visibility and distinctness and care should be taken not to disrupt natural hair colours/patterns that may form part of the species' signalling system.

TATTOOING

Primates as small as callitrichids can be tattooed (Savage *et al.*, 1993). Tattooing may be a painful, and hence stressful, procedure and should therefore be conducted under anaesthetic. The tattoo site and needles must be properly disinfected (e.g. with sealed swabs). This method is of little value for dark-skinned species and consideration should be given to which part of the body is tattooed, in respect of both the longevity and the visibility of the mark. Ear tattoos tend to last longer, but are less visible from a distance than those on the chest or inner thigh, which fade more rapidly and may become obscured by regrowth of hair on the limbs and trunk (Rasmussen, 1991). Tattoo guns that run on DC (battery) power are available and are useful in the field (although often constrained by poor needle speed).

EAR NOTCHES OR PUNCHES

Where it is necessary to take ear notches or punches for DNA samples, the pattern or position of these marks may be used to identify individuals. A drawback is the risk of confusion with natural marks (Rasmussen, 1991). Equipment used for this should be sterilised between individuals and where other sources of DNA may be available (e.g. from hair or faeces) the taking of ear notches or punches may be ethically questionable.

TOE CLIPPING

Formerly, studies of small mammals often removed the terminal joint of a digit as an individual mark. Increasingly, such mutilation is considered unethical, especially in primates, which depend on fine manipulative skills.

EAR TAGS

Coloured ear tags, such as those for livestock, can provide an easily visible marking system. Taylor *et al.* (1988) note that rigid, plastic tags can be difficult to see, often being obscured by fur. They trialled two types of tag on captive anubis baboons (*Papio anubis*) in which the fitting, under anaesthetic, of combined Jumbo Rototag (back) and Allflex tags (front) (NASCO) resulted in high rates of grooming out and infection. No such problems occurred with Duflex sheep/goat tags (Fearing Manufacturing Co. Inc.), which incorporate a slow-release, antibacterial coating (chlorohexidine).

MICROCHIPS

Microchips (passive integrated transponder, or PIT, tags) encased in glass, frequently with a bio-compatible coating, are injected subcutaneously using a syringe or injector 'pistol'. In primates they are typically injected into the lateral thigh area; thin-skinned areas such as the hands and feet should be avoided (Rasmussen, 1991). These microchips (approximately 12 mm × 2 mm in size and weighing about 0.06 g) function as radio-frequency transmitters and contain no batteries, but are activated and read using a special microchip reader held less than 20 cm away (depending on the reader). Each microchip has a unique code, which is detected by the reader, and will last for around one million activations. A practical difficulty is the need to get the reader sufficiently close to the chip, especially as the chip may migrate short distances under the skin (Wolfensohn, 1993). However, this system works well in combination with other more transient marks.

COLLARS OR BELTS

Brightly coloured collars or belts, secured using rivets or cable ties (be careful to ensure no sharp ends or edges protrude) can be efficient markers. Collars made of butyl material are commercially available (Telonics Inc.). Plastic discs fixed directly to collars, as used in some studies, can be fragile (Jones & Bush, 1988a), whereas anodised aluminium discs (e.g. pet tags) are more durable. Take care when considering the use of loose-hanging metal tags as they may act as bells and could compromise the ability of tagged animals to avoid predators as well as impair other important behaviour such as in social interaction and the ability to forage or predate effectively.

RADIO-TRACKING

Radio-tracking can facilitate the habituation of study animals by increasing contact time through the rapid location of collared individuals (Campbell & Sussman, 1994). It can also prove a useful tool to measure decrease in flight distance with habituation (Rasmussen, 1998; Chapter 2). It may be that more rapid habituation through radio-tracking serves only to counter the timidity caused by capture (Jones & Bush, 1988a; Campbell & Sussman, 1994; Gursky, 1998; Chapter 7), but all these effects merit note as potentially confounding variables, and the link between radio-tracking and enhanced habituation is generally a bonus to most field studies (Karesh *et al.*, 1998).

Fedigan *et al.* (1988) conducted a comparative study of the determination of home ranges in spider monkeys by radio-tracking and direct observation. They found that not only did radio-tracking ensure a greater probability of encountering any given animal, but it also resulted in larger home range estimates than were derived from observational data. Using radio-tracking, home ranges were ascertained in just two to three months, whereas it took one or more field sessions of eight months to achieve the same results by observation alone. Fedigan *et al.* (1988) emphasised that radio-tracking led researchers to areas of the animals' ranges that they would not otherwise have discovered, a point also made by Campbell & Sussman (1994).

Although, in primates, radio-tracking has primarily been used to investigate aspects of ecology such as seasonal changes in home range (Li *et al.*, 1999), substrate use (Harcourt & Nash, 1986), polyspecific associations (Buchanan-Smith, 1990) and activity patterns (Curtis *et al.*, 1999), it has also proved an extremely valuable tool for studying mating systems and social organisation (Bearder & Martin, 1980; Buchanan-Smith, 1991; Radespiel, 2000). It is a particularly useful tool for studying primates that are nocturnal or live in habitats that afford poor visibility. More specialised studies have examined locomotion (Sellers *et al.*, 1998), predation (Wiens & Zitzmann, 1999), urine marking or washing (Charles-Dominique, 1977; Harcourt, 1981), and calling frequency (Gautier & Gautier-Hion, 1982), and monitored translocated animals (Richard-Hansen *et al.*, 2000) and solitary males (Jones & Bush, 1988b).

The basic equipment needs for a radio-tracking study include: a selection of transmitters (tuned to transmit on different frequencies) with batteries, transmitter mounts, a receiver (plus batteries), a set of earphones, a coaxial cable (with suitable antenna and receiver connectors) and an antenna. Generally, and especially when working in remote areas, it is prudent to have duplicates of much of this list, as back-ups. When ordering transmitters ensure that they are set to transmit at frequencies different from those of any other researchers working within transmitter or dispersal range.

TRANSMITTERS

In the past it was common for researchers to build their own transmitters, as evidenced by practical advice in such guides as Macdonald & Amlaner (1980) and Kenward (2001); however the reliability and economies associated with larger-scale production have made commercial purchase increasingly attractive (see 'List of suppliers and useful Internet sites' at the end of this chapter).

Standard VHF/UHF transmitters emit a pulse on a tuned radio-frequency, which can be detected using an antenna and receiver and used to locate the source of the signal, generally through triangulation. Amongst many trade-offs in transmitter design, an important distinction between VHF and UHF transmitters is that higher frequencies are associated with reduced trans-mission range and high attenuation by environmental features such as veg-etation. An important consideration when selecting the frequency of your transmitters is the associated constraint on the size of both transmitting and receiving antennae. Kenward (2001) discusses, in some detail, the relationship between frequency, wavelength and receiving and transmitting antennae. In summary, the choice of a higher frequency (e.g. up to 230 MHz) will allow the use of shorter, more convenient antennae, but this is compromised by the reduced range of high frequency signals, which are more susceptible to absorption by vegetation and attenuate more rapidly. Lower frequencies (e.g. 142–150 MHz) therefore have an advantage in terms of signal range in areas where vegetation may be damp and dense, such as in rain forest. However, increasing the frequency of the transmitter to 216–230 MHz would require a transmitting antenna of only about two-thirds of the length of that for a lower frequency transmitter for the same efficiency (Kenward, 2001).

Transmitter mounting

External transmitters, with their enclosed batteries, need to be attached to the study animal. Body size and anatomy will determine the best attachment method, for example on a neck collar, harness or waistband. Obvious wel-fare priorities necessitate careful attention to potential growth and seasonal weight changes. Devices designed to drop off on a given time-scale, or to expand, offer some solutions.

Collars are the most frequently used means for attaching transmitters. Nylon webbing often proves to be the most practical and cost effective material, having a good strength to weight ratio, and being non-absorbent and relatively cheap. The transmitter antenna can either be a tuned loop of wire or a brass strip, which can form the basis of the collar, or of whip form protruding from the assembly, wrapped around inside the transmitter cas-ing or attached to the perimeter of the collar, belt or harness. Tuned loops may be more efficient and are certainly more useful for species that might significantly damage a whip antenna, but they suffer from size (diameter) constraints of what is easily tuneable and can become detuned during use or stop transmitting altogether if the loop is broken (e.g. by a predator) (Kenward, 2001). Collars can be fastened using pop rivets or nylon locking nuts through metal backing plates or a similarly secure system and care should be taken to ensure that no protuberances are left on the inside of the collar that may harm the animal. Collars need to be fitted loosely enough not to affect feeding or comfort but tight enough not to come off; a common guide is to be able to place two fingers under the fixed collar. However, collar attachment may be a serious difficulty for animals without a defined neck, such as the slow loris (*Nycticebus coucang*) (Barrett, 1985). Some collars are designed to expand with growth or seasonal weight change, using degradable foam on the inside or pleats held with degradable or forcible thread.

In their study of the social organisation of the mohol galago, Bearder & Martin (1980) used leather belts with PVC (polyvinyl chloride) tubing covering the antenna wire. The transmitter was mounted onto the belt with acrylic that could be broken into to replace the battery. Charles-Dominique (1977) used belt-mounted transmitters to study urine marking in galagos (*Galago alleni*); two differently shaped transmitter assemblies were matched to the anatomy of males and females. They emitted two different pulse rates depending on whether a pair of exposed wires that formed a switch was dry (slow pulse) or wet (fast pulse), with a slowing of the rate as the switch gradually dried. Whilst this study displayed pioneering ingenuity, the bulk of the transmitters (exceeding 9.25% of the animal's body weight) would not meet current welfare standards. Another innovative design was used to study frequency of vocalisations using a microphone (collar) and transmitter (harness) in a male crowned guenon (*Cercopithecus pogonias*), a female redtail guenon (*C. ascanius*) and their four hybrid offspring (Gautier & Gautier-Hion, 1982). Harness-mounted transmitters were also used by Savage *et al.* (1993) in their study of cotton-top tamarins.

Although it is a common method of tagging ectothermic vertebrates, implantation is less widely used for studies of mammals except where physiological measures are a focus. Clearly, implantation requires veterinary skill and attention to infection (Chapters 8 and 18). However, once successfully implanted the transmitter may be less problematic for the subject than an externally attached device.

Transmitter retrieval

For both scientific and ethical reasons it is essential to minimise the effects of radio-tracking (as with other techniques) on the subject, and even minimal effects should be carefully measured to reduce the risk of confounded results. An obvious question is whether it is possible to retrieve transmitters from study animals, either at the end of the study or at the end of the transmitters' life. Transmitter retrieval is not often reported in the literature, although Gursky (1998) states that, in her study of spectral tarsiers (*Tarsius spectrum*), she was able to retrieve 9 out of 16 collars. It is important to make a cost–benefit analysis of the impact of leaving the transmitter in place versus the risks associated with retrieval.

Various devices can be built into a collar or harness, with the aim of causing it to drop off after a timely interval. However, there will be trade-offs, such as the extra bulk of these detachment devices and the risk of premature detachment. Many researchers choose to refurbish and re-use collars and transmitters (e.g. replace the batteries), a service offered by most manufacturers and which may represent a considerable saving over the purchase of new equipment.

When it is considered that attaching a transmitter weighing 3% of an animal's body weight is the equivalent of an 80 kg person carrying an additional 2.4 kg it is clear that care should be taken to detect, minimise and monitor any adverse effects of transmitters. Gursky (1998) found no significant difference in the behaviour of spectral tarsiers fitted with collars weighing 5% or 7% of their body weight. A reduction in the weight of the assembly needs to be balanced against reduced transmitter life and/or range owing to minimising the primary, flexible weight component, the battery.

Transmitter reliability

The life of the battery and therefore of the transmitter is rarely, without additional cost, guaranteed by the manufacturer, with low temperatures, in particular, having an adverse effect on battery life.

To prevent the battery running down when not in use, transmitters are frequently shipped and stored with a detachable magnet fixed to the outside to prevent the closure of a reed switch that would otherwise complete the power circuit and activate the transmitter. On receipt of the transmitters, and periodically during storage before use, it is prudent to check that the magnets are preventing transmission by testing whether the transmitter can be detected using tracking equipment with the magnet in place. If the magnets or reed switches fail, the batteries may be exhausted when the transmitter is required for use. It is also important to double check and record the frequency of an activated transmitter before its use and not simply to rely on the transmitter's label.

RECEIVERS

Receivers are generally the most expensive element of a radio-tracking system. There are two main types of receiver: standard models (approximately US$ 650–850), and more specialised, programmable ones (over US$ 2000).

Standard receivers are pre-tuned to a specific frequency range. Transmitters must match the frequency range of your receiver. Three of the most commonly used receivers are the Telonics TR-4 (Telonics Inc.), the AVM LA12-Q (AVM Instrument Co. Ltd) and the Mariner M-57 (Mariner Radar Ltd). As always,

Table 10.1. *Comparison of the weight, size and power requirements of three commonly used receivers*

Receiver	Size (mm × mm × mm)	Weight (including batteries) (g)	Power
Mariner M-57	220 × 80 × 110	2500	8 × 'C' cells[a]
Telonics TR-4	170 × 90 × 45	425 (including 2 batteries)	1 or 2 × 9 V batteries[b]
AVM LA12-Q	150 × 150 × 60	750	6 × NickelCadmium (Ni-Cad) AA cells[c]

[a] No 12 V or mains operation.
[b] 12 V DC, 220 and 110 V AC adapters available.
[c] 110 and 220 V DC adapters available.

there are trade-offs when choosing between receivers; for example, Kenward (2001) observes that Mariner M-57s can survive being run-over by a Landrover, but they are correspondingly heavier than the Telonics and AVM alternatives and are the least flexible in terms of powering options (Table 10.1).

High humidity can seriously damage receivers and batteries and, when not in use, receivers should be dried, batteries removed and both stored with silica gel.

RECEIVING ANTENNAS

Although more complex antenna systems are available (e.g. aircraft-mounted and static systems; Kenward, 2001) most field studies require portable antennae. The most basic type of antenna, the omni-directional, is typically of whip form and may, for example, be attached by a magnet to the roof of a vehicle. It is particularly useful for detecting the general proximity of a transmitter, without providing any information about its bearing from the observer.

The directional, hand-held antenna (e.g. multi-element Yagi, 2-element Adcock or Telonics 'H' and loop) is the most practical for studying primates in forest or woodland habitats. A common variant is the multi-element Yagi gain-type antenna, which consists of a main driven element (connected via a coaxial cable to the receiver) with a number of parasitic directing elements in front of it and a reflector element behind, all arranged in a specific mathematical relationship. While the spacing relationship between the elements is defined, the number of elements can vary; increasing the number of elements increases the directionality of the antenna (Burger, 1991). Collapsible versions of the Yagi are available, and are a good option for use in dense vegetation. A pair of Yagis, typically of four or more elements, can be used in tandem, in what is called a Null-Peak System. This is significantly more accurate than a single hand-held antenna, but carries the disadvantage of bulk.

Macdonald & Amlaner (1980) outline some of the problems associated with using directional antennae and include some tips for their solution, as does Kenward (2001). Kenward (2001) suggests a procedure for determining the bearing of the transmitter from the observer using a Yagi antenna:

Step 1: Find the strongest signal that is not either a 'back bearing' or a reflection.

Step 2: Sweep the antenna to one side until there is no more signal reception

Step 3: Swing the antenna back towards the strongest signal until it is *just* detectable and mark this bearing using a landmark or compass.

Step 4: Repeat this process (steps 2 and 3) on the other side of the strongest signal and again mark the bearing of the die-off of the signal.

Step 5: The bearing of the transmitter for triangulation is calculated as being the mid-point between those marked in steps 3 and 4. It is always desirable, where possible, to confirm the exact location derived through triangulation by making visual contact with the subject. It may be possible to locate an animal by simply following the bearing of the transmitter signal, correcting as you go, until you find it. However, particularly if the study animal is nervous, poorly habituated or in habitat where stealth is difficult, the subject may be continually displaced ahead of the tracking researcher. In these circumstances a triangulated fix may provide the only genuine measure of the animal's movements.

ENVIRONMENTAL EFFECTS

Radio signals can be affected by the tracking environment including reduction of signal intensity through absorption by vegetation. Fedigan *et al.* (1988) found, in Santa Rosa National Park (Costa Rica), that changes in vegetation and climate between the dry and wet season resulted in a decrease in reception range from 500 to 150 m, causing a suspension of radio-tracking during the wet season. False bearings can be picked up owing to the reflection of the signal by natural topographical features (e.g. cliffs, hills, gullies, rocks, etc.) as well as by woodland, individual trees and artificial features such as buildings. However, reflected signals can prove useful where the direct signal may be totally blocked (e.g. by features described above). Kenward (2001) points out that diffraction of the signal is a particular problem in woodland or forest where it can be difficult to get an accurate bearing owing to interference effects around tree trunks. He also notes that, while, in woodland or forest holding an antenna so that the elements are vertical may, to an extent, improve the accuracy, it is still preferable to hold it horizontally because of the amount of vertical reflection and diffraction produced by trees.

SATELLITE TRACKING

The obvious feature distinguishing satellite tracking from traditional radio-tracking is that it does not require the continuous presence of the researcher. This facilitates the tracking of animals that move considerable distances or that use seasonally inaccessible areas. Two types of device are commonly used for satellite tracking, the GPS and ARGOS (Advanced Research and Global Observation Satellite) system. The GPS uses a ground-based receiver that downloads information from a specialised constellation of low orbit satellites. The receiver requires data from at least three satellites, simultaneously, to calculate its position. The ARGOS system works in reverse to GPS, in that a ground-based transmitter sends information to another specialised satellite constellation. Here only one satellite is required to calculate a fix, which is in turn transmitted to a ground-based station for retrieval (Chapter 4).

GPS receivers produce positional accuracy of 5–15 m, which can be further improved by the use of differential correction (differential GPS (DGPS), Chapter 4). DGPS typically produces sub-metre accuracy and is available built into some receivers along with other options such as temperature and mortality sensors. ARGOS generates variable accuracy that is typically of the order of hundreds of metres, but can be as great as kilometres, making it impractical for animals with small ranges.

GPS receivers store data on board that must be accessed directly by connection to the collar (facilitated by an automatic drop-off mechanism, or recapture), wireless ground-to-ground download (via a VHF link or the mobile phone (Global System for Mobile Communication, or GSM) network), or satellite uplink. Direct connection risks the loss of data owing to inability to relocate the receiver. However, this can be solved by incorporating a drop-off mechanism and a standard VHF transmitter, allowing the recovery of the dropped collar. VHF links work over a range of 0.1–2 km, ground to ground, and up to 5 km from the air depending on the topography and vegetation obscuring the line of sight from transmitter to receiver. An additional disadvantage of downloading data via a standard radio-link is that it may take a considerable time (although this can be shortened with data compression); a typical example might be 30 minutes to download two months' of data. To achieve this within the specified distance, and at a particular time, may be excessively challenging in the field and in almost all of these cases the system (extra batteries, VHF transmitters, drop-off mechanism, etc.) for allowing data retrieval adds to the overall weight of the collar.

Downloading data via GSM requires no ground-based personnel, but works only where there is mobile phone coverage. Satellite uplinks, like the ARGOS system, require no ground-based reception or personnel, but are limited by the number of fixes that can be sent at a time.

GPS and ARGOS systems differ in their positional sampling rate; GPS can store up to one fix per second, whereas, on average, ARGOS can resolve just one per day. This has an influence on value for money, so that if all the costs of typical tracking studies are taken into account, GPS costs in the order of US\$ 1–5 per fix, compared to ARGOS costs of around US\$ 23 per fix (GPS with satellite uplink costs the same as ARGOS). This compares to approximately US\$ 1–20 per fix for standard radio-tracking (S. Ellwood, personal communication, 2002).

Although GPS receiver-based animal tags, and ARGOS transmitters, can now weigh as little as 30 g (plus the collar/belt/harness assembly), bringing this technology within the range of primates weighing as little as about 1.5 kg, GPS reception and ARGOS transmission can be seriously attenuated by thick vegetation. This can mean that in tropical forest neither method is suitable owing to poor fix acquisition rate.

BIOTELEMETRY

Biotelemetry can be achieved via either implantable or external devices, the latter sometimes fitted to a jacket or harness. Among external devices are those that include mercury tilt switches that transmit at different pulse frequencies depending on their orientation and are used to monitor activity and posture (Kenward, 2001). Thus far, the use of implantable transmitters has been minimally applied to wild primates. Implants are often put into the peritoneum, requiring deep anaesthesia, aseptic surgical conditions and the use of analgesics and antibiotics to control pain and potential infection (Chapters 8 and 18).

The necessary restriction on the size of implants constrains battery life. Rasmussen (1991) also cautions that more frequent sampling results in diminished battery life. Some parameters, such as temperature, vary only gradually and so need not be sampled as often as, for example, is likely to be necessary for brain wave activity or heart rate. The sampling of more than one parameter, requiring a multi-channel device, will need a greater signal bandwidth, and therefore more battery power. The choice of transmission frequency is also important as higher frequencies attenuate more rapidly through tissue and therefore have reduced reception range as compared with lower frequencies (Rasmussen, 1991).

REGULATIONS

Radio-tracking studies typically fall under both general (access and research) and specific regulations (governing radio-frequency use and invasive procedures). Flaunting regulations may affect not only your ability to undertake research but also that of other researchers.

Whereas there is a strategy to harmonise frequency band allocation across the European Union by 2008, there is little consistency elsewhere, though many countries allocate 150 kHz to 2 MHz to wildlife radio-tracking (Kenward,

2001). Whilst VHF and UHF transmitters are very low energy, have limited range and are consequently unlikely to interfere with, or suffer interference from, other radio-transmitter users, nevertheless it is important to consult both the authorities and others who may have used radio-tags in your study country.

It is unacceptable to attempt any invasive procedure, such as the implantation of biotelemetry devices, in a trial-and-error manner. Training and supervision are morally, and generally legally, essential.

ACKNOWLEDGEMENTS

Many thanks to Stephen Ellwood for his guidance on the section in this chapter concerning satellite tracking.

REFERENCES

Amlaner, C.J. Jr (1978). Biotelemetry from free-ranging animals. In *Animal Marking: Recognition Marking of Animals in Research*, ed. B. Stonehouse, pp. 205–28. London: Macmillan Press.

Amlaner, C.J. Jr, & Macdonald, D.W. (eds.) (1980). *A Handbook on Biotelemetry and Radio Tracking*. Oxford: Pergamon Press.

Barrett, E. (1985). The ecology of some nocturnal arboreal mammals in the rain forest of peninsular Malaysia. Ph.D. thesis, University of Cambridge.

Bearder, S.K. & Martin, R.D. (1980). The social organization of a nocturnal primate revealed by radio tracking. In *A Handbook on Biotelemetry and Radio Tracking*, ed. C.J. Amlaner & D.W. Macdonald, pp. 633–48. Oxford: Pergamon Press.

Buchanan-Smith, H.M. (1990). Polyspecific association of two tamarin species, *Saguinus labiatus* and *Saguinus fuscicollis*, in Bolivia. *Am. J. Primatol.* **22**, 205–14.

(1991). A field study on the red-bellied tamarin, *Saguinus labiatus*, in Bolivia. *Int. J. Primatol.* **12**, 259–75.

Burger, W. (1991). Receiving antenna 'accuracy'. *Telonics Q.* **4**(2), 2–3.

Campbell, A.F. & Sussman, R.W. (1994). The value of radio tracking in the study of neotropical rainforest monkeys. *Am. J. Primatol.* **32**, 291–301.

Casperd, J. (1992). Primate radiotracking and biotelemetry. *Primate Eye* **47**, 18–23.

Charles-Dominique, P. (1977). Urine marking and territoriality in *Galago alleni* (Waterhouse, 1937 – Lorisoidea, Primates): a field study by radio-telemetry. *Z. Tierpsychol.* **43**, 113–38.

Charles-Dominique, P. & Bearder, S.K. (1979). Field studies of lorisid behaviour: methodological aspects. In *The Study of Prosimian Behavior*, ed. G.A. Doyle & R.D. Martin, pp. 567–629. London: Academic Press.

Curtis, D.J., Zaramody, A. & Martin, R.D. (1999). Cathemerality in the mongoose lemur, *Eulemur mongoz*. *Am. J. Primatol.* **47**, 279–98.

Drea, C.M. & Wallen, K. (1999). Low status monkeys 'play dumb' when learning in mixed social groups. *Proc. Natl. Acad. Sci., USA* **96**, 12965–9.

Fedigan, L.M., Fedigan, L., Chapman, C. & Glander, K.E. (1988). Spider monkey home ranges: a comparison of radio telemetry and direct observation. *Am. J. Primatol.* **16**, 19–29.

Gautier, J.-P. & Gautier-Hion, A. (1982). Vocal communication within a group of monkeys: analysis by biotelemetry. In *Primate Communication*, ed. C.T. Snowdon, C.H. Brown & M.R. Petersen, pp. 5–29. Cambridge: Cambridge University Press.

Gursky, S. (1998). Effect of radio transmitter weight on a small nocturnal primate. *Am J. Primatol.* **46**, 145–55.

Halloren, E., Price, E. & McGrew, W. (1989). Technique for non-invasive marking of primate infants. *Lab. Primate Newsl.* **28**(4), 13–15.

Harcourt, C. (1981). An examination of the function of urine washing in *Galago senegalensis*. *Z. Tierpsychol.* **55**, 119–28.

Harcourt, C. & Nash, L. (1986). Species differences in substrate use and diet between sympatric galagos in two Kenyan coastal forests. *Primates* **27**(1), 41–52.

Ingram, J.C. (1978). Primate markings. In *Animal Marking: Recognition Marking of Animals in Research*, ed. B. Stonehouse, pp. 169–74. London: Macmillan Press.

Jones, W. & Bush, B. (1988a). Darting and marking techniques for an arboreal forest monkey, *Cercopithecus ascanius*. *Am. J. Primatol.* **14**, 83–9.

(1988b). Movement and reproductive behaviour of solitary male redtail guenons (*Cercopithecus ascanius*). *Am. J. Primatol.* **14**, 203–22.

Karesh, W.B., Wallace, R.B., Painter, L.E., Rumiz, D., Braselton, W.E., Dierenfeld, E.S. & Puche, H. (1998). Immobilisation and health assessment of free-ranging black spider monkeys (*Ateles paniscus chamek*). *Am. J. Primatol.* **44**, 107–23.

Kenward, R. (2001). *A Manual for Wildlife Radio Tracking*. London: Academic Press.

Li, B., Chen, C., Ji, W. & Ren, B. (1999). Seasonal home range changes of the Sichuan snub-nosed monkey (*Rhinopithecus roxellana*) in the Qinling Mountains of China. *Folia Primatol.* **71**, 375–86.

Macdonald, D.W. & Amlaner, C.J. Jr (1980). A practical guide to radio tracking. In *A Handbook on Biotelemetry and Radio Tracking*, ed. C.J. Amlaner Jr & D.W. Macdonald, pp. 143–59. Oxford: Pergamon Press.

Megna, N. (2001). Marking monkeys – Nyanzol-D (discussion). *Lab. Primate Newsl.* **40**(3), 5.

Millspaugh, J.J. & Marzluff, J. (eds.) (2001). *Radio Tracking and Animal Populations*. San Diego: Academic Press.

Radespiel, U. (2000). Sociality in the gray mouse lemur (*Microcebus murinus*) in northwestern Madagascar. *Am. J. Primatol.* **51**, 21–40.

Rasmussen, K. (1991). Identification, capture and biotelemetry of socially living monkeys. *Lab. Anim. Sci.* **41**, 350–4.

(1998). Changes in the range use of Geoffroy's tamarins (*Saguinus geoffroyi*) associated with habituation to observers. *Folia Primatol.* **69**, 153–9.

Richard-Hansen, C., Vié, J. C. & de Thoisy, B. (2000). Translocation of red howler monkeys (*Alouatta seniculus*) in French Guiana. *Biol. Conserv.* **93**, 247–53.

Savage, A., Giraldo, L. H., Blumer, E. S., Soto, L. H., Burger, W. & Snowdon, C. T. (1993). Field techniques for monitoring cotton-top tamarins (*Saguinus oedipus oedipus*) in Columbia. *Am. J. Primatol.* **31**, 189–96.

Sellers, W. I., Varley, J. S. & Waters, S. S. (1998). Remote monitoring of locomotion using accelerometers: a pilot study. *Folia Primatol.* **69**, 82–5.

Taylor, L. L., Easley, S. P. & Coelho, A. M. Jr. (1988). Ear tags for long-term identification of baboons. *Lab. Primate Newsl.* **27**(2), 8–10.

Wiens, F. & Zitzmann, A. (1999). Predation on a wild slow loris (*Nycticebus coucang*) by a reticulated python (*Python reticulatus*). *Folia Primatol.* **70**, 362–4.

Wolfensohn, S. E. (1993). The use of microchip implants in identification of two species of macaque. *Anim. Welfare* **2**, 353–9.

List of suppliers and useful Internet sites

Equipment and utilities for wildlife research :
   <www.animal-info.net/36equip2.html>.
Directory of biotelemetry equipment manufacturers :
   <www.biotelem.org/manufact.html>.

Some manufacturers

AVID plc, Oak Hall Manor, Sheffield Park, Uckfield, East Sussex TN22 3QY, UK. <www.avidplc.com> (microchips and readers).

Advanced Telemetry Systems, Inc., 470 First Avenue North, PO Box 398, Isanti, MN 55040, USA. <www.atstrack.com> (radio-telemetry and GPS systems, data collection computers, antennas etc.).

AVM Instrument Co. Ltd, 1213 South Auburn St., Colfax, CA 95713, USA. www.avminstruments.com (transmitters, receivers, antennas etc.).

Biotrack Ltd, 52 Furzebrook Road, Wareham, Dorset BH20 5AX, UK. <www.biotrack.co.uk> (radio transmitters and receivers).

Fearing Manufacturing Co. Inc. (now Destron Fearing Corporation), 490 Villaume Avenue South, St Paul, MN 55075-2445, USA. <www.destron-fearing.com> – lists worldwide partners (eartags, microchips and readers).

J. Belmar Inc. Andover, MA, USA (manufacture Nyanzol-D).

Lotek Wireless Inc., 115 Pony Drive, Newmarket, Ontario, Canada, L3Y 7B5. <www.lotek.com> (radio-telemetry and GPS systems, archival tags, etc.).

Mariner Radar Ltd, Bridleway, Campsheath, Lowestoft, Suffolk NR32 5DN, UK. (radio-receivers, also available through Biotrack Ltd (see above)).

NASCO, Janesville Ave, PO Box 901, Fort Atkinson, WI 53538-0901, USA. <www.enasco.com> – on-line catalogue including outside USA (eartags, applicators, etc.).

NAVSYS Corp., 14960 Woodcarver Road, Colorado Springs, CO 80921, USA. <www.navsys.com> (GPS systems).

North Star Science and Technology, Technology Center Building. Rm 4.036, 1450 S. Rolling Rd, Baltimore, MD 21227, USA. <www.northstarst.com> (GPS systems and radio-transmitters).

OxLoc Ltd, University Field Laboratory, Wytham, Oxford OX2 8QJ, UK. <www.oxloc.com> (GPS systems).

Redken Laboratories Inc., 575 Fifth Ave, New York City, NY 10017 USA <www.redken.com> (hairdye).

Sirtrack Ltd, Private Bag 1403, Goddard Lane, Havelock North, New Zealand. <sirtrack.landcareresearch.co.nz/> (radio-telemetry and GPS systems, etc.).

Telonics Inc., 932 E. Impala Ave, Meza, AZ 85204-6699, USA. <www.telonics. com>. Website includes *Telonics Quarterly* on-line (radio-telemetry and GPS, accessories, etc.).

TVP Positioning AB (Televilt), Bandygatan 2, SE-71134 Lindesberg, Sweden. <www.positioning.televilt.se> (radio-telemetry and GPS systems, etc.).

11 • Feeding ecology and seed dispersal

Department of Anthropology, University of California, Davis, California, USA

INTRODUCTION

The study of frugivore feeding ecology is an important part of tropical biology because frugivores are among the predominant groups of tropical vertebrates in terms of species diversity and biomass. Fluctuating resource availabilities throughout the year dictate, for most frugivores, variable diets that may include leaves, fruits, flowers and seeds, as well as animal matter. Identifying food plant taxa and parts eaten is a central part of this research, and these tasks can be quite complicated when one is studying canopy-living animals in regions with diverse flora. The primary sources of information on these diets are accurate, detailed observation of feeding behaviours, the collection and classification of feeding remains, and faecal contents analysis.

Frugivores such as primates are often described as important ecological interactors (mutualists) because of their role as seed dispersers. To date, however, few studies have documented clearly the actual importance of these animals to the fleshy fruited plants on which they feed. Studying plant and animal strategies simultaneously is no simple task, but doing so provides new perspectives on botany, feeding ecology, digestive physiology, co-evolution and plant–animal interdependence, which may provide valuable tools for conservation. Particularly in the tropics, the study of endozoochory remains one area in which the dedicated natural historian still has much to reveal to the world of science. For these reasons, frugivory and seed dispersal remain topics ripe for exploration. Here I describe some of the methods and equipment required for animal observation, plant sample identification, and the analysis of faecal contents and feeding remains. I also give some practical advice on studying ranging, seed dispersal, gut passage times, seed germination and seedling survivorship in the field.

BASIC MATERIALS

Aside from the field biologist's basic kit of field clothes, notebook, compass and binoculars, the most important piece of gear for studying seed dispersal

Field and Laboratory Methods in Primatology: A Practical Guide, ed. Joanna M. Setchell and Deborah J. Curtis. Published by Cambridge University Press. © Cambridge University Press 2003.

is the simple plastic sandwich-bag – or many such bags. These are ideal for collecting dung and seed specimens. Washing and reusing a soiled bag is not worth the added mess in your pockets, or (more importantly) the risk of disease transmission (Chapters 1 and 8), so bring a plentiful supply. Buy them by the hundreds and tie them into bundles in your luggage. I recommend the cheaper bags that use small twists of wire for closure, because self-seal bags are much more expensive. Specimen bags and vials sold by biological supply houses are even more expensive and I have not found them to be substantially better. Permanent markers are required for labelling the bags with the date and time the sample was collected. Keeping markers on a string tied to your belt or daypack may slow their inevitable disappearance in the field, but, again, erring on the side of abundance is best.

Another important piece of equipment is flagging tape to mark the trees in which your animals feed (another use for the markers). Because tree size is an important determinant of habitat selection, and since trunk diameter correlates closely with fruit production, a measuring tape is necessary for measuring the diameters of feeding trees, while a range finder allows the accurate estimation of canopy heights and distances (see Chapter 3). You will also need a good pair of calipers. For dietary description, the fruits, fruitlets and seeds of food plants can be measured with these on three axes from which volume can then be calculated with the formula:

$$\frac{4\pi}{3} \, (\text{length}/2)(\text{width}/2)(\text{breadth}/2)$$

If animals remove mouthfuls from larger fruits, or fruitlets from compound fruits, then measuring the sizes of the portions removed may be just as important as measuring the sizes of entire fruits. Graduated cylinders may be useful for measuring the volumes of irregularly shaped fruits and fruit parts using water displacement (for more on measuring food mechanics and nutritional content, see Chapters 12 and 13).

Finally, a plant press and plant dryer, as well as paper bags of various sizes, will allow you to preserve specimens of food plants, particularly the fruits, which are typically rare in herbarium collections. Two companies that supply many of these tools are the Ben Meadows Company and Forestry Suppliers (see 'Useful Internet sites' at the end of this chapter). Food plant specimens are pressed between layers of newspaper or blotting paper and corrugated cardboard and left in a dryer for a period of days until they are desiccated and non-perishable. Plant dryers typically use light bulbs or lamps as a heat source. Herbaria and university botany departments may have spare presses and can offer advice on constructing homemade dryers. Take a good supply of newspaper or blotting paper to the field for these purposes.

SAMPLE IDENTIFICATION AND COLLABORATORS

Tropical botany remains a specialised field, with few guidebooks and an immense diversity of taxa to learn. Some of the field biologist's most

important collaborators are the botanists who can help to identify plant specimens. If you are fortunate enough to work at a field site with local guides
who know the indigenous flora, then the work will be made much easier. Be
aware that local names may vary from region to region and, in particular, the
same local name can refer to completely different plants. Local names provide
very useful information and facilitate communication with local guides, but
Latin names are standard. You must therefore be prepared to collect multiple
herbarium specimens of food plants for identification. These may be sent to
a competent botanist in the country in which you work (usually affiliated to
a prominent university), the national herbarium, and/or to a large international herbarium such as those at Kew Gardens (UK), INPA (Instituto Nacional
de Pesquisas da Amazônia, Brazil), the Muséum National d'Histoire Naturelle
(France), the Smithsonian Institution or the Missouri Botanical Garden (USA).
Tropical plant taxonomy is a science of few textbooks, but three invaluable
resources used by botanists worldwide are Gentry (1996), Letouzey (1986) and
the library of edited volumes produced by the Foundation Flora Malesiana
(van Steenis, 1948–2001).

ANIMAL OBSERVATION

In the study of frugivory and seed dispersal it is particularly important to
be with the animals when they awaken at dawn (or dusk in the case of nocturnal taxa) because a large proportion of dispersed seeds are voided as the
animals begin to stir upon waking. Dung should be collected as soon as it
hits the ground, so a flashlight may be useful for collecting as many specimens as possible before the animals begin their daily activities. In order to
have the most complete list of food plants, it is a good idea to collect dung
from as many animals as possible (a sort of scan sample of the group's diet).
If the animals depart before you have finished collecting dung and seeds
(a scenario which you can expect to experience quite often), then it is time to
cut your losses and follow, because tracking a particular focal animal provides
the most accurate data. Ideally, you will be able to follow an individual from
the time of food ingestion to the time of defecation in order to track gut
passage time and dispersal distance – two key measures of dispersal ability. It
may be necessary to follow a group for several days in a row in order to determine these parameters. If animals are not easily recognisable as individuals,
then tagging or marking them may be useful (Chapter 10).

Detailed observation of fruit and seed handling is obviously a key part of
this research. Focal animal time sampling (Altmann, 1974; Martin & Bateson,
1993) is the standard data collection method. However, observant *ad libitum*
recording of feeding behaviours is also very important. Keep track of the
exact timing of feeding bouts for later determination of probable gut passage
times. By recording the times and locations of each feeding bout and defecation over the course of many continuous hours you will, with luck, be able to
later deduce the rhythm of gut passage. It may be necessary to mark feeding

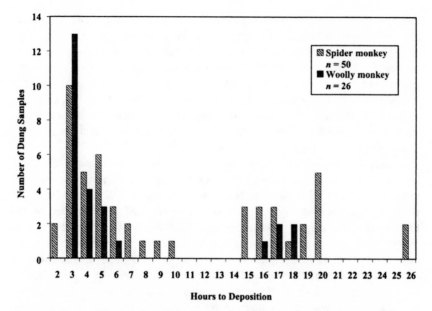

Fig. 11.1. Passage times of seeds dispersed by spider monkeys (*Ateles belzebuth belzebuth*) and woolly monkeys (*Lagothrix lagotricha poeppigii*) in the wild. *n*, number of dung samples collected.

trees quickly during follows and return to them later for measurement, botanical specimen collection, and mapping (Chapters 3 & 4). Be sure to plan for such botanical data collection in addition to time spent following animals.

Figure 11.1 shows how complicated it often is to determine gut passage time in the wild. The two modes display the different passage times of seeds ingested at different times of day. Seeds in the second mode, at roughly 1700 hours, were ingested primarily in the afternoon and deposited under sleeping sites the following morning. The passage times and dispersal distances of frugivorous spider monkeys (*Ateles belzebuth belzebuth*) and woolly monkeys (*Lagothrix lagotricha poeppigii*) in this study in eastern Ecuador did not differ significantly. Mean seed dispersal distances were 243 m and 246 m, respectively (Dew, 2001).

FAECAL CONTENTS ANALYSIS

Dung samples contain a wealth of biological information. Specimens can be manipulated through their collection bags for items such as large seeds, and they can be dissected or washed through a sieve to identify small seeds, fibre, plant parts and insect parts (Chapter 14). The rapidly passed dung of many frugivores can emerge from the animal with even delicate structures such as flower parts still intact. Counting and identifying gut-passed seeds is important, although estimating or recording simple presence/absence data may be more practical for plants with large numbers of tiny seeds, such as figs. Seeds should be examined closely for signs of damage, as the slightest nick in a seed coat can kill the plant embryo inside. Washing dung samples through a sieve may catch masticated fragments of seed coats that might otherwise be missed evidence of granivory. These methods can also be useful in the examination of stomach contents, which, sadly, may be all-too-readily available sources of dietary data (Chapter 9). Along with dung samples, these can be preserved in alcohol or dried in a food dehydrator for later analysis in a nutritional laboratory.

Be prepared to find a confusing array of seed taxa in dung, many of which will match no plant you witnessed being eaten by a focal animal. One of the first items of business will be to begin building a seed reference collection. With this you will be able to estimate the number of fleshy fruited plant species eaten and dispersed even if you cannot identify all of them taxonomically. The simplest method is to keep examples of each seed morphotype in individual plastic bags labelled with collection dates and specimen numbers. Seed specimens may also be stored in jars of alcohol in order to preserve their original appearance. Seeds of some plant species can vary greatly in size and in shape, so it is wise to be conservative in 'lumping' apparent morphotypes rather than 'splitting' them until their identities can be verified with fruit specimens.

FOOD PLANTS AND FEEDING REMAINS

Tracking the changing availabilities of different plant resources is a key component of feeding ecology research (Chapter 3). A central part of dietary detective work involves closely examining fruits and other plant parts eaten and dropped by animals as they feed. It is important to note the stages of ripeness and the particular portions of the fruits, leaves, flowers, seeds, or pith swallowed by the animals, in contrast to those portions that are rejected. Plant secondary compounds may be concentrated in different plant structures at different times and these factors influence animal food selection (Chapter 13). Since food items often come from vines or epiphytes located in the obscure portions of tall tree crowns, collecting decent botanical specimens may be impossible in a frustratingly large number of cases. Fallen fruits and other foods may be preserved in alcohol as well as in the dryer, and such specimens can be particularly useful if leaves or other plant parts are impossible to collect. Locally available ethanol has been known to serve this purpose well. Colour photographs of fruits are also recommended for identification.

RANGING, SEED DISPERSAL AND GUT PASSAGE

A well-mapped trail system is very helpful for measuring ranging and dispersal distances. Sleeping sites, food plants, and defecation sites can then be mapped with regard to their distances and positions relative to trail points (Chapter 4). Determining gut passage rates in the field is much more difficult than measuring them in a captive environment. When trying to identify the parent plant of a dispersed seed, one hopes to find a situation in which only one individual of a particular food plant species was visited, and that plant was visited only once during a particular day or period. The work is made more difficult when animals visit multiple individuals of the same fruiting plant species, or when the focal animal returns to a food plant multiple times during a follow. Of the various seed species collected from dung during a particular follow there may be few or none that can indicate passage time. When animals repeatedly return to the same feeding tree, however, it may be possible to measure dispersal distance without knowing exact passage time. As genetic testing techniques improve, it should become easier to identify the parent plants of individual seeds and seedlings using laboratory methods.

Frugivorous primates' gut passage rates are typically in the range of a few hours, and, for most primate species, once a focal animal has been followed for one or more days you can have some confidence that most voided seeds were ingested during the follow period (Fig. 11.1; Milton, 1984; Julliot, 1996; Yumoto et al., 1999; Stevenson, 2000; Dew, 2001). If all fruit sources have been marked and mapped during a follow then passage rate and dispersal distance can be determined from the analysis of these bouts. Start conservatively by first analysing ingestion and defecation patterns of the second or third day of a continuous follow. Sleeping sites are the best place to start backtracking through your data to determine the origins of dispersed seeds.

Large-bodied animals such as pongids, some cercopithecine monkeys, and animals with highly folivorous diets may have considerably longer passage times than most frugivores (Milton, 1984; Lambert, 1999). When studying such species, and cathemeral animals, following seeds from ingestion to deposition may be impossible. In such cases, you may be able to estimate minimum and maximum dispersal distances by analysing ranging data. It is simpler and also quite valuable to ascertain the proportions of dispersed seeds that fall in the 'hot zones' of high seed mortality under or near fruiting crowns with those that fall 15+ m away from any fruiting crown (Howe et al., 1985; Schupp, 1988; Terborgh et al., 1993). Likewise, you can describe dispersal sites with regard to canopy cover, substrate, microhabitat, and other factors likely to affect germination and survival (Rogers et al., 1998). Describing range usage is also important in studying seed dissemination patterns, as if a study species uses the same resting sites repeatedly, seed distribution may be heavily clustered.

GERMINATION AND SEED SURVIVORSHIP TRIALS

Intact seed passage through the gut typically shortens the time to germination (Traveset, 1998). However, plant species vary widely in time to germination, and this information is known for few tropical plants. One is likely to collect relatively few gut-dispersed seeds per plant species, and without large sample sizes the comparison of germination rates is statistically risky, but it is also important to know whether or not the gut-dispersed seeds of particular plants actually germinate following ingestion by particular frugivores. Seed germination trials therefore remain important in seed dispersal studies, and small sample sizes should not preclude you from collecting germination data.

If large quantities of gut-dispersed seeds are available from some plants, then comparing the germination of gut passed versus unpassed seeds may be possible for those taxa. This will require collecting large quantities of unswallowed seeds that have not yet been infested by granivores such as weevils (Bruchidae). Since the fruits fallen under a parent plant are often infested, this can be a tricky task.

The usual method for studying seed germination is to bury seeds in containers of damp sand or soil. Improvised local materials may be most practical in field conditions, but some studies have used more precise laboratory set-ups with Petri dishes and moistened paper (Dew & Wright, 1998; Traveset, 1998). A laboratory building or screened area free from rodents and other pests is the best place to conduct undisrupted germination trials. Be sure to dissect all seeds at the end of the study to determine whether viable embryos remain. Seeds of some plants, such as *Spondias mombin* (Anacardiaceae), may not germinate for six months or more after deposition (Dew, 2001).

SEED GERMINATION AND SURVIVORSHIP IN THE WILD

A growing number of studies have examined the survivorship of gut-dispersed seeds under natural or semi-natural conditions (Janzen, 1982; Chapman, 1989; Estrada & Coates-Estrada, 1991; Feer, 1999; Andresen, 2001; Dew, 2001). Such studies can help to determine the roles of secondary dispersers such as dung beetles and cache-hoarding rodents. Seeds can be placed along transects in marked locations and checked periodically for germination success removal and/or mortality (Blate *et al.*, 1998; Andresen, 1999; Dew, 2001). Seed predation has been studied by enclosing sub-sets of seeds within wire cages that prevent their removal (Silman, 1996; Asquith *et al.*, 1997; Brewer & Rejmanek, 1999; Wenny, 2000). Outside of such enclosures, you run into the difficulty of marking individual gut-dispersed seeds. Studies by Pierre-Michel Forget and others have pioneered a successful tracking method of marking large seeds with lengths of nylon thread (Forget *et al.*, 2000; Wenny, 2000). Survival analysis is the standard statistical method for comparing time to germination and seed survivorship (Cox & Oakes, 1984; Lieberman, 1996).

CONCLUSION

Dietary ecology remains a field full of questions, particularly in the tropics. Thorough study can reveal the interdependence between species in new ways.

Feeding ecologists excel at knowing their study animals well and scrutinising behaviours in very fine detail. Knowledge of food characteristics, plant morphology, physiology and phenological patterns also provide valuable contributions to science. With thoughtful investigation, the interactions between frugivorous animals and the plants that sustain them may provide indispensable information to the fields of community ecology and conservation biology. Hopefully, the lessons learned will make a difference in the preservation of these spectacular species and their habitats.

ACKNOWLEDGEMENTS

Many thanks are due to Peter Rodman, whom I was honored to have as my graduate advisor. I also thank Pat Wright, Katharine Milton, Colin Chapman, and Tony Di Fiore for their advice and friendship. My research has been funded by The Explorers Club, The National Science Foundation, The Douroucouli Foundation, and Primate Conservation Incorporated.

REFERENCES

Altmann, J. (1974). Observational study of behavior: sampling methods. *Behaviour* **49**, 227–65.

Andresen, E. (1999). Seed dispersal by monkeys and the fate of dispersed seeds in a Peruvian rain forest. *Biotropica* **31**, 145–58.

(2001). Effects of dung presence, dung amount and secondary dispersal by dung beetles on the fate of *Micropholis guyanensis* (Sapotaceae) seeds in Central Amazonia. *J. Trop. Ecol.* **17**, 61–78.

Asquith, N.M., Wright, S.J. & Clauss, M.J. (1997). Does mammal community composition control recruitment in neotropical forests? Evidence from Panama. *Ecology* **78**, 941–6.

Blate, G.M., Peart, D.R. & Leighton, M. (1998). Post-dispersal predation on isolated seeds: a comparative study of 40 tree species in a Southeast Asian rainforest. *Oikos* **82**, 522–38.

Brewer, S.W. & Rejmanek, M. (1999). Small rodents as significant dispersers of tree seeds in a neotropical forest. *J. Veg. Sci.* **10**, 165–74.

Chapman, C.A. (1989). Primate seed dispersal: the fate of dispersed seeds. *Biotropica* **21**, 148–54.

Cox, D.R. & Oakes, D. (1984). *Analysis of Survival Data*. London, New York: Chapman & Hall.

Dew, J.L. (2001). Synecology and seed dispersal in woolly monkeys (*Lagothrix lagotricha poeppigii*) and spider monkeys (*Ateles belzebuth belzebuth*) in Parque Nacional Yasuni, Ecuador. Ph.D. thesis, University of California, Davis.

Dew, J.L. & Wright, P.C. (1998). Frugivory and seed dispersal by four species of primates in Madagascar's eastern rainforest. *Biotropica* **30**, 425–37.

Estrada, A. & Coates-Estrada, R. (1991). Howler monkeys (*Alouatta palliata*), dung beetles (Scarabaeidae) and seed dispersal: ecological interactions in the tropical rain forests of Los Tuxtlas, Mexico. *J. Trop. Ecol.* **7**, 459–74.

Feer, F. (1999). Effects of dung beetles (Scarabaeidae) on seeds dispersed by howler monkeys (*Alouatta seniculus*) in the French Guianan rain forest. *J. Trop. Ecol.* **15**, 129–42.

Forget, P.-M., Milleron, T., Feer, F., Henry, O. & Dubost, G. (2000). Effects of dispersal pattern and mammalian herbivores on seedling recruitment for *Virola michelii* (Myristicaceae) in French Guiana. *Biotropica* **32**, 452–62.

Gentry, A. H. (1996). *A Field Guide to the Families and Genera of Woody Plants of Northwest South America (Colombia, Ecuador, Peru), with Supplementary Notes on Herbaceous Taxa.* Chicago: University of Chicago Press.

Howe, H. F., Schupp, E. W. & Westley, L. C. (1985). Early consequences of seed dispersal for a neotropical tree (*Virola surinamensis*). *Ecology* **66**, 781–91.

Janzen, D. H. (1982). Removal of seeds from horse dung by tropical rodents: influence of habitat and amount of dung. *Ecology* **63**, 1887–990.

Julliot, C. (1996). Seed dispersal by the red howler monkey (*Alouatta seniculus*) in the tropical rain forest French Guiana. *Int. J. Primatol.* **17**, 239–58.

Lambert, J. E. (1999). Seed handling in chimpanzees (*Pan troglodytes*) and redtail monkeys (*Cercopithecus ascanius*): implications for understanding hominoid and cercopithecine fruit-processing strategies and seed dispersal. *Am. J. Phys. Anthropol.* **109**, 365–86.

Letouzey, R. (1986). *Manual of Forest Botany. Tropical Africa*, vols I, II, III. (Translated by R. Huggett.) Centre de Technique Forestier Tropical, Nogent-sur-Marne, France.

Lieberman, D. (1996). Demography of tropical tree seedlings: a review. In *The Ecology of Tropical Forest Seedlings*, ed. M. D. Swaine, pp. 141–8. Paris: Parthenon Publishing Group.

Martin, P. R. & Bateson, P. P. G. (1993). *Measuring Behaviour: An Introductory Guide.* Cambridge, New York: Cambridge University Press.

Milton, K. (1984). The role of food-processing factors in primate food choice. In *Adaptations for Foraging in Nonhuman Primates*, ed. P. S. Rodman & J. G. H. Cant, pp. 249–79. New York: Columbia University Press.

Rogers, M. E., Voysey, B. C., McDonald, K. E., Parnell, R. J. & Tutin, C. E. G. (1998). Lowland gorillas and seed dispersal: the importance of nest sites. *Am. J. Primatol.* **45**, 45–68.

Schupp, E. W. (1988). Seed and early seedling predation in the forest understory and in treefall gaps. *Oikos* **51**, 71–8.

Silman, M. R. (1996). Regeneration from seed in a neotropical rain forest. Ph.D. thesis, Duke University, Durham, NC.

Stevenson, P. R. (2000). Seed dispersal by woolly monkeys (*Lagothrix lagothricha*) at Tinigua National Park, Colombia: dispersal distance, germination rates, and dispersal quantity. *Am. J. Primatol.* **50**, 275–89.

Terborgh, J., Losos, E., Riley, M. P. & Riley, M. B. (1993). Predation by vertebrates and invertebrates on the seeds of five canopy tree species of an Amazonian forest. *Vegetatio* **107–108**, 375–86.

Traveset, A. (1998). Effect of seed passage through vertebrate frugivores' guts on germination: a review. *Perspect. Plant Ecol. Evol. Syst.* **1**, 151–90.

van Steenis, C. G. G. J. (ed.) (1948–2001). *Flora Malesiana, Being an Illustrated Systematic Account of the Malaysian Flora*, vols. 1–13. Dordrecht: Kluwer Academic.

Wenny, D. G. (2000). Seed dispersal of a high quality fruit by specialized frugivores: high quality dispersal? *Biotropica* **32**, 327–37.

Yumoto, T., Kimura, K. & Nishimura, A. (1999). Estimation of retention times and distances of seed dispersed by two monkey species, *Alouatta seniculus* and *Lagothrix lagotricha*, in a Colombian forest. *Ecol. Res.* **14**, 179–91.

Useful Internet sites

Ben Meadows Company: <www.benmeadows.com>.
Forestry Suppliers: <www.forestry-suppliers.com>.

12 • Dietary analysis I: Food physics

PETER W. LUCAS[1], DANIEL OSORIO[2], NAYUTA YAMASHITA[3],
JONATHAN F. PRINZ[4], NATHANIEL J. DOMINY[5] AND BRIAN W.
DARVELL[6]

[1]*Department of Anatomy, University of Hong Kong, Hong Kong, People's Republic of China*
[2]*School of Biological Sciences, University of Sussex, Brighton, UK*
[3]*Department of Cell and Neurobiology, Keck School of Medicine, University of Southern California, Los Angeles, California, USA*
[4]*Wageningen Centre for Food Sciences, Wageningen, The Netherlands*
[5]*Department of Ecology and Evolution, University of Chicago, Chicago, Illinois, USA*
[6]*Dental Materials Science, Faculty of Dentistry, University of Hong Kong, Hong Kong People's Republic of China*

INTRODUCTION

Both this chapter and the next focus on measurements of the physical and chemical attributes of potential foods that primates select or reject. The major reason for analysing the diets of primate in this manner is to understand the basis for their food choice. Observing primates as they feed quickly raises questions in the observer's mind about the possible foraging strategies that the animals might be following in order to survive. How do primates distinguish food from what is otherwise scenery? Can we measure the attributes of potential foods in the form that primates are actually sensing them? What do primates get out of the foods they choose and are their choices, based on sensory capabilities, optimal in terms of nutrients? Tests of hypotheses that address these questions will require objective dietary analysis. It is important to tailor your measurements to the questions being asked.

The physiochemical characteristics of foods may form important sensory cues for their detection, selection and subsequent processing by primates, but all these characteristics are affected to some degree by specimen storage. Physical characteristics, such as colour, geometry and mechanical properties, are ruined. So it is often important and sometimes vital to make measurements almost immediately, while the specimen is fresh. The alternative of drying specimens for later chemical analysis not only involves a substantial time lag between fieldwork and subsequent access to a laboratory but also can lead to inaccurate results. Thus a major function for Chapters 12 and 13 is to describe the practical application of a new range of field techniques that can produce results *in situ*. A full field laboratory would be the ideal situation, but in reality many students can only dream of such facilities. Thus,

Field and Laboratory Methods in Primatology: A Practical Guide, ed. Joanna M. Setchell and Deborah J. Curtis. Published by Cambridge University Press. © Cambridge University Press 2003.

wherever possible, we also include field options that are possible on a smaller budget. Otherwise, for chemistry, we describe briefly what can be achieved 'back home' in the laboratory. Overall, fieldworkers should think broadly and prepare well before setting out because many food properties can now be estimated synchronously with the period of observation. It is wise though to remember that, while many things can be done, only one thing is essential – the accurate observation of primate feeding.

The current high rates of tropical deforestation, coupled with the vulnerability of primates to hunting even in intact forests, means that many first studies of primate ecology will also be the last for that particular species and location. Primatologist researchers therefore have an unusually strong obligation to target their measurements at future generations of primatologists, rather than simply their thesis examiners or journal referees. The most important step towards making present-day measurements 'upwardly compatible' with future questions is to document precisely how samples were selected, collected, stored, processed and analysed. Too much published data is unusable because the methods are inadequately described. Our approach here is meant to be positive and pragmatic: we want to encourage primatologists to gain much more information from their field studies while still on site.

BASIC EQUIPMENT

'In the field' does not just mean 'in the shade': a roofed enclosure is needed. A power supply from solar panels, generators or (rather rarely available) a connection to the mains, is required for a majority of the tests we describe. The tests can be conducted with rechargeable batteries that require intermittent access to a power supply for recharging and everything can be run off, or recharged using, solar panels. Basic equipment can be expensive and we present lower cost options whenever possible. We assume that most researchers will have access to a laptop computer in the field. Adequate space for operating the equipment should also be taken into account.

Researchers interested in testing mechanical properties of foods will need to invest in a portable tester (see e.g. Darvell *et al.*, 1996). Although toughness tests can be conducted without a computer, other tests that measure a force–displacement gradient require one with the appropriate software loaded. We are not aware of alternatives that cost less than that described by Darvell *et al.* (approximately US$ 10 000 including everything other than a laptop), but other testers have been used successfully in the field (Overdorff & Strait, 1998). The accurate measurement of colour requires a spectrometer, which can also be used for quantifying leaf chemistry, as described in Chapter 13.

Additional equipment is described in the relevant sections. Finally, careful laboratory notes for both mechanical and chemical tests are essential, including details of all tests conducted.

TESTS

All tests follow directly from observations of the exact plant part that is consumed. Mechanical properties, colour and nutrients change throughout the
developmental cycle of a plant, so the accurate identification of the developmental stage of a food is critical.

FOOD GEOMETRY

Aspects of food size and shape are important determinants of primate acceptance, affecting visibility at long range. Depending on the manual dexterity
and incisal form of a primate, the size and shape of food items also affect the
method of ingestion. Dimensional measurements with mechanical calipers
are straightforward and a portable analytical balance (accurate to ±0.01 g; e.g.
a 'jeweller's balance' such as Acculab PP2060D, approximately US$ 500) can be
used to obtain food weight. In addition, digital cameras, either still or video
(see Chapter 16), provide simple electronic records of basic form and colour.
Downloading these images to a computer allows the use of image analysis
software for quantifying shape. The advent of high resolution digital cameras, which need to be calibrated, may make this a viable means of recording
all the macrosurface features of foods, including colour (Parrága *et al.*, 2002).

FOOD COLOUR

The detection of food colour at long distances (say 10 m or more) is likely
to provide an important sensory cue for primates (Dominy *et al.*, 2001). At
shorter ranges, colour may well be useful for identifying the developmental
stage of fruit and leaves. The colour vision of primates is very variable, both
within and between species. Although it has long been assumed that such
differences are important for foraging primates, there were almost no field
data to support this until recently. Now simple field techniques have begun
to change this state of affairs, and we expect many advances in the next few
years.

The reflectance of an object is fairly easy to record with a small computer-
based spectroradiometer. There are two aspects to the coloration of most objects: a diffuse, spectrally selective reflection due to chemical pigmentation
and a more directional, non-spectrally selective specular reflection (shininess)
that affects brightness and saturation, but not hue. For recording reflectance
spectra, it is conventional to use a viewing geometry that minimises specular reflection, both because such measurements are more reproducible and
because pigment reflectance probably relates more closely to the principal
attraction for a primate. Illumination can be either directional or diffuse.
Diffuse sources (i.e. integrating spheres) are widely used for colorimetry in
industry (and can be used in biology; see Lucas *et al.*, 1998) – they are very

accurate, but expensive (e.g. Minolta CM2600D, which records the spectrum into the ultraviolet (UV) range, approximately US$ 10 000). Directional lighting is far more usual in work on animal colour vision and is cheaper. Lucas *et al.* (2001) describe a set-up that includes a fibre optic spectrometer (Ocean Optics USB2000-UV-VIS: USB connection being the cheapest option, approximately US$ 2650), a fibre-optic light source (Ocean Optics PX-2 pulsed xenon lamp, approximately US$ 770) and a home-made lightproof specimen box that arranges the lighting at a fixed angle to the specimen surface. A microscope illuminator may be a cheaper alternative to this light source. Orienting the specimen is easy for flat leaves, difficult for curved shiny fruit. The reflectance must be referred to a white standard. For this, compacted barium sulphate powder is cheap, while commercial standards are robust but expensive (Ocean Optics WS-1 diffuse reflection standard, approximately US$ 360).

From these spectra, it is also relatively straightforward to predict the responses of the primate cone photoreceptors, since the spectral sensitivities of a considerable number of species have now been determined. This tells us the information that the eye makes available to the brain. With care, it can be predicted whether a pair of colours will be discriminable for a given animal – for example that of a food item as seen against a leaf background, or adjacent fruits or leaves at different developmental stages (as shown in the examples in Fig. 12.1). The spectra of food items alone are meaningless: food, non-food and background – all have to be recorded. Typical backgrounds are foliage and bark, but earth and sky are others.

Several papers have presented and reviewed analytical methods for measuring colour (Regan *et al.*, 2001; Kelber *et al.*, 2003), but Fig. 12.1 illustrates the general approach. Object reflectance, referenced to a standard, needs to be 'lit' as it would be for a foraging primate. Natural lighting, especially in a forest, is neither highly directional nor completely diffuse and varies in its spectral composition with direction. Figure 12.1 shows examples of two illuminants recorded in forests, as well as their effect on non-specular reflection: this is small but noticeable. Results on chromatic signals are for a trichromatic catarrhine or platyrrhine and assume the retinal photoreceptor responses obtained from psychophysical investigations of humans. Where receptor sensitivities for other primates differ significantly from humans, they can be estimated from photoreceptor spectral absorbance functions, taking account of light absorbance within ocular media (e.g. the lens and macular pigment).

Diurnal vision in primates is restricted to wavelengths between 400 and 700 nm. Although primates cannot see into the UV region (350–400 nm), many other animals can, and if questions are asked about food selection by primates *vis-à-vis* their competitors, then it may be desirable to measure UV reflectance (Bennett *et al.*, 1994; Altshuler, 2001). In nocturnal primates, the tapetum increases sensitivity of the retina in dim light, but it also fluoresces, permitting utilisation of a portion of the UV spectrum when this is present (Martin, 1990). A pulsed xenon lamp produces UV light, but a simple and very cheap

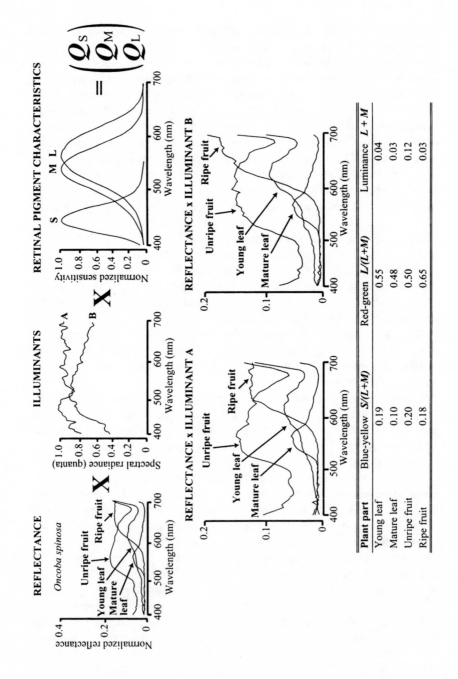

REFLECTANCE

Oncoba spinosa

Normalized reflectance

Young leaf
Mature leaf
Unripe fruit
Ripe fruit

Wavelength (nm)

ILLUMINANTS

Spectral radiance (quanta)

A
B

Wavelength (nm)

RETINAL PIGMENT CHARACTERISTICS

Normalized sensitivity

S   M L

Wavelength (nm)

$$\begin{pmatrix} Q_S \\ Q_M \\ Q_L \end{pmatrix}$$

REFLECTANCE x ILLUMINANT A

Unripe fruit
Ripe fruit
Young leaf
Mature leaf

Wavelength (nm)

REFLECTANCE x ILLUMINANT B

Unripe fruit
Ripe fruit
Young leaf
Mature leaf

Wavelength (nm)

Plant part	Blue-yellow $S/(L+M)$	Red-green $L/(L+M)$	Luminance $L + M$
Young leaf	0.19	0.55	0.04
Mature leaf	0.10	0.48	0.03
Unripe fruit	0.20	0.50	0.12
Ripe fruit	0.18	0.65	0.03

presence/absence alternative might be simply to use UV light-emitting diodes (UV LEDs) of the type used to detect forged bank notes.

Given that primate species differ in their colour vision, human perception cannot be relied on to describe spectral stimuli. However, one low budget alternative used in several published studies, and which is semi-quantified while retaining some subjectivity, is the use of Munsell colour cards (approximately US$ 100), matching these to food and background samples. These are limited, particularly in range of brightness, but are far superior to the simple descriptive colour categories commonly found in the ecological literature. However, even for those with no funds whatsoever, use of these basic colour terms has certainly shown trends in the fruit choice of different primate species that correspond to inter-specific variations in colour vision (Dominy, 2003).

Fig. 12.1. Colour processing. The estimation of photoreceptor responses and chromatic signals from reflectance and illumination spectra.

Top row: Relative quantal flux (i.e. light intensity) as a function of wavelength incident on the eye is given by multiplying a reflectance spectrum (left, referenced to a white standard) by the illumination spectrum (centre) at each wavelength across the visible spectrum. Illuminant A, from Yakushima, Japan, is unshaded sunlight, while B is slightly shaded by foliage and is from Kibale, Uganda. Relative quantal catches, of short (S) medium (M) and long (L) wavelength sensitive cone photoreceptors ($Q_{S,M,L}$) are given by multiplying the incident flux by the photoreceptor's spectral sensitivity (right). Spectral sensitivities plotted here are from a human observer (Smith & Pokorny, 1972; Wysecki & Stiles, 2000).

Middle row: Quantal flux for a given set of objects varies substantially under different natural illumination spectra. In the visual system, this variation is substantially 'discounted' by colour constancy mechanisms. We can model the consequences of colour constancy by dividing the estimated receptor quantal catches of a given receptor by the quantal catch to some standard. This gives estimated outputs for short, medium and long wavelength cones: S, M and L, respectively. This standard might be the illumination spectrum itself, or some 'average background'. This algorithm can be thought of as a model of photoreceptor adaptation (Osorio & Vorobyev, 1996; Kelber *et al.*, 2003).

Lower row: Chromatic signals derived from receptor outputs can be modelled in various ways, but we have calculated $S/(L+M)$ and $L/(L+M)$ after Macleod & Boynton (1979). An advantage of this method is that $S/(L+M)$ represents the blue-yellow chromatic signal available to a dichromatic mammal, while the red–green parameter, $L/(L+M)$, is available only to trichromats. Bivariate plots of $S/(L+M)$ and $L/(L+M)$ are useful to show separations but this does not predict their discriminability (Osorio & Vorobyev, 1996; Kelber *et al.*, 2003). The parameter $(L + M)$ represents luminance.

SURFACE TEXTURE OF FOODS

Many potential plant food items have irritant thorns, spines or hairs that can deter primates completely or slow down their feeding rate. Leaves in particular can contain amorphous opaline silica or calcium oxalate crystals, the regular ingestion of which can wear teeth. Silica is surprisingly common on the surfaces of leaves eaten by primates. The Moraceae, a family ubiquitous in primate diets, provides a prime example (Fig. 12.2). Estimation of the abrasiveness of leaves may be desirable. A recent laboratory method to ascertain the wear effect of grasses has been described by Hammond & Ennos (2000) and adapted to a field setting by J. F. P and N. J. D (Fig. 12.2). Scratching could be analysed by examining tooth microwear, but a simpler solution is probably just to take the dried food back to the laboratory and quantify the density of features under a scanning electron microscope (as in Fig. 12.2; Lucas & Teaford, 1995).

MECHANICAL PROPERTIES OF FOODS

The physical resistance of foods to being ingested, chewed and swallowed probably forms an important factor in their acceptance or rejection by primates. To simplify the understanding of this, this section treats foods as though they were real solids. Of course, they are not and there is an extensive theory of cellular mechanics that deals with foods, such as plant tissues, wherein a trellis of cell walls traps fluid (Ashby & Gibson, 1997).

The most important properties of solid foods are toughness, Young's modulus (see below) and the yield stress. Young's modulus and yield stress can be measured under force control (using weights as 'dead' loads, and measuring the effect on specimen dimensions), but toughness can be measured only under displacement control (i.e. the dimensions of the specimens are systematically changed, while monitoring the effect on the force). All these properties are directional and the amount of information required to fully characterise a food object depends on its heterogeneity. This level of complexity, though, is not the place to start, certainly not in the field, and any measure of one or more of these food properties could go a long way towards explaining food choices. In fact, groupings of two of these properties can explain the general basis of plant defence against fracture (Lucas *et al.*, 2000), defences that a primate must overcome in order to eat these items. The square root of the ratio of toughness to Young's modulus describes what is usually referred to in lay terms as 'toughness', while the square root of their product is a quantity that most would denote as 'hardness'. Note though that any specific analysis typically requires more information than this, with a variable mix of friction (briefly mentioned in the 'Other possibilities' section), indentation (controlled by the yield stress) and fracture. No one has yet attempted measurements of all these in the field.

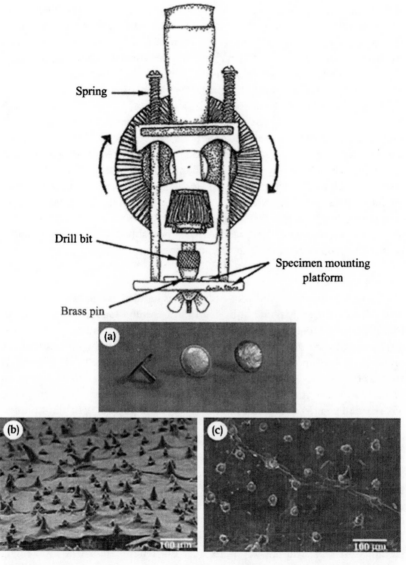

Fig. 12.2. Abrasion testing. A toy mini-drill, modified so that a brass pin, shown in (a), inserted in the drill bit rotates against a food specimen to check its abrasiveness. A standard load is imposed by two springs. The specimen-mounting platform has a holder for clamping food specimens such as leaves. The pin has been pre-coated with nail varnish. The drill can be turned a set number of revolutions (e.g. three) and the pin removed and inspected for scratches to the varnish, as seen in (a). Pins can be saved for quantitative wear analysis. (b) and (c) are scanning electron micrographs of *Poulsenia armata* (Moraceae) leaves (from Barro Colorado Island, Panama), showing the surface armour that does the damage.

It is important to note that we know much less about how the mechanical properties of foods are sensed by primates than we do about food attributes such as colour. Even in humans, the branch of food science that deals with mechanics (texture studies) has no generally accepted methodology. Increasingly, however, the influence of materials science is being felt and we strongly advocate this fundamental approach. However, the learning curve for mechanical tests is steep and you will need to read and practice. Vincent (1992) is invaluable as a 'cook book' for methods and Ashby & Jones (1996, 1998) are superlative general introductions to the science behind materials testing. Darvell *et al.* (1996) have described a portable tester, which now has provision for jigs for many types of mechanical test (Lucas *et al.*, 2001). Supplementary equipment includes calipers, a thickness gauge (particularly useful for measuring leaf thickness) and tools for making test specimens such as cork borers (perhaps US$ 500 in all). The best way to preserve plants in the field in the state in which they were eaten is probably to place them in plastic bags with damp paper towels to preserve moisture.

Young's modulus

The Young's or elastic modulus of a material is defined as the ratio of stress to strain during the early phase of its deformation under load. It is a measure of the stiffness of a food and calculated from the early (greatest) slope of a force–displacement diagram. Young's modulus can be measured in bending, compression, or tension tests, but we recommend the compression of cylinders of foods as the easiest (described in Fig. 12.3c). In any type of test, there may be an initial 'toe' at the start of the test where some slack is taken up in tension or where small geometrical errors are being corrected (Fig. 12.3). In plant tissues, ignore the toe when measuring the gradient, but remember that many soft animal tissues have curved stress–strain relationships even at low loads. Knowledge of limits to the dimensional ratios of specimens in tests will avoid errors. In a bending test, the length of the specimen beam should be more then 10 times its depth, while in a cylindrical compression test, the height:diameter ratio should be optimised at about 2 (and definitely less then 10) to avoid buckling. Both compression and tension tests are subject to specimen 'end effects'. In a tension test, compression of the specimen by the grips may affect the modulus estimate (the answer is to keep the length:thickness ratio high), while, in a compression test, friction of the food against the compressing platform does the same. A strip of Teflon™ (polytetrafluoroethylene)-coated adhesive tape attached to the compression plates can help to prevent this.

Toughness

Toughness, the energy expended per unit area of crack growth, is probably the vital mechanical element in the resistance of most foods to consumption by primates. In plant tissues, toughness arises from the structure and dimensions of the cell walls, while the characteristics of the cuticle are all-important in insects.

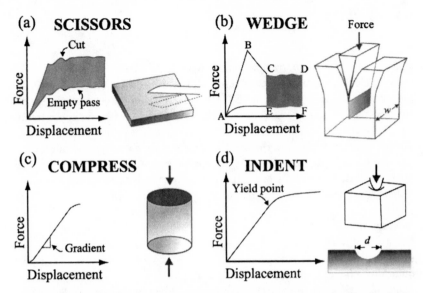

Fig. 12.3. Mechanical testing. (a) A scissors test gives toughness estimates for foods that are difficult to shape or in the form of rods or sheets. Application of a force (e.g. to the handles) causes the scissor blades to act like double wedges, producing a crack that tracks along the fine dotted line as the scissors are closed. This produces the force–displacement trace marked as 'cut'. The area below this curve gives the work done during the test. However, friction, largely work done by the scissors blades against each other, needs to be deducted: this is achieved simply by running an empty pass over the identical path used to cut the specimen. The work done during the cut, minus the work done by the blades against themselves gives the work done on the specimen (shaded in the figure). Dividing this work by the fractured cross-sectional area (using calipers, etc.) gives the toughness estimate.

(b) A sharp wedge is forced through a block of tissue to measure toughness. The loading curve moves through points ABCD. From A, the force builds with indentation until cracking at B. This is unstable at first but settles and, from C, further movement of the wedge between C and D makes the crack run just ahead of the wedge tip. Reversing the wedge back to position A, then running it through the cracked specimen produces a friction curve (often negligible) labelled AEF. The area bounded by CDFE gives the work done to produce the crack. The physical crack area (shaded) is given by the depth of wedge penetration (distance CD) multiplied by the width, w. Of the block dimensions, only the width needs to be cut accurately.

(c) A cylinder of tissue is compressed between plates to obtain the force–displacement gradient at small deformations. The force–displacement gradient at small deformations is converted into Young's modulus by first dividing the force by the original cross-sectional area (the area of the black top of the cylinder) and then the displacement by the original cylinder height.

Indentation is shown in (d). Indentors can be made in various shapes. A hemispherical indentor is shown, but a Vickers pyramid is probably best. The hardness of a specimen is the force divided by the area of indentation, where the area is measured in the plane of the specimen, in this case, from measuring the diameter, d.

Scissors tests on floppy specimens shaped like sheets or rods, or wedge tests on blocks, are simple yet effective (described in Fig. 12.3a, b). In order to be accurate, all the work done in the test must contribute to crack growth. Usually, this requires a specimen to be notched, the length of the notch to be accurately measured and for the crack to grow from the end of the notch. This palaver is avoided by the scissors and wedge tests. For the majority of primate foods, a portable device is sufficiently rigid to withstand deformation during a test. Some foods (e.g. seeds), however, have to be taken home and tested with a laboratory universal tester.

Indentation

Indentation tests are some of the oldest in the mechanical canon and have many uses in field biology. They generally involve a conical, spherical or pyramidal spike being pressed into an object. The force that produces a permanent mark of given surface area is called the hardness of the object (Fig. 12.3d). This use of the term, however, is not the same as that used in lay language, which is often closer to the square root of (Young's modulus × toughness) as mentioned above. However, hardness tests involve contacts similar to those of teeth on food particles, and this imitative aspect has led to the use of simple and cheap instruments like durometers in ecology. It is also relatively straightforward to make your own indentation device. There are problems, however, in the interpretation of the data. Hardness is not a material property in itself, but reflects the yield stress, the stress at which the mark becomes permanent (Fig. 12.3d). In most real solids, hardness values hover around three times that of the yield stress (Ashby & Jones, 1996) and this relationship probably applies to the scratching of tooth enamel by plant silica described above. However, in cellular solids such as most foods, this will not be so. The lightest touch with an indentor produces a recoverable dent. In a true solid, this is an indirect measure of Young's modulus, but edible tissues are not true solids and the meaning is somewhat different. Resistance to the lightest touch in thin-walled plant tissues is likely to be due to the turgidity of cellular contents, a feature ingeniously exploited by the Outwater ball tonometer to measure the internal cell pressure (Lintilhac *et al.*, 2000). Tonometers may have applications in primatology, since moisture is probably important in food choice. On heavier loading, turgid cells will burst and the tissue is likely to collapse in on itself. When this happens, hardness is about equal to yield stress (Wilsea *et al.*, 1975). However, the cells of many ripe fruits slide across each other and tissues such as those of seed shells are often very dense woods (Lucas *et al.*, 2000), leading to contacts that result in less volume reduction.

Although food hardness is an undeniably important aspect of primate ecology, measurements are rarely related to fundamental properties and vary from the descriptive to the quantitative. For example, Boubli (1999) describes fruits as hard (like a walnut), medium (like an avocado), or soft (like a grape). However, for those without funds, even this level of resolution proved informative for understanding the ecological specialisations of black-headed uakaris (*Cacajao melanocephalus*); primates adapted to forests with unusually well-protected fruits. Other studies, making quantified measures with a variety of instruments, have shown the importance of food mechanics to dietary divergence and species co-existence (Kinzey & Norconk, 1993; Yamashita, 1996, 2000; Overdorff & Strait, 1998).

One specific application of indentation tests is to estimate the bite forces that primates produce. When they feed, primates often drop food parts, either deliberately (e.g. fruit peel) or accidentally. Examination of these food fragments can reveal a substantial amount of information about how a food was ingested. In particular, when mammals feed on harder foods, indentations produced by tooth cusps can usually be seen on or around the fracture line. Peak forces are usually produced during these indentations: Lucas *et al.* (1994) and Hill *et al.* (1995) provide a methodology for establishing the magnitude of forces from an examination of these indentations.

Very hard foods, such as woods or the cuticle of beetles, are low in water and may survive with physical properties intact for investigations in a laboratory 'back home'. Microhardness testers are built into microscopes, and are very accurate.

OTHER POSSIBILITIES

In addition to reflectance, it may be worth measuring the transmission of light through young leaves. Their translucence is very obvious when they are seen against the light, which is how some primates have been observed to feed on them. However, because leaves scatter light, the accurate measurement of transmission needs an extended light source (e.g. a light box or uniformly lit white paper) rather than a spectrophotometric set-up such as that described in Chapter 13.

Fruits eaten by higher primates, i.e. anthropoids with spatulate incisors, are often characterised by outer tissues, which botanists call rinds, that peel cleanly from the underlying flesh at ripeness. The work done in peeling is work done against adhesion (Kendall, 1994). The work of adhesion is a quantity analogous to toughness and can be measured fairly easily with a mechanical tester. The stickiness of a food is also an adhesion problem. It is important: many primate foods contain latex, but we cannot offer any obvious test of its effectiveness. The surfaces of food objects have been very underinvestigated. The shininess of object surfaces (i.e. the specular component of object reflectance) could be very important for attracting the attention of a primate to a fruit and may be worth measuring systematically. Surfaces are important for mechanical action too. Friction helps hands to grip branches that contain potential food items, but when teeth are used to remove tree bark or soils (geophagy), friction simply raises the forces that need to be exerted to obtain fracture and this adds to tooth wear. Field devices that measure the coefficient of friction may be needed for the analysis of such actions. Resistance to being gripped also lies in the domain of friction, depending not only on the surface of the animal or plant part being gripped, but on the surfaces of the hands or teeth that are gripping them. Tooth enamel is very smooth, but intra-oral friction is not always low: one of the major actions of tannins in plant foods is to reduce the lubricating effect of saliva (the more so in primates that have tannin-specific binding proteins), allowing tannin levels to be detected by an increase in friction between oral tissues (Prinz & Lucas, 2000). This characteristic is often referred to as 'astringency' in food texture studies.

This chapter has only scratched the surface of food physics. Almost ignored, this subject offers enormous potential for explaining many aspects of primate food choice. We have limited the discussion to solid foods, but for the sake of completeness, gums and flower nectar are, of course, viscous liquids and the ripening of fruit flesh often involves considerable flow. Investigations of these foods require viscometers of some sort, a large variety of which are available. We have not yet tried any of these in the field.

ACKNOWLEDGEMENTS

We thank the Research Grants Council of Hong Kong and the Croucher Foundation of Hong Kong for their support. Camilla Pizano drew the abrasion tester shown in Fig. 12.3; Johnny Leung arranged it.

REFERENCES

Altshuler, D. L. (2001). Ultraviolet reflectance in fruits, ambient light composition, and fruit removal in a tropical forest. *Evol. Ecol. Res.* **3**, 767–78.

Ashby, M. F. & Gibson, L. J. (1997). *Cellular Solids Structure and Properties.* Cambridge: Cambridge University Press.

Ashby, M. F. & Jones, D. R. H. (1996). *Engineering Materials 1*, 2nd edition. Oxford: Butterworth Heineman.

(1998). *Engineering Materials 2*, 2nd edition. Oxford: Butterworth Heineman.

Bennett, A. T. D., Cuthill, I. C. & Norris, K. J. (1994). Sexual selection and the mismeasure of color. *Am. Nat.* **144**, 848–60.

Boubli, J. P. (1999). Feeding ecology of black-headed uacaris (*Cacajao melanocephalus melanocephalus*) in Pico da Neblina National Park, Brazil. *Int. J. Primatol.* **20**, 719–49.

Darvell, B. W., Lee, P. K. D., Yuen, T. D. B. & Lucas, P. W. (1996). A portable fracture toughness tester for biological materials. *Meas. Sci. Technol.* **7**, 954–62.

Dominy, N. J. (2003). Color as an indicator of food quality to anthropoid primates: ecological evidence and an evolutionary scenario. In *Anthropoid Origins*, ed. C. R. Ross & R. F. Kay. New York: Kluwer Academic, in press.

Dominy, N. J., Lucas, P. W., Osorio, D. & Yamashita, N. (2001). The sensory ecology of primate food perception. *Evol. Anthropol.* **10**, 171–86.

Hammond, T. A. & Ennos, A. R. (2000). Mechanical testing of the abrasiveness of grass leaves. In *Plant Biomechanics 2000*, ed. H.-C. Spatz & T. Speck, pp. 535–40. Stuttgart: Georg Thieme Verlag.

Hill, D. A., Lucas, P. W. & Cheng, P. Y. (1995). Bite forces used by Japanese macaques (*Macaca fuscata yakui*) on Yakushima Island, Japan, to open aphid-induced galls on *Distylium racemosum. J. Zool.* **237**, 57–63.

Kelber, A., Vorobyev, M. & Osorio, D. (2003). Animal colour vision – behavioural tests and physiological concepts. *Biol. Rev.*, **87**, 81–118.

Kendall, K. (1994). Adhesion: molecules and mechanics. *Science* **263**, 1720–5.

Kinzey, W. G. & Norconk, M. A. (1993). Physical and chemical properties of fruit and seeds eaten by *Pithecia* and *Chiropotes* in Surinam and Venezuela. *Int. J. Primatol.* **14**, 207–27.

Lintilhac, P. M., Wei, C., Tanguay, J. J. & Outwater, J. O. (2000). Ball tonometry: a rapid, non-destructive method for measuring cell turgor pressure in thin-walled plant cells. *J. Plant Growth Reg.* **19**, 90–7.

Lucas, P. W. & Teaford, M. F. (1995). Significance of silica in leaves eaten by long-tailed macaques (*Macaca fascicularis*). *Folia Primatol.* **64**, 30–6.

Lucas, P. W., Peters, C. R. & Arrandale, S. (1994). Seed-breaking forces exerted by orang-utans with their teeth in captivity and a new technique for estimating forces produced in the wild. *Am. J. Phys. Anthropol.* **94**, 365–78.

Lucas, P. W., Darvell, B. W., Lee, P. K. D., Yuen, T. D. B. & Choong, M. F. (1998). Colour cues for leaf food selection by long-tailed macaques (*Macaca fascicularis*) with a new suggestion for the evolution of trichromatic colour vision. *Folia Primatol.* **69**, 139–52.

Lucas, P. W., Turner, I. M., Dominy, N. J. & Yamashita, N. (2000). Mechanical defences to herbivory. *Ann. Bot.* **86**, 913–20.

Lucas, P. W., Beta, T., Darvell, B. W., Dominy, N. J., Essackjee, H. C., Lee, P. K. D., Osorio, D., Ramsden, L., Yamashita, N. & Yuen, T. D. B. (2001). Field kit to characterize physical, chemical and spatial aspects of potential foods of primates. *Folia Primatol.* **72**, 11–15.

Macleod, D. I. A. & Boynton, R. M. (1979). Chromaticity diagram showing cone excitation by stimuli of equal luminance. *J. Opt. Soc. Am.* **69**, 1183–6.

Martin, R. D. (1990). *Primate Origins and Evolution.* Princeton, NJ: Princeton University Press.

Osorio, D. & Vorobyev, M. (1996). Colour-vision as an adaptation to frugivory in primates. *Proc. Roy. Soc. Lond. Series B*, **263**, 593–9.

Overdorff, D. J. & Strait, S. G. (1998). Seed handling by three prosimian primates in southeastern Madagascar: Implications for seed dispersal. *Am. J. Primatol.* **45**, 69–82.

Parrága, C. A., Troscianko, T. & Tolhurst, D. J. (2002). Spatio-chromatic properties of natural images and human vision. *Curr. Biol.* **12**, 483–7.

Prinz, J. F. & Lucas, P. W. (2000). Saliva tannin interactions. *J. Oral Rehabil.* **27**, 991–4.

Regan, B. C., Julliot, C., Simmen, B., Vienot, F., Charles-Dominique, P. & Mollon, J. D. (2001). Fruits, foliage and the evolution of primate colour vision. *Phil. Trans. R. Soc. Lond.* B **356**, 229–83.

Smith V. C. & Pokorny, J. (1972). Spectral sensitivity of colorblind observers and the cone pigments. *Vis. Res.* **12**, 2059–71.

Vincent, J. F. V. (1992). *Biomaterials – A Practical Approach.* Oxford: IRL Press.

Wilsea, M., Johnson, K. L. & Ashby, M. F. (1975). Indentation of foamed plastics. *Int. J. Mech. Sci.* **17**, 457–60.

Wysecki, G. & Stiles, W. S. (2000). *Color Science.* New York: John Wiley.

Yamashita, N. (1996). Seasonality and site specificity of mechanical dietary patterns in two Malagasy lemur families (Lemuridae and Indriidae). *Int. J. Primatol.* **17**, 355–87.

(2000). Mechanical thresholds as a criterion for food selection in two prosimian species. In *Plant Biomechanics 2000,* ed. H.-C. Spatz & T. Speck, pp. 590–5. Stuttgart: Georg Thieme Verlag.

13 • Dietary analysis II: Food chemistry

PETER W. LUCAS[1], RICHARD T. CORLETT[2], NATHANIEL
J. DOMINY[3], HAFEJEE C. ESSACKJEE[4], PABLO
RIBA-HERNANDEZ[5], KATHRYN E. STONER[6] AND NAYUTA
YAMASHITA[7]

[1]*Department of Anatomy, University of Hong Kong, Hong Kong SAR, People's Republic of China*
[2]*Ecology and Biodiversity, University of Hong Kong, Hong Kong SAR, People's Republic of China*
[3]*Department of Ecology and Evolution, University of Chicago, Chicago, Illinois, USA*
[4]*Physiological Laboratory, University of Liverpool, Liverpool, UK*
[5]*School of Biology, University of Costa Rica, San Jose, Costa Rica*
[6]*Ecology Institute, Universidad Nacional Autonomade México, Michoacan, Mexico*
[7]*Department of Cell and Neurobiology, Keck School of Medicine, University of Southern California, Los Angeles, California, USA*

INTRODUCTION

Chapter 12 introduced dietary analysis and discussed physical aspects of potential foods as they might influence feeding behaviour. Here, we deal with chemical aspects of potential foods. Attempts to explain the influence of chemical factors on primate nutrition, and the dietary factors that promote or deter the uptake of nutrients, are limited by our understanding of how the primate gut operates. We are still not sure what the optimal dietary requirements are for humans (Challem, 1999). Research is developing both on theoretical (e.g. Jumars, 2000a,b) and practical levels (Minekus *et al.*, 1999), but the effective rate of uptake is not simply a question of enzymatic action. The quantity of plant fibre that a primate ingests is a major influence on the rate of passage of food through the gut and thus digestibility. A variable gut population of micro-organisms and parasites also play a large positive or negative role, as do specialisations in the stomach or large intestine. The situation is even less clear when it comes to chemical compounds that act as feeding deterrents, toxins or anti-nutritional factors. These have largely been bred or processed out of the agricultural products on which humans feed, so they have received relatively little attention in food science. Currently, these chemicals are assessed by crude measures, such as total phenolics, which, of necessity, ignore the wide range of variation within each class of chemicals in the nature and intensity of the biological effect.

Field chemistry tests in primatological studies have typically been basic, employing 'presence/absence' tests for compounds such as alkaloids

Field and Laboratory Methods in Primatology: A Practical Guide, ed. Joanna M. Setchell and Deborah J. Curtis. Published by Cambridge University Press. © Cambridge University Press 2003.

(Waterman, 1984), cyanogenic glycosides (Glander *et al.*, 1989) and phenolics (Simmen & Sabatier, 1996). There is nevertheless a great deal of scope for such tests: Mowry *et al.* (1996), for example, is noteworthy for the innovative use of Nestlé Lactogen milk-mix to screen for tannins. Others have used semi-quantitative tests, like pH papers to assess acidity (Ungar, 1995) and refractometers to measure sugars in flowers or fruits (White & Stiles, 1985). However, the advent of technology such as portable optical fibre spectrometers and dry pH meters has opened the door to a much wider range of quantitative field tests, for example colorimetric assessment of chemical reactions for several food ingredients. These tests utilise fresh, rather than dried, material, allowing results to be expressed as concentrations (akin to how they might be sensed by a primate) rather than on a dry weight basis (which relates more to nutritional gain) and, very importantly, to give rapid results. Though quantified, these new tests still do not provide what is, in the end, necessary: they do not tell you *exactly* what is in the food item. Instead, they generally refer test results to a standard, describing the results as 'equivalents', a term assuming that the extract contains components with a chemical behaviour equivalent to the standard. Should you want to know more, bring the food home for more detailed laboratory analysis. Bear in mind that the acceptance/rejection of potential foods depends on sensory perceptions that are not one-to-one with nutritional quality and it should not be assumed that a detailed understanding of the latter would tell you anything about foraging decisions.

We limit this chapter to the analysis of plant parts because that is the area in which most research has been directed, and divide it into 'field' techniques, which require a minimal set-up, and 'laboratory techniques', for which we provide only a point of reference. Recent laboratory studies on the nutritional composition of invertebrates (Barker *et al.*, 1998) and exudates (Smith, 2000) indicate that field-adapted techniques would also be valuable, particularly as insects may provide some important amino acids (Hladik, 1977). Analysis of vertebrate meat consumed by primates is straightforward because the major nutritional components are highly digestible and the variation between species is small in comparison with plant materials (see e.g. Ntiamoa-Baidu, 1997). Invertebrates, such as insects, are more problematic because they are less digestible and much more variable in composition, both between species and between life cycle stages (Bell, 1990).

BASIC EQUIPMENT

As with the food physics tests described in Chapter 12, an adequate dust-free space is essential. Electricity, or the means to recharge batteries, is also required. Depending on the tests to be performed, a stovetop or hotplate and refrigeration are necessary.

A method of reducing the particle size of plant parts prior to testing is also essential. A tissue homogeniser (e.g. Tissue Tearor, Dremel, approximately

US\$ 600) is ideal, but a pestle and mortar could also be used, although this combination is slow and difficult to control. You will also need a spectrometer (for advice, see Chapter 12) and cuvette holder (e.g. Ocean Optics CUV-UV, 1 cm path with lenses, approximately US\$ 400). General requirements for accurate field spectrometry start with good water quality. Rainwater may be used, but it requires filtration. Both ceramic or membrane filters are good. Low-cost deionisers are available to produce dH_2O (approximately US\$ 500 for a range of choices), invaluable if you wish to measure ion concentrations in food samples later (e.g. calcium ion concentrations in figs; O'Brien *et al.*, 1998). Bottled water is also a possibility, but it should be distilled and should not contain mineral additives. It is best to use quartz cuvettes, although they are expensive (approximately US\$ 50).

The chemicals required for individual tests are given in the relevant sections. Bear in mind that some solvents may be difficult to locate in some countries, in which case they will need to be brought in. Only small quantities are needed, and these are difficult to price. Additional equipment includes a good analytical balance ($\pm$0.01 g; Chapter 12), adjustable micropipettes (e.g. Gilson, approximately US\$ 200 each), glassware, plasticware, gloves and related accessories such as plastic Eppendorff microtubes (1.5 ml) and two or three microtube racks. Altogether about US\$ 1000 should be budgeted for consumables.

FOOD PRESERVATION AND EXTRACTION

Most researchers have dried plant food for transport to a laboratory. Drying samples at high temperature ($>50\,^{\circ}C$) or in direct sunlight is never a good idea because many constituents of plant parts are irreversibly altered by such treatment. If necessary, we recommend the low temperature drying techniques of Wrangham *et al.* (1998). For small amounts of material, rapid drying with silica gel or molecular sieves in the dark at room temperature is probably best. Silica gel can also be used to keep material dry. Field extraction from fresh material is usually preferable to laboratory extraction from dried material and there is evidence for considerable change in some components, even with great care (e.g. tannins: Hagerman, 1988; Orians, 1995). We have used 50% methanol : 50% dH_2O (hereinafter called 50% methanol) because most nutrients are sufficiently soluble in it to provide a balanced extract, but it must be borne in mind that methanol is highly toxic (see below for general precautions).

To extract from fresh material, cut the food sample into pieces of about 1 mm in particle length (incidentally, approximately the degree of comminution that chewing achieves). Weigh accurately about 0.1 g of sample, and transfer it into a homogenising tube with 5 ml increments. Add 50% methanol to a final volume of 5 ml and lower the tip of the homogeniser into the suspension. Operate at low speed first before shifting to high speed. Administer for

30 seconds to 2 minutes using an up–down dabbing motion, and then allow the homogenate to stand. The longer it stands in the solvent, the greater the extraction is likely to be. It is important to standardise the time period for all extractions – we chose 1–2 minutes for our work. Withdraw the supernatant into a 5 ml syringe. Fit the syringe with a Luerlock and a plastic filter holder containing a glass fibre filter (1.6 µm pore size, Type 1, Millipore, USA). Engage the syringe plunger, press gently to direct the extract through the filter and into a 1.5 ml Eppendorff microcentrifuge tube, filling this up to about 10% from the top. Label the tube. In the case of liquid samples like nectar, the sample can be extracted from flowers using a pre-weighed syringe, and poured into an Eppendorff tube with 5 ml of 50% methanol. The extracts produced should be refrigerated if at all possible. All chemical tests depend on these extracts, so at minimum they can be stored and the remainder of the tests conducted back in a laboratory. If plant tissues are dried then a different extraction procedure will be required, and spectrophotometric measurements will be less accurate.

General safety tips: Wear gloves when handling chemicals. Store chemicals away from direct sunlight and label reagent bottles with both the reagent name and date. In field sites, it is very likely there will be people around who will not understand what you are doing or know the danger posed by some of these chemicals, so be very vigilant and lock up chemicals when they are not in use. Do not dispose of chemicals carelessly. If you store chemical waste, be careful not to produce an inflammable mass. Keep your equipment (e.g. spatulas) clean and dry. Remember some important contamination precautions – never pipette directly from your stock reagent. Dispense what is needed into a separate aliquot. The bottom line is: be clean, be thorough and be safe.

FIELD TESTS

Three examples of field tests, for protein, phenolics and tannins, are presented here in some detail to show exactly what can be achieved in the field. Other possibilities are suggested below. Only the protein test can be conducted without a fridge, as some of the stock and buffer solutions employed in the phenolics and tannin tests require low temperature storage.

PROTEIN

Proteins react with a dye called Coomassie Brilliant Blue, producing a strong blue colour, the absorbance of which is measured spectrophotometrically at 595 nm. The test is very simple and a typical standard curve uses bovine serum albumin (BSA), a protein chosen simply because it gives a good linear relation between absorbance and concentration. Results are reported as percentage BSA equivalents. This method follows Read & Northcote (1981) and Sapan *et al.* (1999). To prepare a stock solution of Coomassie Blue (100 ml) weigh 0.33 g of Coomassie Blue G (Sigma). Dissolve the dye in 2:1 (volume × volume) phosphoric acid (85%) and ethanol (100%). Filter through a Whatman No. 1 paper and store at environmental temperature. The stock solution needs to be protected from light, for example by wrapping the reagent bottle in aluminium foil. You may be able to take ready-made stock solution with you to the field. To prepare the diluted working protein reagent (100 ml), dispense 3 ml of the concentrated stock solution into a 100 ml measuring cylinder. Next, add 8 ml of phosphoric acid (85%), and 3.75 ml of ethanol (100%). Make up the final volume to 100 ml with dH_2O. Filter twice through a Whatman No. 1 paper and store at room temperature. The resulting solution is reddish-brown in colour. You will need just 1 ml of working reagent per test, so prepare only the amount that you will need for a day or two of work.

To perform the tests, you need Eppendorff tube racks, a 200 µl and a 1 ml adjustable micropipette with disposable tips, 50% methanol, working protein reagent, Eppendorff tubes containing extracts, an equal number of empty Eppendorff tubes plus one extra for the blank and spectrometry equipment (Lucas *et al.*, 2001). First, align the extracts that you plan to analyse along one row of an Eppendorff rack. Place an empty Eppendorff tube opposite each extract, reserving one for a blank (a solution containing everything except the extract, which is replaced by 50% methanol). Using a new, sterile pipette tip for each extract, dispense 50 µl of each extract into the respective Eppendorff tube (the blank receives only 50 µl of 50% methanol). Add 1 ml of the working staining reagent into each. Cap and mix the tubes by gently inverting them. Allow the tubes to stand for 5 minutes at room temperature. A blue colour will develop almost immediately. To perform the spectroscopic measurement you first need to measure the electronic activity of the detectors in dark conditions. This is called the 'dark current' and should be deducted from every further measurement taken. The incident light then needs to be measured with the light source switched on and everything in place except a solution of sample or blank. Then measure the absorbance of the blank by taking 1 ml from the microtube, filling the cuvette and placing it in the transmittance stage. For this test, the absorbance for the blank can be as high as 0.5. Empty the cuvette completely using the pipette. It is possible to evacuate every noticeable drop by doing this, which avoids having to wash the cuvette after each absorbance reading. Invert the cuvette over absorbent paper and shake and wipe the pipette clean. Each sample is then measured identically to the procedure for the blank. Any sample absorbance above 1.0 needs dilution to·get an accurate reading. To ensure the quality of the results, we suggest periodically retesting the blank.

PHENOLICS

These compounds are the parent group of tannins, distinguished by the presence of six-membered carbon rings with hydroxyl (OH^-) groups, that are presumed to have a variety of defence functions in plants (Waterman & Mole, 1994). In the following test, the phenolics present in a plant sample oxidise potassium ferricyanide to produce ferrous ions. These ions then react with ferric chloride in HCl to produce a Prussian Blue complex, the absorbance of which can be measured spectrophotometrically at 700 nm. The stabiliser is an attempt to prevent precipitation of Prussian Blue by increasing the viscosity of the solution. It is only partially successful and you should adopt strict time protocols. The standard is gallic acid, a simple phenolic acid with small molecular weight. Too small a molecule to precipitate protein, gallic acid is an example of a phenolic compound not considered a tannin. We do not advise running more than 50 samples at a time, as it requires considerable practice to build up to that level. The amount of phenolics is determined at 700 nm. The test follows Hagerman (1998), converted to a microassay (Lucas *et al.*, 2001). It should be noted that this measures the reducing power of phenolics, not their overall concentration (Appel *et al.*, 2001). First, you need to make 100 ml of 0.1 M HCl. This chemical is very corrosive and dangerous. [**Never add water to concentrated acid. Always add acid to water.**] Measure 100 ml of dH_2O in the large measuring cylinder and transfer into a 500 ml reagent bottle. Next, carefully add 830 μl of concentrated HCl. Cap the bottle and store at room temperature. Secondly, make 0.02 M $FeCl_3$, in 0.1 M HCl (100 ml). Weigh out 0.32 g of $FeCl_3$ and transfer to a large reagent bottle containing 100 ml of 0.1 M HCl. Label the bottle, for example as 'CL', date it and store at room temperature. Thirdly, make 0.0016 M $K_3Fe(CN)_6$ (100 ml). Measure 100 ml of dH_2O in the large measuring cylinder and transfer to a 500 ml reagent bottle. Weigh out 0.53 g of $K_3Fe(CN)_6$ and transfer to the 100 ml of dH_2O in the large reagent bottle. Label the bottle, for example as 'CN', date it, and store at room temperature. To make the stabiliser (1% gum-arabic solution), measure 80 ml of dH_2O in the large measuring cylinder and add it to the conical flask. Weigh out 1.00 g of gum-arabic and transfer it to the flask. Heat the solution to boiling so as to dissolve the gum-arabic completely. Filter the solution through Whatman No. 1 paper into another flask. This will take time since the gum solution is very viscous. Once filtration is complete, make the final volume of filtrate up to 100 ml with dH_2O. Store in a small reagent bottle at 4 °C. It should be usable for about a week.

Set up the extracts again along the back row of an Eppendorff rack, leaving the space furthest to the left free. Load the entire front row with empty Eppendorff tubes. The first tube will be for the blank. Add 300 μl of dH_2O to each tube. Add 10 μl of 50% methanol to the 'blank' Eppendorff tube. This is a very small volume to pipette: make sure the pipette tip touches the surface of the water in the microtube while the pipette plunger is fully depressed. Retract the pipette before releasing the plunger. Remove the pipette tip and reserve it for methanol only. Add 10 μl of the sample extract from the tube in the back row to the empty tube placed in front of it using a new tip for each extract. Add 100 µl of CN solution to each of the front row mixtures, working evenly from left to right. Set a stopwatch to 15 minutes, put a new tip on the 200 µl pipette and start the watch. Add 100 µl of CL solution to each of the front row mixtures, working at an even pace from left to right. When the stopwatch goes off, add 1 ml of stabiliser to each tube on the front row of the rack, again working at the same even pace from left to right. The solutions should now be relatively stable for long enough to read the absorbance. If the Prussian Blue complex precipitates in spite of the stabiliser, or the absorbance is >1.3, we advise repeating the assay with a diluted sample. A two-fold dilution with 50% methanol is normally sufficient. Remember to take the dilution factor into account for the determination of the final concentration.

TANNINS

Tannins precipitate proteins. A simple method based on this mechanism, devised by Hagerman (1987), works well for field estimation. Plant extracts, loaded into wells of protein-containing agarose gel, combine with the protein to form a whitish precipitate. The following chemicals are required for the preparation of the buffer solution reagents: dH_2O, 2 M sodium hydroxide, glacial acetic acid and ascorbic acid, while agarose and BSA form the gel. Note BSA is very prone to fungal contamination in the absence of refrigeration. In addition to items listed in the 'Protein' and 'Phenolics' sections, you need a pH meter (a dry pH meter is best, e.g. Sentron NV, Netherlands; the model we used is no longer available), 500 ml measuring plastic cylinder, thermometer, 25 ml wide mouth pipette with a pipette filler and eight or more plastic Petri dishes, the lids of which are fitted with 5 mm diameter acrylic pins (as in Fig. 3 of Lucas et al., 2001). These pins will make wells during gel setting and are used for the loading of plant extracts. You will also need standard lids without pins. On the undersurface of the Petri dishes, just beside the future locations of each well, label the wells anticlockwise 1–6 with a waterproof marker pen. To make the buffer, weigh 5.3 g of ascorbic acid and add this to a 500 ml glass beaker. Dispense 400 ml of dH_2O to the beaker and swirl to dissolve. Add 2.43 ml of glacial acetic acid and mix well with a spatula. Make a 2 M solution of NaOH by dissolving 8 g of NaOH in 100 ml of dH_2O and use this to bring the pH to 5.0. Transfer the pH 5.0 solution into the cylinder and make up the volume to 500 ml with dH_2O. This buffer can be stored for one week at 4 °C. To make an agarose solution sufficient for eight Petri dishes, dispense 200 ml of the buffer into a 250 ml conical flask. Weigh out 2.0 g of agarose and 0.02 g of BSA. Add the agarose to the buffer and boil with constant stirring until completely dissolved. Monitor the cooling of the agarose solution to 40–45 °C with a thermometer (this temperature range is critical: agarose will gel/solidify at temperatures lower than 40 °C, while denaturation of BSA occurs at temperatures higher than 45 °C). Add the BSA and swirl gently to dissolve it completely.

Now working rapidly yet smoothly, withdraw 25 ml of solution using a pipette equipped with a filler (see Fig. 3 of Lucas *et al.*, 2001) and dispense this into each dish. It is very important to avoid air bubbles and you need to practise. The trick is to empty the pipette gently on the side of the Petri dish. Cover each dish with the modified lids. Allow the agarose to gel at room temperature. When cool, remove the modified lids cautiously to expose the wells while avoiding cracks. Cover with a standard Petri dish lid. Store upside down to avoid condensation in the wells. Be certain that the wells are free from condensation before the samples are loaded. You can carefully aspirate the wells with a Pasteur pipette, taking care not to puncture them, or alternatively store them upside down for a short period to allow moisture to drain out and the wells to dry. Within two hours, you can start loading 40 μl of your chosen sample extract into each well. Move clockwise in doing this, while keeping record of which extract goes in which numbered well. Replace the standard lid, seal the dishes with clingfilm or Parafilm M® (to preserve moisture and to limit microbial contamination) and incubate for 96 hours at room temperature (although the precipitation ring tends to develop within 36 hours). We express tannin concentrations as equivalents to six-point standard curves based on crude quebracho tannin (gift from Dr A. E. Hagerman). Although the ring edge is sometimes difficult to discern, we use calipers to measure ring diameters. Methods for differentially staining the gel to yield rings for water-soluble and condensed tannins have been described (Hagerman *et al.*, 1997), but these appear too complex for fieldwork. Remis *et al.* (2001) describe staining their gels with Prussian Blue reagent and measuring ring size with NIH Image software (see 'Useful Internet sites', at the end of this chapter). This could be arranged in the field with a digital camera mounted on a tripod and a ruler for scale.

LABORATORY CHEMISTRY

The advantages of laboratory over field analyses are the much wider range of chemical components that can be measured with potentially greater accuracy. Fast and accurate instrumental techniques for analysis have replaced much traditional wet chemistry over the last two decades. It is also possible to look into commercial analysis services, for example those that assess animal forage and feeds. These will do some routine analyses faster, cheaper and more accurately than an inexperienced researcher and/or an underequipped laboratory. A huge range of non-routine analyses is also available commercially to the researcher with sufficient funds.

FIBRE

In food science, 'fibre' includes all the digestion-resistant components of plant materials. The most useful measure of fibre content is neutral detergent fibre (NDF) – that which remains after the sample is boiled for an hour in neutral detergent solution. NDF includes cellulose and lignin, plus most hemicelluloses and is quick, cheap and easy to determine with the appropriate equipment. The standard procedure leaves a variable amount of digestible starch that can be removed by pre-treatment with amylase. Acid detergent fibre (ADF) is basically NDF without hemicelluloses. Specialised laboratories can run the sequence NDF–ADF–lignin, allowing a segregation of all components of the cell wall (Wrangham *et al.*, 1998). In human food science, the standard measure of fibre content is dietary fibre (DF): the residue after sequential digestion with amylase, protease and amyloglucosidase to remove starch and protein. This procedure is designed to mimic human digestive physiology and has no obvious advantage over NDF for primate foods.

PROTEIN AND AMINO ACIDS

Crude protein has traditionally been estimated from the nitrogen content of samples measured by the Kjeldahl method. This is still widely used, but requires special glassware and produces corrosive fumes and toxic waste products. Micro-Kjeldahl equipment requires less than 0.25 g of sample material and is semi-automated. The method does not determine the total nitrogen content if the sample contains significant amounts of heterocyclic N, N–N or N–O linkages (such as in nitrate) and is being replaced in many laboratories by automatic nitrogen analysers, which use combustion methods to determine true total nitrogen. The commonly used Dumas combustion method gives a 25% higher nitrogen reading than the Kjeldahl method (Simonne *et al.*, 1998).

Total nitrogen has traditionally been converted into protein by multiplying by a factor of 6.25, derived from studies with animal proteins. Numerous studies have shown that this overestimates protein in plant materials because a variable amount of nitrogen is in non-protein forms (Conklin-Brittain *et al.*, 1999), some of which are nutritionally valuable (e.g. free amino acids), while others may have a neutral or negative impact on nutrition (e.g. nitrogen-based plant secondary chemicals). A similar problem occurs with the analysis of insects, since the chitin component of the exoskeleton contains 6.9% nitrogen (Bell, 1990). Simply using a lower conversion factor may be adequate for a rough estimate of the protein content in the diet (Conklin-Brittain *et al.*, 1999). Amino acid analysers can measure their individual concentrations in proteins, which is important in assessing protein quality (plant proteins often have a seriously unbalanced amino acid composition).

NON-STRUCTURAL CARBOHYDRATES

Extraction methods for soluble carbohydrates are a compromise between completeness of extraction and avoidance of hydrolysis. Prolonged boiling maximises extraction efficiency but can result in large changes in composition, while cold extraction is usually incomplete. There are wet chemistry methods available for all aspects of carbohydrate analysis (Chaplin & Kennedy, 1994) but it takes considerable skill to apply these accurately to complex, unknown mixtures of carbohydrates and other organic materials. High pressure liquid chromatography (HPLC) has largely replaced these methods for quantitative analysis. Different combinations of columns, solvents, detectors and standards are needed for different carbohydrates, so an analysis targeted at glucose, fructose and sucrose, for instance, may overlook less common sugars in the sample. However, it is those three sugars that plants generally produce in bulk and which seem to play an exceedingly important role in primate food choice, particularly for fruits. Sucrose is the transport sugar, the splitting of which produces equal quantities of fructose and glucose. However, sucrose is not always present in fruits, particularly those associated with some bird groups that lack a sucrase enzyme. Primates can detect sucrose often at lower concentrations than the other sugars and this may be related to fruit consumption. Sugars increase during fruit development and, in leaves, they may increase with sun exposure (Ganzhorn, 1995) and time of day (Ganzhorn & Wright, 1994). Indeed, it has been claimed that chimpanzees (*Pan troglodytes*) may prefer young leaves to mature leaves owing to significantly higher levels of glucose (Reynolds *et al.*, 1998). Even ants can be relatively sugar rich (Hladik, 1977). While methods of sugar quantification vary, many primatologists turn to HPLC (Simmen & Sabatier, 1996; Reynolds *et al.*, 1998). Recently other methods, such as gas chromatography (Molnár-Perl *et al.*, 1994) and near-infrared spectroscopy (Schulz *et al.*, 1998), have been developed to quantify sugars (and acids, carotenoids and other constituents).

With the exception of many fruits, plant parts usually contain more starch than sugars. Starch is often assumed to be as tasteless and unattractive to other primates as it is to humans. However, many mammals have a salivary amylase that can break down starch extremely rapidly to release tasty sugars, provided that the starch is not present in an insoluble granular form (e.g. Laska *et al.*, 2001).

LIPIDS

These are defined by their solubility in organic solvents and are usually extracted from plant materials in a Soxhlet apparatus (Gunstone, 1996). Crude fat (or ether extract) is defined as the lipid extracted with anhydrous ether (which is flammable and explosive). Other solvents, such as petroleum ether or hexane, are less dangerous, but give slightly different results. Crude fat is a complex mixture of substances including plant cuticles, which are not digested (even by ruminants) and could potentially be analysed from the faeces to indicate plant composition in the diet. Methods for measuring fatty acids are given by Chamberlain *et al.* (1993) and Simmen & Sabbatier (1996).

Detailed lipid analysis is probably only worth-while if fresh material is available because plant lipids are very sensitive to oxidation under normal storage and preparation conditions. Seeds can be stored intact as long as they remain alive. Thin-layer chromatography is used to separate the total lipids into the major classes of compounds. Fatty acid composition is then determined by gas chromatography.

SECONDARY COMPOUNDS

Laboratory determinations of terpenoids have been reported by Wrangham *et al.* (1998). For accurate identification, these must be extracted and quantified individually, usually by gas–liquid (GLC) or gas chromatography. Saponins are another widespread and potentially interesting group of compounds known for their toxicity. Individual saponins can be identified and quantified by HPLC coupled with mass spectrometry (Marston *et al.*, 2000). Waterman & Mole (1994) describe a simple spectrophotometer-based haemolysis technique that is only suitable if the extract is tannin free. Alkaloids are so heterogeneous as to lack even a satisfactory definition. They are less widespread in plant material than the other two major classes of secondary chemicals, but can be crucial in feeding decisions when present. Although there are more or less reliable general tests to screen for the presence of alkaloids (Harborne, 1998), they must be quantified individually, using GLC or HPLC.

OTHER POSSIBILITIES

The only limit on possible field tests is the time needed to develop them. Lipids elude field investigation because of the problem of extraction. They can certainly be stained, however, in freehand thin sections. There are now commercial microscopes that connect to a computer and which thus permit quantification in two dimensions using image analysis software. A quantitative test for alkaloids and cyanide, for example, seems possible, the latter by adapting a small commercial kit (cyanide test kit, Cole–Parmer). The ninhydrin test (Lucas *et al.*, 2001) can be refined to give accurate overall levels of amino acids. Others have described methods for measuring mineral content (Silver *et al.*, 2000; Remis *et al.*, 2001). Ion concentrations can be measured with ion-specific electrodes that can be attached to pH meters. Vitamin C (Milton & Jenness, 1987) and the antioxidant potential of these and other compounds have also received attention (Leitão *et al.*, 1999).

Terpenoids as a whole raise many problems, particularly the low molecular weight volatile monoterpenes. The latter seem to be released through the plant cuticle and could be sensed by smell. The quantification of scent chemistry is now practical, using headspace techniques for field collection

(Pettersson & Knudsen, 2001). A further development, the 'electronic nose' lies on the horizon (Dominy *et al.*, 2001).

ACKNOWLEDGEMENTS

We thank the Research Grants Council of Hong Kong and the Croucher Foundation of Hong Kong for their support. Dr A. E. Hagerman (University of Miami, Oxford, OH, USA) gave valuable advice concerning field chemistry.

REFERENCES

Appel, H. M., Governor, H. L., D'Ascenzo, M., Siska, E. & Schultz, J. C. (2001). Limitations of Folin assays of foliar phenolics in ecological studies. *J. Chem. Ecol.* **27**, 761–78.

Barker, D. Fitzpatrick, M. P. & Dierenfeld, E. S. (1998). Nutrient composition of selected whole invertebrates. *Zoo Biol.* **17**, 123–34.

Bell, G. P. (1990). Birds and mammals on an insect diet: a primer on diet composition analysis in relation to ecological energetics. *Studies Avian Biol.* **13**, 416–22.

Challem, J. J. (1999). Toward a new definition of essential nutrients: is it now time for a third 'vitamin' paradigm? *Med. Hypoth.* **52**, 417–22.

Chamberlain, J., Nelson, G. & Milton, K. (1993). Fatty acid profiles of major food sources of howler monkeys (*Alouatta palliata*) in the neotropics. *Experientia* **49**, 820–4.

Chaplin, M. F. & Kennedy, J. F. (1994). *Carbohydrate Analysis: A Practical Approach.* Oxford: IRL Press.

Conklin-Brittain, N. L., Dierenfeld, E. S., Wrangham, R. W., Norconk, M. & Silver, S. C. (1999). Chemical protein analyses: a comparison of Kjeldahl crude protein and total ninhydrin protein from wild, tropical vegetation. *J. Chem. Ecol.* **25**, 2601–22.

Dominy, N. J., Lucas, P. W., Osorio, D. & Yamashita, N. (2001). The sensory ecology of primate food perception. *Evol. Anthropol.* **10**, 171–86.

Ganzhorn, J. U. (1995). Low-level forest disturbance effects on primary production, leaf chemistry, and lemur populations. *Ecology* **76**, 2084–96.

Ganzhorn, J. U. & Wright, P. C. (1994). Temporal patterns in primate leaf eating: the possible role of leaf chemistry. *Folia Primatol.* **63**, 203–8.

Glander, K. E., Wright, P. C., Seigler, D. S., Randrianasolo, V. & Randrianasolo, B. (1989). Consumption of a cyanogenic bamboo by a newly discovered species of bamboo lemur. *Am. J. Primatol.* **19**, 119–24.

Gunstone, F. D. (1996). *Fatty Acid and Lipid Chemistry.* London: Blackie Academic & Professional.

Hagerman, A. E. (1987). Radial diffusion method for determining tannin in plant extracts. *J. Chem. Ecol.* **13**, 437–49.

(1988). Extraction of tannin from fresh and preserved leaves. *J. Chem. Ecol.* **14**, 453–61.

(1998). *The Tannin Handbook*. Website: <miavx1.muohio.edu/~hagermae>.

Hagerman, A. E., Zhao, Y. & Johnson, S. (1997). Methods for determination of condensed and hydrolyzable tannins. In *Antinutrients and Phytochemicals in Foods*, ed. F. Shahadi, pp. 209–22. Washington, DC: American Chemical Society.

Harborne, J. B. (ed.)(1998). *Phytochemical Methods: A Guide to Modern Techniques of Plant Analysis*, 3rd edn. London: Chapman and Hall.

Hladik, C. M. (1977). Chimpanzees of Gabon and chimpanzees of Gombe: some comparative data on the diet. In *Primate Ecology: Studies of Feeding and Ranging Behaviour in Lemurs, Monkeys and Apes*, ed. T. H. Clutton-Brock, pp. 481–501. London: Academic Press.

Jumars, P. A. (2000a). Animal guts as ideal chemical reactors: maximizing absorption rates. *Am. Nat.* **155**, 527–43.

(2000b). Animal guts as non-ideal chemical reactors: partial mixing and axial variation in absorption kinetics. *Am. Nat.* **155**, 544–55.

Laska, M., Kohlmann, S., Scheuber, H., Salazar, L. T. H. & Luna, E. R. (2001). Gustatory responsiveness to polycose in four species of nonhuman primates. *J. Chem. Ecol.* **27**, 1997–2011.

Leitão, G. G., Mensor, L. L., Amaral, L. F. G., Floriano, N., Limeira, V. L. G., Menezes, F. S. & Leitão, S. G. (1999). Phenolic content and antioxidant activity: a study on plants eaten by a group of howler monkeys (*Alouatta fusca*). In *Plant Polyphenols 2: Chemistry, Biology, Pharmacology, Ecology*, ed. G. G. Gross, R. W. Hemingway & T. Yoshida, pp. 883–95. New York: Kluwer Academic.

Lucas, P. W., Beta, T., Darvell, B. W., Dominy, N. J., Essackjee, H. C., Lee, P. K. D., Osorio, D., Ramsden, L., Yamashita, N. & Yuen, T. D. B. (2001). Field kit to characterize physical, chemical and spatial aspects of potential foods of primates. *Folia Primatol.* **72**, 11–15.

Marston, A., Wolfender, J. L. & Hostettmann, K. (2000). Analysis and isolation of saponins from plant material. In *Saponins in Food, Feedstuffs and Medicinal Plants*, ed. W. Oleszek & A. Marston, pp. 1–12. Dordrecht: Kluwer Academic.

Milton, K. & Jenness, R. (1987). Ascorbate content of neotropical plant parts available to monkeys and bats. *Experientia* **43**, 339–42.

Minekus, M., Smeets-Peeters, M., Bernalier, A., Marol-Bonnin, S., Havenaar, R., Marteau, P., Alric, M., Fonty, G. & Huis in't Veld, J. H. (1999). A computer-controlled system to simulate conditions of the large intestine with peristaltic mixing, absorption of fermentation products and a high-density microflora. *Appl. Microbiol. Technol.* **53**, 108–14.

Molnár-Perl, I., Tisza, S. & Sass, P. (1994). Optimization of the simultaneous quantification of acids and sugars as their silyl(oxime) derivatives by GC/MS. *Acta Hort.* **368**, 291–309.

Mowry, C. B., Decker, B. S. & Shure, D. J. (1996). The role of phytochemistry in dietary choices of Tana River red colobus monkeys. *Int. J. Primatol.* **17**, 63–84.

Ntiamoa-Baidu, Y. (1997). *Wildlife and Food Security in Africa*. Rome: FAO.

O'Brien, T. G., Kinnaird, M. F., Dierenfeld, E. S., Conklin-Brittain, N. L., Wrangham, R. W. & Silver, S. C. (1998). What's so special about figs? *Nature*, **392**, 668.

Orians, C. M. (1995). Preserving leaves for tannin and phenolic glycoside analyses: a comparison of methods using three willow taxa. *J. Chem. Ecol.* **21**, 1235–43.

Pettersson, S. & Knudsen, J. T. (2001). Floral scent and nectar production in *Parkia biglobosa* Jacq. (Leguminosae: Mimosoideae). *Bot. J. Linn. Soc.* **135**, 97–106.

Read, S. M. & Northcote, D. H. (1981). Minimization of variation in the response to different proteins of the Coomassie Blue G dye-binding assay for protein. *Anal. Biochem.* **116**, 53–64.

Remis, M. J., Dierenfeld, E. S. Mowry, C. B. & Carroll, R. W. (2001). Nutritional aspects of Western lowland gorilla (*Gorilla gorilla gorilla*) diet during seasons of fruit scarcity at Bai Hokou, Central African Republic. *Int. J. Primatol.* **22**, 807–36.

Reynolds, V., Plumptre, A. J., Greenham, J. & Harborne, J. (1998). Condensed tannins and sugars in the diet of chimpanzees (*Pan troglodytes schweinfurthii*) in the Budongo Forest, Uganda. *Oecologia* **115**, 331–6.

Sapan, C. V., Lundblad, R. L. & Price, N. C. (1999). Colorimetric protein assay techniques. *Biotechnol. Appl. Biochem.* **29**, 99–108.

Schulz, H., Drews, H.-H., Quilitzsch, R. & Krüger, H. (1998). Application of near infrared spectroscopy for the quantification of quality parameters in selected vegetables and essential oil plants. *Near Infrared Spectrosc.* **6**, A125–A130.

Silver, S. C., Ostro, L. E. T., Yeager, C. P. & Dierenfeld, E. S. (2000). Phytochemical and mineral components of foods consumed by black howler monkeys (*Alouatta pigra*) at two sites in Belize. *Zoo Biol.* **19**, 95–109.

Simmen, B. & Sabatier, D. (1996). Diets of some French Guianan primates: food composition and food choices. *Int. J. Primatol.* **17**, 661–93.

Simonne, E. H., Harris, C. E. & Mills, H. A. (1998). Does the nitrate fraction account for differences between Dumas-N and Kjeldahl-N values in vegetable leaves? *J. Plant Nutr.* **21**, 2527–34.

Smith, A. C. (2000). Composition and proposed nutritional importance of exudates eaten by saddleback (*Saguinus fuscicollis*) and mustached (*Saguinus mystax*) tamarins. *Int. J. Primatol.* **21**, 69–83.

Ungar, P. S. (1995). Fruit preferences of four sympatric primate species at Ketambe, northern Sumatra, Indonesia. *Int. J. Primatol.* **16**, 221–35.

Waterman, P. G. (1984). Food acquisition and processing as a function of plant chemistry. In *Food Acquisition and Processing in Primates*, ed. D. J. Chivers, B. A. Wood & A. Bilsborough, pp. 177–211. New York: Plenum Press.

Waterman, P. G. & Mole, S. (1994). *Analysis of Phenolic Plant Metabolites*. Oxford: Blackwell.

White, D. W. & Stiles, E. W. (1985). The use of refractometry to estimate nutrient rewards in vertebrate-dispersed fruits. *Ecology* **66**, 303–7.

Wrangham, R. W., Conklin-Brittain N. L. & Hunt, K. D. (1998). Dietary response of chimpanzees and cercopithecines to seasonal variation in fruit abundance. I. Antifeedants. *Int. J. Primatol.* **19**, 949–69.

Useful Internet sites

Computer-linked microscopes : see for example <www.Pocketscope.com>.
NIH Image software : <rsb.info.nih.gov/nih-image/>.

14 • Collecting arthropods and arthropod remains for primate studies

CLAIRE M. P. OZANNE AND JAMES R. BELL
Centre for Research in Ecology and the Environment, School of Life and Sports Sciences, University of Surrey Roehampton, London, UK

INTRODUCTION

Arthropods, along with other invertebrates, make up 95% of the global fauna, with 1.5 million described species and at least a further 6 million that remain undescribed. These organisms, which include insects, arachnids and myriapods, play a significant role in the life histories of a range of primate species, not least in their diet. Body size plays an important part in determining the degree of insectivory (Kay, 1984); however, an example of insect eating can be found within all lineages of the primate phylogeny. Prosimians have long been described as insectivores, some having morphological adaptations for insect feeding (e.g. the extractive finger of the aye-aye, *Daubentonia madagascariensis*) and others adopting behaviours such as aerial snatching (e.g. galagos, *Galago demidovii*), or stalking specific arthropod prey (e.g. angwantibos, *Arctocebus calabarensis*) (Charles-Dominique, 1974). Some Loridae, like the potto (*Perodicticus potto*), have evolved specialised diets in which the arthropod prey emit distasteful chemicals (e.g. formic acid from ants: *Crematogaster* spp.) or have the potential to inject venom (e.g. large scolopendrid centipede: *Spirostreptus* spp.) (Charles-Dominique, 1974). Chimpanzees (*Pan troglodytes*) use tools to help to eat insects such as bees (*Apis* spp.), and the honey they produce (Tutin & Fernandez, 1992). Even folivores that appear to be entirely vegetarian inadvertently eat thousands of insects a day (e.g. mountain gorillas, *Gorilla gorilla beringei*, and langurs, *Semnopithecus entellus*). However, apart from the large patas monkey (*Erythrocebus patas*), non-prosimian primates are unable to process enough arthropod material for these animals to form a major part of a higher primate's diet (Isbel, 1998).

The time invested in catching some agile insects can be high compared to picking fruit or foliage off a shrub, for example. However, when arthropods are caught, they are a good source of proteins and lipids, providing an energy level equivalent to that of vertebrate tissue (1.93 kcal/g – see Kay, 1984), but they have other non-dietary uses too. For example, the millipede (*Orthoporus dorsovittatus*) secretes two insect-repellent chemicals known as benzoquinones

Field and Laboratory Methods in Primatology: A Practical Guide, ed. Joanna M. Setchell and Deborah J. Curtis. Published by Cambridge University Press. © Cambridge University Press 2003.

and is used by wedge-capped capuchin monkeys (*Cebus olivaceus*) to anoint themselves, keeping mosquitoes and botflies away during the rainy season (Valderrama *et al.*, 2000). However, not all interactions are beneficial and a significant number of arthropods can cause disease either by means of vectoring micro-organisms or by actual damage (e.g. bites, stings, and myiasis). In this chapter we confine ourselves to methods associated with collecting terrestrial arthropods (i.e. invertebrates with jointed legs such as grasshoppers and spiders), because there is only limited evidence in the literature of primates eating other non-arthropod invertebrates (although gastropods (slugs and snails) are an occasional exception). An additional point of interest is that there are a small number of crab-eating primates (e.g. long-tailed macaques (*Macaca fascicularis*), Payne & Francis, 1985; mandrills (*Mandrillus sphinx*), Hoshino, 1985). Crabs are firmly included within the Arthropoda, and, while some crabs are aquatic, others may be almost entirely land-bound. However, although methods included in this chapter may suit the collection of crabs, we have not specifically tailored our methods to include the crab fauna. On this basis, if students wish to collect crabs, some methods will need refinement, such as the addition of lures and the redesign of traps to make them escape-proof.

ARTHROPOD TAXONOMY AND IDENTIFICATION

Arthropods are diverse, numerous and in the tropics are often poorly known both taxonomically and ecologically. These arthropods are morphologically complex (e.g. the difference between two species may be the position of a single hair) but can usefully be identified to the ordinal level (e.g. Coleoptera, beetles) or placed in a morpho-taxon group (e.g. beetle-like; see Oliver & Beattie, 1996). Researchers new to arthropod studies can familiarise themselves with general body forms using colour identification guides (e.g. Chinery, 1994), later moving on to more descriptive taxonomic works when more detailed anatomical analyses are necessary (e.g. McGavin, 2001). Technical nomenclature is widely used in arthropod taxonomy and an entomological dictionary, such as that of Gordh & Headrick (2001), is useful, although many entomology textbooks will contain a sufficient glossary for basic identification work. Students uncertain of the phylogenetic relationships of the Arthropoda should visit the Tree of Life (see Other useful Internet sites), entering at the level of Metazoa – i.e. multi-cellular organisms that develop from embryos. In basic terms, arthropods lack backbones and are thus split from the Chordata, which includes primates, on the basis of embryogeny and intial development. They are then split from other animals without backbones (e.g. worms) because of their jointed legs.

The types of arthropod selected by primates include the larger-bodied animals such as Orthoptera (grasshoppers, crickets and katydids), Dictyoptera (cockroaches and preying mantises), Phasmida (stick insects), Odonata (dragonflies), Lepidoptera (moths and butterflies), Coleoptera (beetles), Hymenoptera (bees, ants and wasps), Isoptera (termites), Arachnida (spiders and scorpions), Myriapoda (centipedes and millipedes), and Hemiptera (cicadas, shield bugs). For an example of a detailed list of insectivory, refer to Niemitz (1984), who gives an insight into the catholic diet of the genus *Tarsius*.

GETTING STARTED – EQUIPMENT LIST

Two books that have been specifically written for expedition work, being compact, ring-bound and full of information on arthropods, are McGavin (1997) and Anonymous (1996). Both should be tucked into your rucksack before you leave.

ESSENTIALS

Collecting net, paint brushes (for delicate specimens), paper and pencil, forceps, plastic storage jars and vials, binoculars, plastic bags for foliage samples, hand lens (10 ×–30 ×), white sorting tray (to separate organisms from debris), pooter or aspirator (used to suck arthropods up without swallowing them; ideally, take both a manual and battery powered aspirator), and alcohol (methanol/ethanol for long-term storage, or alternatively use isopropyl alcohol (2-propanol or 'rubbing alcohol') for field use, but change to a long-term storage solution on your return): all concentrations must be greater than 70% in an aqueous solution to prevent rotting.

IDEALLY

Binocular microscope (70 ×–100 ×), calico bags, branch clippers, Winkler bags, Petri dishes, and cold packs.

WITH LOTS OF THOUGHT, PREPARATION AND GOOD FUNDING

Knockdown apparatus: mister/fogger, (for example: misters Hurricane-Major™ and the Stihl™, foggers Swing-Fog™ and Dyna-Fog™); chemical (natural pyrethrum or synthetic pyrethroid); collecting trays; rope/twine; petrol and two-stroke oil; and refrigerator.

KILLING, SETTING AND STORAGE OF SPECIMENS

If arthropods are not collected into alcohol they can be killed rapidly using a killing jar, either a polyethylene or glass container (500 ml) labelled 'POISON', with a few drops of ethyl acetate (the chemical in nail varnish remover) as the killing agent. In the long term, most arthropod specimens are best stored in alcohol (ideally 70–100% solution) in tubes or vials with tight lids. Primate

GPS or map reference	17° 36.945 S 145° 47.748 E
Site name	Palmerston Nat. Pk. QLD
Capture technique, e.g. malaise trap	Pyrethrum mist
Sample code, e.g. q1 (quadrat 1)	Planted Trans. 1, 0 m
Date	2 Dec 2000
Name of collector	C. Ozanne

Fig. 14.1. Example label for samples and specimens.

faecal samples should be dried, stored and kept cool in any air-tight container, with silica gel if needed. Labelling of samples and specimens is essential either with a paper and pencil or printed on acid free paper (Fig. 14.1).

Some insects require pinning, such as butterflies and moths. However, a short-term solution is to place them flat in small paper envelopes. Further curatorial information can be found in Fitton (2002) or McGavin (1997).

METHODS FOR DIRECT OBSERVATION

Where primates are visible, it is often possible to observe their insectivory using binoculars or a camera, although the prey may be concealed and diffi-cult to identify. Some authors have been able to determine insects to species level by comparing the remains dropped during or after feeding to reference collections, for example Orthoptera preyed upon by saddle-back tamarins (*Saguinus fuscicollis nigrifrons*) and moustached tamarins (*Saguinus mystax mystax*) (Nickle & Heymann, 1996). This method could be applied even when primates are feeding high in the canopy.

If primates are feeding on small arthropods on ground vegetation or in the leaf litter, the location can be noted and samples collected using forceps, fine paintbrushes or aspirators. If they are nest-dwelling insects, samples can be collected by making a small opening in the side of the nest and dropped into alcohol. Using an aspirator to collect ants may leave a bitter taste in your mouth owing to the release of formic acid by defensive soldiers – use a battery-powered aspirator if possible.

Determination of primate insect-feeding to a taxon-specific level seems a viable, but largely uncharted means of gaining dietary information. Certain assumptions could be made about the types of insect eaten using primate behavioural information. For example, field notes might read 'primate seen sitting and picking over leaf litter for extended periods with little investment in searching effort. Cannot see insects, but they are held between thumb and forefinger'. These notes would suggest that the insects are abundant, small, probably social and likely to either be ant- or termite-like. Another field notebook example might be 'primate seen stalking along tree trunk and moving slowly. Caught insect by snatching action'. These notes indicate that the target prey had a good sense of its environment and is a fast flyer, jumper or walker. The likely candidates here are cicadas, dragonflies, grasshoppers or crickets – generally animals with larger eyes or other acute sensory organs. Other ethologies could, for example, be related to venomous, leaf-eating or nest-building arthropods.

METHODS FOR INDIRECT OBSERVATION

Where direct observation of feeding is not possible, a number of indirect techniques can be used to give detailed information about the arthropods that are available at the feeding site.

ARTHROPOD BY-PRODUCTS – SECONDARY SIGNS

Arthropods often betray their presence, for example by leaving webs (spiders and caterpillars), frass (insect excreta), nests (e.g. wasps and bees), mounds (ants and termites), holes (e.g. solitary wasps and antlions), galleries (e.g. wood-boring beetles) or chewed leaves (e.g. caterpillars and leaf cutter bees), or by making sounds (e.g. cicadas, grasshoppers and crickets). These signs can be used to detect the type of arthropod selected for consumption or use by a primate. For example, if a primate is seen feeding from a hole in a tree trunk, take a picture of the hole and try to extract the concealed arthropod, but never put your hand where you cannot see. If the arthropod is determined not to expose itself, carefully collect some small ants in a pooter (aspirator) and blow them into the hole. All arthropods have strong reactions to ants and will take the shortest exit to escape their presence. Likewise, take recordings of audible songs of crickets or grasshoppers that are selected by a primate (Chapter 15). These recordings will help to identify the animals later.

FAECAL SAMPLES

Faecal sampling provides a measure of the type of arthropod food that has been eaten, and is particularly useful if one is studying the feeding habitats of a primate species that is difficult to observe directly. Arthropods are often encased in chitin (sclerotised epidermis), and it is this that passes through the gut of primates, later appearing as fragments in the faeces. However, identifying the remains of arthropods can be difficult: where adult beetles are often covered in chitin, the larvae (grubs) have a spartan covering, often in the region around the head and legs, and fly larvae (maggots) may be almost without any identifiable chitin at all. Furthermore, within the faeces, arthropod body parts may be jumbled up, incomplete and obscured by plant debris.

Despite these constraints, many studies have successfully studied the arthropod diet of primates, restricting the identification of the arthropod fragments to the ordinal (e.g. beetles: Coleoptera) or sub-ordinal level (e.g. true bugs: Heteroptera and Homoptera), rarely choosing to define family or species unless the body part is readily identifiable (e.g. the pedicel or waist of an ant (Formicidae), see Bolton (1994). Identification of arthropod fragments is made easier with Gentry *et al.* (1991), a two-volume clinical manual for food quality control containing colour photographs of invertebrate fragments. An alternative budget option is to use a set of invertebrate guides, which will introduce the arthropod form to the ordinal level (e.g. Tilling (1987) and Chinery (1994) are widely used in Britain), and combine these with Shiel *et al.* (1997), which gives a detailed account of arthropod fragments in bat droppings. This should be US$ 300 cheaper and sufficient for most ordinal studies. It is prudent to make a reference collection of 'whole' arthropods collected from the strata of vegetation used by a primate (see below for the optimum sampling protocol). These whole animals can then be compared with fragments found in the faeces.

Counting arthropod fragments can lead to problems in analysis if anything other than presence/absence data are required. Estimating density or mass is fraught with problems of pseudo-replication and independence: as an individual's body parts become fragmented in the gut, these parts (e.g. legs) can be mistaken for separate individuals, yielding overestimates. Counting only 'heads' can dampen these problems, but it is possible that arthropods may be decapitated before consumption (e.g. locusts, mantids, etc.). Additionally, larval stages do not often have separate head body segments (e.g. maggots) and these may be missed altogether. In both scenarios, it is simple to demonstrate that data generated from faecal samples can be considered only 'estimates of consumption', which are far from absolute. Lastly, the choice of statistical tests may also help. If there is believed to be a large, but unquantifiable, error term, estimates can be made coarser by choosing a non-parametric test (e.g. Spearman rank correlation) over parametric ones (e.g. Pearson product moment correlation).

SAMPLING ARTHROPODS FROM THE SUB-CANOPY AND CANOPY

Primates may feed actively on arthropods in the canopy of understorey and overstorey trees. The two most effective quantitative sampling methods for plant canopies are branch clipping and pyrethrum knockdown (for a more extensive treatment, see Ozanne, 2003a; Basset *et al.*, 1997).

Branch clipping involves cutting off a section of foliage into a bag that is then drawn shut to prevent the escape of mobile arthropods. Arthropods can be shaken off the foliage and identified live on site or the bag filled with a chemical such as CO_2 or ethyl acetate (Basset, 1992). In the latter case arthropods would be picked or brushed carefully off the foliage and taken to the laboratory for storage and sorting. The bag (e.g. plastic, muslin or mesh) and clippers are attached to fixed or telescopic poles (e.g. canes, branches, tent poles), allowing them to be pushed up into the canopy from the ground (Basset & Höft, 1994). Normally the bag is drawn over the foliage, then the branch is cut and the bag drawn shut with a cord that can be operated from the ground (for construction see Johnson, 2000). The amount of foliage clipped should be standardised where possible. This method would be particularly useful for studies of primates such as the patas monkey, which feeds on arthropods in the swollen thorns of *Acacia* trees (Isbel, 1998) or Japanese macaques (*Macaca fuscata*), which bite into woody plant galls to extract aphids (Hill *et al.*, 1995) and for primates that feed on insects rolled in dead leaves, for example saddle-back tamarins (Nickle & Heymann, 1996).

Branch clipping is a fairly comprehensive sampling method for most arthropods, particularly sedentary animals, which can be standardised to plant biomass or area units (Schowalter *et al.*, 1981). However, clipping under-samples large mobile (e.g. dragonflies) and some small aerial (e.g. midge clouds) insects (Cooper & Whitmore, 1990; Johnson, 2000) and over-estimates inhabitants of branch tips.

Pyrethrum knockdown involves delivering pyrethrum or synthetic equivalents into the canopy in droplet form using one of two techniques, fogging or misting. In both cases pyrethrum and synthetics are shortlived and non-toxic to vertebrates (excluding fish): they cause loss of motor function to arthropods, which then fall towards the ground into collecting trays. Foggers produce a thermal cloud of pyrethrin that rises upwards and outwards filling the canopy space. This method is useful for high canopies (15 m upwards), but is more difficult to operate and control than misting. Fogging must be carried out in still, dry air and is typically done for 5–10 minutes at a time (Stork & Hammond, 1997; Adis *et al.*, 1998). In contrast, misters use a fan mounted in a back-pack to produce a powerful air current that shears off droplets of chemical, forming a fine mist. The mist reaches a more targeted area of foliage than a fog, but the machine must be hoisted into the canopy in order to reach foliage above 12–15 m. Misting is typically carried out for 1–5 minutes, depending on the machine, which may be set up for low volume (20–300 l/ha) or ultra-low volume delivery (5–20 l/ha). Cone-shaped collecting trays, constructed from waterproof material, for example vinyl or tent material, are suspended with rope below the canopy. Arthropods are allowed to drop into these for about 2 hours after application and samples are standardised to the area of the tray, usually 1 m².

These techniques are particularly applicable where primates feed on a variety of insects in the mid-canopy (e.g. moustached tamarins; Nickle & Heymann, 1996) and the high canopy and for those species that feed by collecting insects from the leaf and bark surface, for example pied bare-faced tamarin (*Saguinus bicolor bicolour*; Egler, 1992). Fogging and misting give a more accurate representation of arthropod communities than does any other foliage sampling method. However, both techniques are expensive, unless equipment can be borrowed, and are limited to calm, dry weather conditions.

VEGETATION AND GROUND SAMPLING

Low vegetation and the litter layer are rich sources of arthropods exploited by several primate groups. For example, saddle-back tamarins feed on arthropods in the lower parts of the forest canopy and lowland gorillas (*Gorilla gorilla gorilla*) and bonobos (*Pan paniscus*) collect ants and termites from the ground (Tutin & Fernandez, 1992; Bermejo *et al.*, 1994). Low vegetation is most effectively sampled by branch clipping or pyrethrum knockdown (see above), but arthropods may also be collected by beating the foliage with a stick so that the animals drop onto a collecting sheet. This provides only relative count data (catch per unit effort), but mimics the technique used by primates such as saddle-back tamarins, which are particularly adept at capturing prey items they have flushed out by shaking the foliage (Peres, 1992). For sampling ground vegetation up to 25 cm in height (and even understorey shrubs), the most effective technique is to use a suction device such as a petrol driven garden leaf-vacuum modified to include a small muslin bag (Stewart & Wright, 1995; Ozanne, 2003b). Vacuuming an area of vegetation yields samples that are highly representative of the arthropod community, although they tend to oversample owing to an edge effect caused by high suction power and can become inefficient if the net is not emptied on a regular basis.

Litter layer arthropods such as ants are effectively sampled using Winkler extractors (Bestelmeyer *et al.*, 2000). This involves collecting leaf litter (normally from a $1 m^2$ area) and sifting it through a grid mounted in a cotton bag: live animals and fine debris fall through the grid. The sifted material is then placed in a Winkler extractor comprising an inner net bag in which the sample is placed and an outer funnel shaped bag with a collecting vessel at the base. The extractor is hung in an airy location and, as the sample slowly dries out, animals seeking moisture are driven into the second bag and fall or roll down into the jar filled with preservative fluid. This technique would be most appropriate for sampling arthropods that are fed on by non-tool-using primates such as lowland gorillas and vervet monkeys (*Chlorocebus aethiops*) (Yamagiwa *et al.*, 1994; Harrison & Byrne, 2000).

A range of traps for collecting arthropods from vegetation and the ground is available or can be constructed from simple materials (Table 14.1). The advantage of using traps to collect arthropods is that they make use of arthropod activity and can be left to passively sample the habitat while the researcher is absent. However, when movement determines the numbers caught, this detracts from an absolute measure (i.e. per square metre), yielding data that are best described in less desirable units of 'active trappability density'.

On the whole, however, we do not recommend the use of traps for this type of study. All are activity based and draw from a wide area of habitat. Thus it will be difficult to determine how representative are the samples of prey availability for primates.

Table 14.1. *Arthropod trap types, their advantages and disadvantages (for references, see Southwood & Henderson, 2000)*

Trap type	Dominant arthropod groups collected	Pros	Cons
Malaise	Flies, termites (winged), parasitic wasps	Inexpensive	To be set in flight lines; very large catches
Interception	Beetles	Easily constructed; very effective for beetles	Large catches; group specific
Colour and pan traps	Flies, Hymenoptera (bees, wasps), Hemiptera (bugs)	Cheap, simple effective	Location dependent
Baited traps	Bait dependent: spiders, beetles, ants, moths and butterflies, flies	Bait can be designed to capture particular sections of the fauna	Draw insects from a wide area
Emergence traps	Flies, beetles	Only method of collecting	Life-cycle stage specific (adult)
Pitfall traps	Spiders, ants, beetles, millipedes, centipedes, spring-tails, crustaceans	Inexpensive	Must be carefully set
Light traps	Moths, beetles	Very effective	Night activity only

DISCUSSION AND CONCLUSION

There is a wide range of techniques for collecting arthropods and we can include only a small, but representative, sample here. For a more detailed treatment please refer to Sutherland (1996), New (1998) or Southwood & Henderson
(2000). Whatever arthropod sampling technique is used, there are a number of factors that should be taken into account when a study is designed.
First, quantitative sampling methods yield the most flexible data. Arthropod
densities can be quantified per unit area or volume and per unit plant biomass, allowing many analytical tools to be applied to the data sets. Secondly,
arthropod spatial distribution is often highly variable, causing decision-
making about the number and location of samples to be difficult. A pilot
collecting session will probably be needed to determine population distributions, but careful location of collecting events in the habitat where the
primates are feeding should pre-empt major problems. Thirdly, the timing
of collection is crucial. Arthropod communities exhibit diurnal and seasonal
variation, thus sampling should be timed to coincide with primate feeding
periods. Primates may also vary in their selection of arthropods diurnally and
seasonally. For example, *Galago demidovii* becomes more insectivorous as the
night progresses, whilst reducing gum and fruit intake (Charles-Dominique,
1974) and spectral tarsiers (*Tarsius spectrum*) change their arthropod diet between seasons, taking Orthoptera and Lepidoptera in the dry season and
adding a more varied fauna, which includes Coleoptera and Hymenoptera,
during the wet season (Gursky, 2000). Primates will of course disturb arthropods, causing them to disperse or hide, so sampling should not be carried out
immediately after a feeding event. Instead sites may be labelled with flagging
tape and revisited at an equivalent time on another day.

Finally, not all arthropods on which primates feed can be collected from
the habitat in which they live. Commensal insectivory (e.g. amongst moustached and saddle-back tamarins; Peres, 1992) may be a useful way of supple-
menting diet as well as reducing parasite loads and forms an important link
between diet and social behaviour. Studies of primates that seek to include
interactions with arthropods could also consider ways of monitoring insect
and arachnid disease vectors (e.g. ticks and mosquitoes) and could investigate
symbiotic relationships such as those between dung beetles (Scarabaeidae)
and primates. For example howler monkeys (*Alouatta palliata*) in the Mexican
rain forest disperse seeds that dung beetles bury, protecting them against
rodent predation (Estrada & Coatesestrada, 1991).

ACKNOWLEDGEMENTS

We would like to thank the University of Surrey Roehampton for supporting
this work.

REFERENCES

Anonymous (1996). *Biodiversity Assessment: A Guide to Good Practice*. London: HMSO.

Adis, J., Basset, Y., Floren, A., Hammond, P.M. & Linsenmair, K.E. (1998). Canopy fogging of an overstorey tree – recommendations for standardization. *Ecotropica* **4**, 93–7.

Basset, Y. (1992). Host specificity of arboreal and free living insect herbivores in rain forest. *Biol. J. Linn. Soc.* **47**, 115–33.

Basset, Y. & Höft, R. (1994). Can apparent leaf damage in tropical trees be predicted by herbivore load or host-related variables? A case study in Papua New Guinea. *Selbyana* **15**, 3–13.

Basset, Y., Springate, N.D., Aberlanc, H.P. & Delvare. G. (1997). A review of methods for sampling arthropods in tree canopies. In *Canopy Arthropods*, ed. N. Stork, J. Adis & R. Didham, pp. 27–52. London: Chapman & Hall.

Bermejo, M., Illera, G. & Pi, J.S. (1994). Animals and mushrooms consumed by bonobos (*Pan paniscus*) – new records from Lilungu (Ikela), Zaire. *Int. J. Primatol.* **15**, 879–98.

Bestelmeyer, B.T.D., Agosti., D., Alonso, L.E., Brandao, C.R.F., Brown W.L. Jr, Delabie, J.H.C. & Silvestre, R. (2000). Field techniques for the study of ground dwelling ants: an overview, description, and evaluation. In *Ants: Standard Methods for Measuring and Monitoring Biodiversity*, ed. D. Agosti, J. Majer, L.E. Alonso, T. Schultz, pp. 122–44. Washington, DC: Smithsonian Institution Press.

Bolton, B. (1994). *Identification Guide to the Ant Genera of the World*. Harvard, MA: Harvard University Press.

Charles-Dominique, P. (1974). Ecology and feeding behaviour of five sympatric lorisids in Gabon. In *Prosimian Biology*, ed. R.D. Martin., G.A. Doyle & A.C. Walker, pp. 131–50. London: Gerald Duckworth and Co.

Chinery, M. (1994). *Insects of Britain and Northern Europe*. London: Harper Collins.

Cooper, R.J. & Whitmore, R.C. (1990). Arthropod sampling methods in ornithology. *Studies Avian Biol.* **13**, 29–37.

Egler, S.G. (1992). Feeding ecology of *Saguinus bicolour bicolour* (Callitrichidae, Primates) in a relict forest in Manaus, Brazilian Amazonia. *Folia Primatol.* **59**, 61–76.

Estrada, A. & Coatesestrada, R. (1991). Howler monkeys (*Alouatta palliata*), dung beetles (Scarabaeidae) and seed dispersal – ecological interactions in the tropical rain-forests of Los-Tuxtlas, Mexico. *J. Tropical Ecol.* **7**, 459–74.

Fitton, M.G. (2002). *Manual of Insect Curation*. London: Butterworth-Heinemann.

Gentry, J.W., Harris, K.L. & Gentry, J.W. Jr. (1991). *Microanalytical Entomology for Food Sanitation Control*, vols. I and II. Melbourne: Association of Official Analytical Chemists (AOAC) International.

Gordh, G. & Headrick, D.H. (2001). *A Dictionary of Entomology*. London: CABI.

Gursky, S. (2000). Effect of seasonality on the behaviour of an insectivorous primate, *Tarsius spectrum*. *Int. J. Primatol.* **21**, 477–95.

Harrison, K. E. & Byrne, R. W. (2000). Hand preferences in unimanual and bimanual feeding by wild vervet monkeys (*Cercopithecus aethiops*). *J. Comp. Psychol.* **114**, 13–21.

Hill, D. A., Lucas, P. W. & Cheng, P. Y. (1995). Bite forces used by Japanese macaques (*Macaca fuscata yakui*) on Yakushima Island, Japan, to open aphid-induced galls on *Distylium racemosum* (Hamamelidaceae). *J. Zool.* **237**, 57–63.

Hoshino, J. (1985). Feeding ecology of mandrills (*Mandrillus sphinx*) in the Campo Animal Reserve, Cameroon. *Primates* **26**, 248–71.

Isbel, L. A. (1998). Diet for a small primate: insectivory and gummivory in the large patas monkey (*Erthrocebus patas pyrrhonotus*). *Am. J. Primatol.* **45**, 381–98.

Johnson, M. D. (2000). Evaluation of an arthropod sampling technique for measuring food availability for forest insectivorous birds. *J. Field Ornithol.* **71**, 88–109.

Kay, R. F. (1984). On the use of anatomical features to infer foraging behaviour in extinct primates. In *Adaptations for Foraging in Nonhuman Primates*, ed. P. S. Rodman & J. G. H. Cant, pp. 21–53. New York: Columbia University Press.

McGavin, G. C. (1997). *Expedition Field Techniques: Insects and other Terrestrial Arthropods*. London: Royal Geographical Society.

(2001). *Essential Entomology: An Order by Order Introduction*. Oxford: Oxford University Press.

New, T. R. (1998). *Invertebrate Surveys for Conservation*. Oxford: Oxford University Press.

Nickle, D. A. & Heymann, E. W. (1996). Predation on Orthoptera and other orders of insects by tamarin monkeys, *Saguinus mystax mystax* and *Saguinus fuscicollis nigriforns* (Primates: Callitichidae), in north-eastern Peru. *J. Zool.* **239**, 799–819.

Niemitz, C. (1984). Synecological relationships and feeding behaviour of the genus *Tarsius*. In *Biology of Tarsiers*, ed. C. Niemitz. pp. 59–76. Stuttgart: Gustav Fischer Verlag.

Oliver, I. & Beattie, A. J. (1996). Designing a cost-effective invertebrate survey: a test of methods for rapid assessment of biodiversity. *Ecol. Appl.* **6**, 594–607.

Ozanne, C. M. P. (2003a). Sampling forest canopy arthropods. In *Insect Sampling*, ed. S. Leather. Oxford: Blackwell Scientific Ltd, in press.

Ozanne, C. M. P. (2003b). Sampling arthropods from forest understorey vegetation. In *Insect Sampling*, ed. S. Leather. Oxford: Blackwell Scientific Ltd, in press.

Payne, J. & Francis, C. M. (1985). *A Field Guide to the Mammals of Borneo*. Sabah, Malaysia: Sabah Society.

Peres, C. A. (1992). Prey-capture benefits in a mixed-species group of Amazonian tamarins, *Saguinus fuscicollis* and *Saguinus mystax*. *Behav. Ecol. Sociobiol.* **31**, 339–47.

Schowalter, T. D., Webb, W. J. & Crossley, D. A. Jr. (1981). Community structure and nutrient content of canopy arthropods in clearcut and uncut forest ecosystems. *Ecology* **62**, 1010–19.

Shiel, C., McAney, C., Sullivan, C. & Fairley, J. (1997). *Identification of Arthropod Fragments in Bat Droppings*. London: Mammal Society Publication no. 17.

Southwood, T. R. E. & Henderson, P. A. (2000). *Ecological Methods*. Oxford: Blackwell Sciences Ltd.

Stewart, A. J. A. & Wright, A. F. (1995). A new inexpensive suction apparatus for sampling arthropods in grassland. *Ecol. Entomol.* **20**, 98–102.

Stork, N. E. & Hammond, P. (1997). Sampling arthropods from tree-crowns by fogging with knockdown insecticides: lessons from studies of oak tree beetle assemblages in Richmond Park (UK). In *Canopy Arthropods*, ed. N. Stork, J. Adis & R. Didham, pp. 3–26. London: Chapman & Hall.

Sutherland, W. J. (1996). *Ecological Census Techniques: A Handbook*. Cambridge: Cambridge University Press.

Tilling, S. M. (1987). *A key to the major groups of British invertebrates. Field Studies*, **6**, 695–766.

Tutin, C. E. G. & Fernandez, M. (1992). Insect-eating by sympatric lowland gorillas (*Gorilla gorilla gorilla*) and chimpanzees (*Pan t. troglodytes*) in the Lopé Reserve, Gabon. *Am. J. Primatol.* **28**, 29–40.

Valderrama, X., Robinson, J. G., Attygalle, A. B. & Eisner, T. (2000). Seasonal anointment with millipedes in a wild primate: a chemical defense against insects? *J. Chem. Ecol.* **26**, 2781–90.

Yamagiwa, J., Mwanza, N., Yumoto, T. & Maruhashi, T. (1994). Seasonal change in the composition of the diet of Eastern lowland gorillas. *Primates* **35**, 1–14.

List of suppliers and their Internet sites

Britain

Watkins and Doncaster, PO Box 5, Cranbrook, Kent TN18 5EZ, UK. Tel: + 44 (0) 1580 753133; Fax: + 44 (0) 1580 754054; website: <www.watdon.com>.

Marris House Nets, 54 Richmond Park Avenue, Queens Park, Bournemouth, Dorset BH8 9DR, UK. Tel: + 44 (0) 1202 515238; Fax: + 44 (0) 1202 510303; website: <www.pwbelg.clara.net/marris> (nets, rearing cages, etc.).

Australia

Australian Entomological Supplies Pty Ltd, PO Box 250, Bangalow NSW 2479, Australia. Tel: + 61 (0) 2 6684 7188; Fax: + 61 (0) 2 6684 7188; email: austento@nor.com.au; website: <www.entosupplies.com.au>.

North America

BioQuip Products, 17803 LaSalle Avenue, Gardena, CA 90248–3602, USA. Tel: + 1 310 324 0620; Fax: + 1 310 324 7931; e-mail: bioquip@aol.com; website : <www.bioquip.com> (entomological and botanical equipment, books, software).

Other useful Internet sites

Tree of Life: <tolweb.org/tree?group=Animals&contgroup=Eukaryotes>.

15 • Tape-recording primate vocalisations

Anthropological Institute and Museum, University of Zürich, Zürich, Switzerland

INTRODUCTION

Ornithologists have been exploring the possibilities and methodology of tape-recording and archiving animal sounds for many decades. Primatologists, however, have only recently become aware that tape-recordings of primate sound may be just as valuable as traditional scientific specimens such as skins or skeletons, and should be preserved for posterity. Audio recordings should be fully documented, archived and curated to ensure proper care and accessibility. As natural populations disappear, sound archives will become increasingly important.

This article describes how to tape-record non-human primate vocalisations. It provides information on the advantages and disadvantages of various types of equipment, and gives tips for better recordings of primate vocalisations, both in the field and in the zoo. Ornithologists have to deal with very similar problems, and their introductory texts are recommended for further study (e.g. Budney & Grotke, 1997; Kroodsma *et al.*, 1996; see also 'Internet sites with review texts on recording nature sounds' at the end of this chapter). Methods and equipment required for sound analysis are not included, as information on that topic has been presented elsewhere (Charif *et al.*, 1995; Hopp *et al.*, 1998).

As a rule, prices for sound equipment go up over the years. Prices for equipment discussed below should only be used as very rough estimates. Retailers' prices differ widely and comparing prices over the Internet before buying is recommended.

SOUND RECORDERS

An ideal recorder records a signal without alterations, by matching its dynamic and frequency range and preserving all its features. Traditional analogue tape-recorders, both compact cassette and reel-to-reel recorders, are not perfect. They degrade the signals they record by adding hiss, distortion,

Field and Laboratory Methods in Primatology: A Practical Guide, ed. Joanna M. Setchell and Deborah J. Curtis. Published by Cambridge University Press. © Cambridge University Press 2003.

228

frequency response alterations, speed variations, print-through effects and drop-outs. Digital recorders do not have these problems. Within the dynamic range and the frequency limits (owing to the number of bits and sampling frequency they use), they record and reproduce signals with great accuracy, low noise, flat frequency response and no speed variations. Nevertheless, the quality of the signal recorded by good analogue tape-recorders is good enough for most purposes in bioacoustic studies.

ANALOGUE TAPE-RECORDERS (AUDIO)

Avoid the use of noise reduction features such as Dolby or DBX during recording, as they add distortion and limit the high frequencies of many sounds. If possible, use a tape-recorder with adjustable sensitivity for signal intensity. The automatic level control (ALC) found on many portable cassette recorders is often ineffective for recording sounds in the field. Unfortunately, there are very few portable machines with adjustable sensitivity.

The Sony Walkman WM-D6 is small (approximately 0.64 kg, with batteries), relatively affordable (about US$ 360), and works well in rough field conditions. However, it uses bar-type metering or light emitting diodes (LEDs). The LEDs are difficult to read in bright light; in addition they are positioned on the top rather than on the side facing you as you record, making them difficult to see while recording. The mini-jack microphone connectors are the weakest part of the machine and broke repeatedly in some machines used by my students and myself. Their survival can be prolonged by avoiding lateral pressure acting on the microphone plugs while they are connected to the Walkman and by disconnecting the microphone when not using it. Despite these negative comments I have used the WM-D6 in the field for many years and can recommend it. It should be good enough for most purposes, but has no built in speaker.

The heavier Sony TC-D5M and TC-D5 Pro II (approximately 1.7 kg, with batteries) exhibit more useful features. The former comes with $\frac{1}{4}$ inch jack microphone connectors and costs about US$ 700, the latter has Cannon-XLR microphone connectors instead and costs US$ 1000. As I have never had problems with the $\frac{1}{4}$ inch jack connectors, I prefer the cheaper of these two. Both are well designed and very durable (with metal rather than plastic housing, and a leather carrying case). They have both volume unit (VU) and peak metering; the former is an averaging system for determining recording levels, the latter is instantaneous and is important for sound recording where there may be very sudden sound level shifts that can cause overloading. These recorders come with a built-in loudspeaker, which makes it much more comfortable to check your recordings, and it can be used for simple playback experiments. However, these speakers exhibit relatively low sound pressure levels and are not recommended for serious playback experiments with primates.

A single channel Sony, the TCM-5000EV (about US$ 500), is also available (also with built-in speakers) but has mini-jack connectors (see above). It also has a built-in speaker, which can be used as a playback (see above) or monitor speaker. It is intermediate in price and most features between the Walkman and the more expensive machines described above. Unfortunately, it does not allow you to use the advantages of different types of cassette (see below).

Several portable cassette recorders with adjustable sensitivity are also available from Marantz (PMD-201 and 222, and CP 430). These machines offer features similar to their Sony counterparts described above, but are generally less expensive (US$ 350, US$ 420 and US$ 550, respectively). Only VU metering is provided (no LEDs). Carrying cases are not included but are available. However, the extensive use of plastics in the outer shell and the way some (plastic) function keys are exposed makes them less sturdy.

The decision between stereo and mono depends on what your recording will be used for. For most field recordings, stereo is an unnecessary complication. However, if you are interested in recording vocal interactions between several individuals or groups, then a stereo machine is preferable.

DIGITAL AUDIO TAPE (DAT)

The DAT recording system stores sound information in a binary code, thereby making it immune to speed errors, tape noise (hiss) and non-linear frequency response problems. The frequency response of DAT recorders allows very good recordings from low frequency signals, as low as 10 Hz, up to 22 kHz. Unfortunately, they can not be used to record ultrasounds.

The DAT recorder delivers a sound quality slightly better than that of a compact disc (CD) in a small, easy to use and easy to store, long duration, tape-based format. In addition, digital transfer to and from a computer allows duplication, editing and analysis in an entirely digital domain. Digital transfer from DATs to computers is possible using specific soundboards with digital input/output (I/O), which are now produced by several companies. Other features include real-time tape counters and a built-in indexing system, which allow the operator to access the starting points of the various recordings quickly.

DAT recorders are more expensive than analogue tape-recorders and are available from Sony (TCD-D10 US$ 1000), Tascam (DA-P1 US$ 1500), and Fostex (PD-4 US$ 5400). Unfortunately, DATs are generally very sensitive to high humidity and are, therefore, not reliable in some recording environments such as rain forest. Most DAT recorders have a built-in 'dew', or humidity, sensor that shuts the machine down whenever high humidity is detected. Most DAT machines will operate for only 2 hours per battery charge, and only accept rechargeable cadmium batteries, a problem when you are working in areas with no or unreliable access to mainline power.

MINIDISC (MD), DIGITAL COMPACT CASSETTE (DCC), MP3

Avoid working with technologies that remove information from the sound, such as MiniDisc, Digital Compact Cassette and Mpeg Layer 3 (MP3). There is also a forthcoming MP4 protocol as well as rival formats, such as Liquid Audio and VQF (also called TwinVQ, transform-domain weighted interleave vector quantisation). All these technologies use sound compression algorithms, discarding sound details that appear to be non-audible. For instance, MiniDisc technology incorporates input filtering and data compression strategies deliberately designed to distort the signal in subtle ways. This makes the signal more difficult to analyse and impossible for accurate frequency-related measurements. It also leads to perceptible distortion if multiple copies are made. Any such technology will change (i.e. degrade) the sound you are recording and make your recordings unsuitable for sound analysis.

WHAT TO BUY – WHERE IS THE FUTURE?

It is obvious from changes in the product lists of the major producers of audio equipment that both analogue audiocassette recorders and DAT are on their way out. Whereas analogue cassettes will still be around for years to come, DAT technology never built up an equally large user base and may disappear faster. Only a few good portable machines of either technology are still being sold, mainly for specialists, and prices for these machines will probably go up as the market dries up. The future is in digital media. Unfortunately, currently popular or announced technologies such as MiniDisc, MP3 or MP4 are not an option for bioacoustic analyses. Direct-to-hard-disc recording may be the option of the future, but no easily manageable and robust solution is available for long-term use in the rain forest. For the time being, at least, this author is sticking to analogue recording using audiocassettes.

MICROPHONES AND PARABOLIC REFLECTORS

If possible, use a directional microphone or a parabolic reflector (also called a parabola). This improves sound collection in nature considerably by attenuating unwanted ambient noise and sounds from other directions (directional microphones) or by giving emphasis to sounds coming frontally (parabolic reflectors).

Directional microphones are the most useful in bioacoustic field recordings. Various names are in use for microphones with various degrees of directionality. Here, I will simply differentiate between shotgun or ultradirectional microphones and semi-directional microphones.

SHOTGUN OR ULTRADIRECTIONAL MICROPHONES

A shotgun microphone is a cardioid microphone fitted with an interference tube on its frontal face. It is characterised by a flat frequency response, is less sensitive to wind and handling noise but offers a lower sensitivity than a microphone mounted in a parabola. The interference tube cancels off-axis signals, while the in-axis signals reach the microphone's diaphragm without attenuation or gain. As a rule, shotgun microphones work in mono only, so you would need two of them if you wanted to make a stereo recording.

There are many brands of high quality microphones available, but they are relatively expensive. As a simple rule, the degree of directionality in shotgun microphones correlates with their length (and, usually, with their price). I have tested various models of the Sennheiser line of microphones. They are expensive but have proven to be very durable, exhibit low noise figures, and are relatively immune to high humidity.

Particularly useful in the field is a combination of the ME 66 microphone capsule and the K6 pre-amp/power unit using an AA battery (about US$ 180 and US$ 215, respectively). Put together, this system has a length of 32 cm. For better directionality, you can plug the longer ME 67 (about US$ 250) onto the K6, with a combined length of 44 cm. The top-of-the-line Sennheiser shotgun microphones have particularly flat frequency curves but are more expensive (MKH 416: US$ 830, 25 cm; MKH 70: US$ 1550, 41 cm). Longer microphones are not very handy in dense tropical forests – they get in the way and are more easily damaged than shorter microphones.

Directional microphones by Audio-Technica are less expensive than the Sennheisers, and still exhibit good directionality (e.g. AT 835: US$ 330, 23 cm). They also come equipped with a windscreen.

Note that you also need special microphone cables with appropriate connectors for some Sennheiser directional microphones (available separately, about US$ 30). On windy days, a rubber foam windscreen may be necessary for your directional microphone. These function to keep wind turbulence as far away from the microphone surface as possible. The cheapest option for the Sennheiser ME 66 is MZW 66. This is relatively expensive (about US$ 30), although it is basically just a tight fitting glove of rubber foam for your microphone. A similar windscreen of your own manufacture should work reasonably well and costs almost nothing.

SEMI-DIRECTIONAL MICROPHONES

Semi-directional microphones produced for video cameras are less directional than the shotgun microphones, but they are cheaper and smaller. They may suffice if you are tape-recording 'loud-calls' of primates or if you are studying captive animals, where background noise is frequently less of a problem and where the microphone can usually be positioned closer to the vocalising animals. However, the degree of directionality and other quality characteristics vary strongly (Wölfel & Schoppmann, 1994). In order to tape-record gibbon (Hylobatidae) songs, leaf monkey (Presbytini) loud calls and macaque (*Macaca* spp.) vocalisations in the field, I have used the following microphones with good results: JVC MZ-707, and Sennheiser MKE 300 (both below US$ 200). Furthermore, these video-microphones can be plugged directly into the Sony WM-D6C, with no special connecting cable needed (unlike some of the directional microphones from Sennheiser), but I recommend buying a cheap elongation cable (1.5 m, US$ 6). The microphone should never be too close to the cassette recorder while recording; otherwise you might have the noise of the machine on your tape-recordings.

Be careful to select a microphone with a frequency range suitable for your needs. The frequency range of some of the cheaper video-microphones does not exceed 10 kHz, which makes them less suited to tape-recording primates with high pitched vocalisations (e.g. callitrichids).

PARABOLIC REFLECTORS

A parabola focuses incoming sound waves that are parallel to its axis onto a single point, the focus, where a microphone is placed. Whereas a shotgun microphone simply screens off sound from directions other than the sound source, a parabolic reflector can actually amplify the sound from the target direction. In addition, it also acts as a high pass filter, especially for low frequency environmental sounds. The effectiveness is determined by the diameter of the reflector in relation to the wavelength of the sound. Gain and directivity increase proportionally as the diameter:wavelength ratio increases. For wavelengths larger than the diameter of the parabola, the response is predominantly that of the microphone itself. As the wavelengths become smaller than the parabola diameter, gain and directivity increase with the frequency. Common diameters are 45 cm, 60 cm and 90 cm, with directionality starting at about 750, 550 and 375 Hz, respectively. Only the larger dishes can be recommended for primate studies: they are able to cover the frequency range of calls produced by the smaller primate species, at least. Many larger species, however, produce calls that go well below 375 Hz.

It is obvious that the size of a parabolic reflector makes it difficult to use in tropical forests, even if some plastic parabola (Telinga) can be rolled up for travelling. Parabola may be of better use in stationary recording sites or in more open forests and in open habitat.

There are a variety of parabolas available made of metal, fibreglass or clear plastic. Microphones for use in parabolic reflectors typically cost anywhere between US$ 300 and 1500. Some manufacturers, such as Telinga and Saul Mineroff Electronics, sell complete systems that include a parabolic reflector with a microphone and shock-mount system.

If you are working on a tight budget, you might consider building a parabolic reflector yourself. With a large bowl- or umbrella-shaped dish, you can add some directionality to any non-directional microphone. The microphone is positioned in the focal point of the reflector, which, fortunately, is not very narrowly defined. The sound is concentrated in this area, and increases sensitivity.

OTHER EQUIPMENT

Many ornithologists recommend the use of good-quality headphones that allow you to listen while recording, making it easier to aim the microphone, as well as giving a clear idea of the quality of the recording being made. I find it inconvenient, however, to monitor the vocal activities of primates, often calling from several directions at the same time, with earphones on.

I recommend using quality audio-cassettes (e.g. TDK, Maxell). Much cheaper cassettes are available, but they may yield fuzzy recordings and some frequency levels may be underrepresented. Of the available cassettes (types I, II and IV), type II tape offers the best high frequency response and the lowest signal-to-noise ratio. Interesting vocalisations tend to occur exactly when the cassette is full, and you will miss them while swapping cassettes. I therefore prefer cassettes that allow you to record for 90 or 100 minutes (instead of 60 minutes). Shorter tapes, on the other hand, are more time accurate, because the tape is thicker.

Always carry a set of new, non-rechargeable batteries with you, for emergencies. Your animals often decide to vocalise after you have spent hours tape-recording background noise, and all your rechargeable batteries have expired. In such cases, an emergency set of batteries will come in handy.

ULTRASOUND

A number of small primates (e.g. *Cebuella, Galago, Microcebus, Nycticebus*) have been shown to emit ultrasound vocalisations above 20 kHz (Pariente, 1974; Pola & Snowdon, 1975; Glatston, 1979; Zimmermann, 1981; Cherry et al., 1987; Zietemann, 2000). The fundamental frequencies of some note types of *Microcebus murinus* reached maximum values of about 45 kHz (Zietemann, 2000). Because detecting, recording and analysing ultrasound vocalisations requires special equipment, such calls may be more widespread among small primates (and other small mammals) than has been documented so far. Fortunately, much research is being carried out on ultrasound calls of bats, and scientists wishing to study ultrasound vocalisations of primates can benefit greatly from exploring the expertise of, and the technologies being used by, bat researchers.

BAT DETECTORS

Bat detectors or ultrasound detectors were developed to provide researchers with instruments to study bat echolocation, but are also used for research on other small mammals and grasshoppers. Bat detectors are based on both analogue and digital techniques to detect and record ultrasounds and transform them into audible sounds. The detectors available on the market use three main systems: heterodyne frequency shifting, frequency division and time expansion.

Heterodyne detectors and frequency division (or count-down) detectors are real time methods (i.e. you hear the sound from the detector at the same time as it is emitted by the bat). They are useful for quick yes/no statements of bat activity. They allow the recording of an audible representation of an ultrasonic call, not of the full ultrasonic signal structure.

If you want to study the frequency spectrograms, you will need a time expansion detector (Petterson D980: US$ 3500; Ultra Sound Advice: U30 + PUSP: US$ 2867). The method is similar to making a high speed tape-recording of the sound and then playing it back at a lower speed. The ultrasonic signal is sampled at high speed, and then replayed at a lower sample rate, for example one tenth, to be made audible, and stored in this format on a tape-recorder. If the signal is stored at a sample rate 10 times lower than the original, frequencies are reduced by 10 while time is expanded by the same factor. Clearly this is not a real time conversion method, but it does offer a number of benefits over and above these methods. Since the signal is stretched out in time, it is possible to hear details of the sound not audible with other types of detector (e.g. you can actually hear frequency differences in single short pulses). Time expansion preserves all characteristics of the original signal, making time-expanded signals ideal for sound analysis in the laboratory.

DIGITAL DATA ACQUISITION

A second method of preserving all characteristics of the original signal involves recording the ultrasound directly into the memory of a computer-based digital signal acquisition system with fast, high capacity data storage. Modern laptop computers allow for direct high speed sampling of ultrasonic signals using a separate data acquisition card. With this method, the signal is available for immediate sound analysis, and the sound quality is usually better than in analogue recordings.

In modern systems, several seconds of sampled sound are stored in a circularly addressed memory ('circular buffer'). The maximum duration of the stored sound depends on the available memory and the selected sampling rate. The advantage of this system is that you can decide *post hoc* whether you want to keep the current section of your random access memory (RAM) or whether you want to continue sampling. With this convenient method, the recorded sound is constantly being deleted and memory freed while you monitor the sound activity of your study animals, until you register some interesting vocal signals.

No time expansion is required with this method. You need a bat detector with high frequency output (Petterson D980: US$ 3500; Ultra Sound Advice: U30 without PUSP: US$ 668). Unlike the Petterson equipment, the bat detector (U30) and time expansion (PUSP) unit are separate units in the Ultra Sound Advice system. Because you need only the bat detector for digital data acquisition, the Ultra Sound Advice solution is cheaper.

The bat detector should be connected to a high-speed A/D board in a laptop computer equipped with a large RAM and specialised software (e.g. Batsound Pro, 128 MB recommended). A complete solution (without the laptop) is available from Pettersson Elektronik AB (Sweden) for about US$ 6000.

Obviously, ultrasound equipment is more expensive than that required to record audible sound and also less suitable for fieldwork in tropical forests. Of the two recording techniques described above, time expansion recording on audiotape is more robust in humid tropical environments than the laptop-based system.

RECORDING PROCEDURE

Get into the habit of making frequent comments on tape while recording your animals. This will be very helpful later when you analyse your recordings. For instance, if several individuals are vocalising together, it is often impossible to determine who made which vocalisation later on. It is often important to record information about the identity, sex and age of a vocalising animal, and the context in which each vocalisation is produced. It is usually impossible to describe a long vocalisation bout in detail from memory *after* the recording has been made. Therefore, do it as the vocal bout goes on but try to speak during intervals between vocalisations. Of course, it is best not to voice input at the same time as your target animal is calling.

Always narrate 'stop' or some other indication that the recorder is turned off before stopping. The lack of critical voiced information on tape, especially where one recording ends and another begins, is one of the greatest failings of the beginner. If you interrupt a tape-recording ('cut'), say so on the tape. Otherwise, it is often difficult to detect that the recorder was paused and confusion can occur during later analysis. If you continue recording the same or new target subjects following the cut, add some narration to say so.

Make long, uninterrupted recordings. Vocalisations tend to occur when you think they won't. If you are waiting for a vocalisation to occur, it may be worthwhile recording continuously, even while your animals are not vocalising. By doing so, you will catch complete vocalisations or vocal bouts, of which you would otherwise miss the beginning. If no vocalisations occur, rewind your cassette from time to time. If you already have vocalisations on the same side of the cassette on which you intend to record, set the rotation-counter to zero before starting to record again. This will allow you to rewind to the end of your last recording when you have been tape-recording for some time with no vocalisations.

Hold the microphone very firmly and steadily. Even moving your fingers may interfere with recordings.

Record from as close to the subject as possible. Directional microphones and parabolic reflectors are no substitute for proximity. Very soft, quiet vocalisations are lost in ambient noise if you are far away from the animal. Occasionally, closer may not be better. If you encounter a loud sound source, such as calling cicadas, somewhere between you and your target animals, going closer would bring you closer to the cicadas as well. In this case, it may be better to change position.

When the side of an audio-cassette is filled with recordings, always break out the record-enabling chip on the upper left, thin side of the cassette, thus making sure that you will not accidentally record on this side again and delete important material. You may decide to give provisional field numbers to your tapes and give them final inventory numbers when you are back in the laboratory. In any case, it is important that tapes can be identified at any time and referred to in your field notes.

Try to be consistent in your methods and the way that you record data. That way, someone else listening to your recordings later can learn your pattern, in your absence, and retrieve important information.

Recordists frequently overlook the importance of adequate documentation of recordings, often obtained with painstaking effort and sometimes at great expense. Lack of documentation seriously weakens the scientific value of your sound recordings. After the recording has been made, the following information should be narrated on tape:

1. Recordist and equipment configuration: name of the recordist, type of recorder, microphone and parabolic reflector.
2. Identification of target subject and the degree of certainty: name of species, type of contact (heard only, seen), number of target subjects on record, sex and age class of target animal(s) or group composition (if known), approximate distance of subject from microphone, cross-references if same individual can be heard on previous recordings.
3. Time and location: date; time at the beginning and the end of recording, time at important occasions (swapping cassettes, calls of new individuals, etc.), location (accurate locality, district, province, country).
4. Other information: type of habitat, weather conditions, behaviour of subject, identification of other sounds in the background.

Try to make a written edit of your tape as soon as possible after recordings are completed. Keep a journal of your recordings detailing the contents of your tapes. You can also add information that was not narrated onto the tape. Editing tape-recordings takes a lot of time but makes your collection much more usable. It is a good idea, while the day's work is still fresh in your memory, to listen to your tape(s), and make a written record of what you have recorded.

Keep your personal collection of wildlife sounds well documented and in one place. Clearly label both your tapes and the storage box.

ACKNOWLEDGEMENTS

I am grateful to Dr Deborah J. Curtis, Dr Sabine Schmidt, Robert Dallmann and Marina Davila Ross for reading and commenting on earlier versions of this manuscript.

REFERENCES

Budney, G. F. & Grotke, R. W. (1997). Techniques for audio recording vocalizations of tropical birds. *Ornithol. Monogr.* no. 48.

Charif, R. A., Mitchell S. & Clark, C. W. (1995). *Canary 1.2 User's Manual.* Ithaca, NY: Cornell Laboratory of Ornithology.

Cherry, J. A., Izard, M. K. & Simons, E. (1987). Description of ultrasonic vocalizations of the mouse lemur (*Microcebus murinus*) and the fat-tailed dwarf lemur (*Cheirogaleus medius*). *Am. J. Primatol.* **13**, 181–5.

Glatston, A. R. (1979). Reproduction and behaviour of the lesser mouse lemur (*Microcebus murinus*, Miller 1777) in captivity. Ph.D. thesis, University College London.

Hopp, S. L., Owren, M. J. & Evans, C. S. (eds.) (1998). *Animal Acoustic Communication – Sound Analysis and Research Methods.* Berlin: Springer-Verlag.

Kroodsma, D. E., Budney, G. F., Grotke, R. W., Vielliard, J. M. E., Gaunt, S. L. L., Ranft, R. & Veprintseva, O. D. (1996). Natural sound archives: guidance for recordists and a request for cooperation. In *Ecology and Evolution of Acoustic Communication in Birds*, ed. D. E. Kroodsma & E. H. Miller, pp. 474–86. Ithaca, NY: Cornell University Press.

Pariente, G. F. (1974). Importance respective du réperage visuel et auditif (absence d'écholocation) chez *Microcebus murinus*. *Mammalia* **38**, 1–6.

Pola, Y. V. & Snowdon, C. T. (1975). The vocalizations of pygmy marmosets (*Cebuella pygmaea*). *Anim. Behav.* **23**, 825–42.

Wölfel, M. & Schoppmann, J. (1994). Vergleichstest: Richtmikrophone – Lasst Mikros sprechen. *VIDEOaktiv* **8**, 4–7.

Zietemann, V. (2000). Artdiversität bei Mausmakis: Die Bedeutung der akustischen Kommunikation. Ph.D. thesis, Institut für Zoologie, Tierärztliche Hochschule Hannover.

Zimmermann, E. (1981). First record of ultrasound in two prosimian species. *Naturwissenschaften* **68**, 531.

Internet sites with review texts on recording nature sounds

.
<eebweb.arizona.edu/faculty/hopp/sound.html>.
<www.birds.cornell.edu/LNS/>.
<www.unipv.it/cibra/>.

16 • Photography and video for field researchers

NOEL ROWE AND MARC MYERS

Primate Conservation Inc., Charlestown, Rhode Island, USA

INTRODUCTION

Photography and video are means of collecting visual data that can be extremely valuable for research, as well as being useful for presentations, teaching, or to interest more general audiences. Good imaging conveys the quality of light, emotion and the essence of the subject to the viewer. The ability to see what is in the viewfinder and know the moment to capture the image is a skill that comes with practice. Although you look through a viewfinder to frame a shot with both still and moving images, in practice each requires a very different set of skills and thought processes. This chapter covers photography first, moves on to video, then discusses protection of equipment and travelling with photographic gear.

STILL PHOTOGRAPHY

TYPES OF STILL CAMERA

Advances in camera technology have greatly simplified the process of obtaining a sharply focused and well-exposed picture. There are three types of commonly used still cameras: point and shoot cameras (including Advanced Photo System (APS) cameras), digital still, and single lens reflex cameras. Every camera has its advantages and disadvantages, and it is important that you choose one that will record the kind of picture you require and that is within your budget.

Point and shoot cameras (PSC)

APS cameras are the smallest and lightest PSCs, and cost US$ 50–250. They have an autofocus lens, built-in flash and weigh 140–220 g. They use film sizes smaller than 35 mm, which record correspondingly less image information per exposure. There is only a limited choice of print films and there are no slide films.

Field and Laboratory Methods in Primatology: A Practical Guide, ed. Joanna M. Setchell and Deborah J. Curtis. Published by Cambridge University Press. © Cambridge University Press 2003.

Compact PSCs use 35 mm film and are also small, light and easy to carry. They cost US\$ 40–300 and come with an autofocus, single focal length lens (cheaper) or a zoom lens (more expensive). Zooms commonly extend from 38 mm (mild wide angle) to 90 mm (portrait) or 135 mm (mild telephoto). Two drawbacks are that the closest focus is at 60 cm, and that when you zoom out to get a useable-sized image of a distant subject, very little light penetrates to the film and the flash turns on, although the flash illuminates only up to 2.5 m. Some PSCs are weatherproof (designed to withstand splashes of water) and other models are waterproof to a depth of 3 m, both of which are useful in tropical forest and when travelling on water.

PSCs can be useful for documenting camp life and local people, but you will generally not get good pictures of primates. However, on days when you do not want to carry your full photographic equipment with you, take your PSC – it is better than no camera at all. Finally, special versions of PSCs can be set up along trails, at nest holes or at special food sites. These camera traps are waterproof and will fire (usually with flash) when triggered by motion or an infrared sensor. A black-and-white digital version uses infrared light without a detectable flash (see 'Useful Internet sites,' at the end of this chapter).

Digital cameras

These represent the photography of the future, but digital scans of slide film currently contain more usable information than images captured by even the best digital cameras. In a digital camera, the image is captured by a charge-coupled device (CCD), made up of a fine matrix of light-sensitive picture elements (pixels), each of which generates an electric charge proportional to the intensity of the impinging light. These charges are read out successively to the camera's microprocessor, which transforms the image into digital bits that are then stored (Johnson, 2001). The greater the number of unduplicated pixels, the higher (sharper) the picture's resolution will be. When printing a photograph you need to translate pixels into dots per inch (DPI), by dividing the number of pixels by the physical printing size. Good-quality prints need 300 DPI or better, depending on the printer. One megapixel (1280 × 960 pixels) is enough resolution to print a photographic quality 3 inch × 5 inch (8 cm × 13 cm) print (Johnson, 2001).

The viewfinder of a digital camera will not show you exactly what you are recording when taking a close up, so you will need a camera with a liquid crystal display (LCD) to see the true image. LCDs are difficult to see in bright light and use a great deal of battery power. However, they provide a major advantage of digital technology: you can see what you have taken immediately and reshoot if the subject permits. At present, the greatest challenge for digital pictures is image storage. If you take more than a few images (particularly large, high quality images) you will need many, relatively expensive, removable media cards, Internet access, or a computer with a spacious hard drive, and a colour printer. If you are using a computer for storage, you will need all the appropriate cables, card readers and software (Johnson, 2001). As with some PSCs, there is a delay between when you push the shutter button and when the shutter actually opens. Often what the photographer thinks is a good picture has in fact been taken a fraction of a second later and the moment is lost.

The only digital cameras with interchangeable lenses are professional models (US$ 4–5000 without lens). Otherwise, digital cameras are PSCs with a flash, and have varying amounts of zoom. The optical zoom rating should cover the ranges you need. Digital zoom digitally enlarges the central portion of the picture and the quality suffers noticeably, so disregard this specification (Johnson, 2001).

Single lens reflex (SLR)

This is the standard camera for fieldwork. The advantage of an SLR is that what you see in the viewfinder is what is recorded on the film. They have fast, interchangeable lenses, ranging from super wide angle (20 mm) to long telephotos (600 mm). However, they are heavy (0.3–1.3 kg without a lens), bulky, and a long telephoto lens is expensive (US$ 3000–6000), heavy (up to 2.7 kg) and needs a tripod (2.3–3.6 kg) to support it in order to take sharp pictures. We suggest buying a brand name camera body (Nikon, Canon, Minolta, Pentax, or Olympus) with manual focus and exposure, and exposure compensation (enabling you to give more or less exposure than is set automatically by the camera). Autofocus can be useful, but manual focussing is often necessary because the autofocus may not be able to focus correctly on a subject in a tree. If available, choose a model with AA batteries rather than one that depends upon expensive lithium batteries, which can be difficult to obtain. Buy extra batteries and remember that extensive use of flash reduces battery life.

ZOOMS, TRIPODS AND MACROS

A 35–200 mm zoom lens will cover most subjects. The wide-angle (35 mm) is useful for photos of the forest and inside buildings. A serious photographer would want two zooms: a wide angle (18–70 mm or 24–70 mm) and a telephoto (70–200 or 70–300 mm). To save money you may want to buy a Sigma or Tamron lens – just make sure that it is compatible with your camera and that the metering works correctly. As a rule these lenses are comparable to name brands in sharpness, but are not built for durability and hard use.

If you have the money then a longer telephoto will be useful. Canon and Nikon make image stabilised 100–400 mm zooms that allow you to take pictures without a tripod at an exposure of 1/60 second (approximately US$ 1500). It is difficult to hand-hold any telephoto lens over 200 mm steady enough to get a sharp picture. The rule is that to hand-hold a camera without shaking the shutter speed should be equal to the focal length of the camera. If you are using a 200 mm lens then you need to shoot at 1/250 second to get a sharp picture. Most people cannot hand-hold any camera for longer than 1/30 second. A tripod gives the option of using slow shutter speeds to get sharp pictures and more depth of field. Use a cable release to trigger the shutter without touching and possibly shaking the camera (Shaw, 2000). Weight adds stability, so most good tripods are between 2 and 4 kg. A monopod may be an effective compromise and is easier to set up and carry around dense forest.

Many zooms have a close up feature that is not a true macro. If you need to take close-ups, buy a 105 mm or 200 mm macro lens. These are more useful than a 50 mm macro because they allow you to be farther away from the subject, meaning that you will have fewer problems with light availability and are less likely to disturb an animal subject.

FLASH AND NIGHT PHOTOGRAPHY

Many SLRs come with a built-in flash, but these often cause 'red eye' effect and are not useful for close-up work. Buy a separate flash unit that is compatible with the through-the-lens (TTL) metering system of your camera. A flash cord enables you to take the flash off the camera to light close-up subjects. Practise with the flash before you leave for the field to find the best f stop (see below) for the best depth of field, and the best position of the flash to your subject.

Flash can often be used to fill in the detail of a primate in shade or backlit in daylight (Hill & Wolfe, 1993). Night photography presents a special challenge. You will need a more sophisticated autofocus camera and a powerful TTL flash for anything that is not very close and co-operative (Shaw, 2000). Nikon and Canon make combinations of these that use infrared to focus in the darkness, but you still need enough light to aim the camera. A helper holding a flashlight covered with red cellophane is often necessary to illuminate the subject. Nocturnal animals' eyes are usually less sensitive to red light, but the blinding flash of white light is likely to disturb them. An advantage of using a flash is that it freezes motion far more effectively than the 1/125 second shutter speed might suggest. Most nocturnal primates have a tapetum lucidum that reflects red, orange or yellow when a flash is used.

APERTURE (F STOP)

The aperture or f stop of a lens (usually f 2.8 to f 22 on a wide angle or standard lens) determines the depth of field, i.e. how much of the foreground to the background is in focus. The most depth of field is provided by f 22, with almost everything in focus, whereas f 2.8 will mean that perhaps only a primate's eyes are in focus. Telephoto lenses generally give you less depth of field, even if you are using f 22, and the largest aperture available on such lenses is often f 4. Shutter speed works inversely with f stop, so that, for any given amount of light, as one goes up the other must go down to get a good exposure. For instance, to expose a 100 ISO film correctly on a sunny day you can take the picture at f 16 and an exposure of 1/60 second for much depth of field or f 4 at 1/1000 second with little depth of field. The reality of photographing primates is that you have little choice of f stop. You are usually shooting with the aperture of the lens wide open at your lowest f stop (f 2.8, f 4 or f 5.6, depending on the lens) so that you can use the fastest possible shutter speed to get a sharp picture (the slower the shutter speed the more the camera shakes, making the photograph blurred).

Remember that the camera's meter does not know what you are photographing. The meter expects to read an average of a normal scene with the sun behind you, so if it is pointed at a dark primate against the sky it will expose for the bright sky. To get a good picture of the primate you need to increase the exposure compensation by 1.5–2 f stops, but remember to change it back to normal for typical conditions. If the camera has a spot meter it may help to get a better exposure if pointed at the primate with little or no sky. If the picture is important to you and you have enough film, then bracket exposures – take three successive exposures, one as measured by the camera and one each over- and underexposed by one f stop at the same shutter speed – to make sure that you have a good exposure. Some cameras have an automatic bracket function, a useful feature, particularly when using slide film.

FILTERS

Three filters can improve the performance of colour film. An ultraviolet (UV) filter can reduce haze, and provides useful protection for the lens. A polarising filter will minimise reflections and consequently enhance the colours in the photograph, making leaves greener and the sky bluer. This filter requires up to 1.5 stops extra exposure. A warming filter, 81A, can take some of the blue out of people photographed in shade (Shaw, 2000).

CHOICE OF FILM

There are at least 89 different films on the market. The first choice is slide (transparency) film, not print (negative) film. Slide film holds the most visual information, is the most colour accurate (important for identification of animals, plants, flowers and fruits), and what you see is what the camera recorded, whereas a print is an interpretation by a printer or machine. Slides can be duplicated, printed, projected and scanned, and are preferred by stock agencies and photograph libraries. The one disadvantage of slides is that the exposure in the camera has to be correct (within $\frac{1}{2}$ an f stop) for the slide to be useable, otherwise it will be too dark (underexposed) or washed out (overexposed). If you must, err towards underexposure with transparency film.

Print film has much more latitude in exposure. I (N.R.) have bracketed exposures and could barely discern any difference in the print. In this case, if you are not sure of the exposure, overexpose the film by 1 f stop. Prints are great to give to people in the local community, and can often be developed and printed satisfactorily in the habitat country, whereas slides are more exacting. Modern scanners digitise well-exposed negatives, prints or slides with comparable results.

Take more film than you think you will need. If you bracket your exposures, you will get only 12 pictures rather than 36 on a roll. Buying film and mail-in processing envelopes from companies that advertise in most photography magazines is usually the cheapest option although you may have no chance of recovering the prints if they are lost in the post. In developing countries, film tends to cost more, choice may be limited, and storage conditions (high temperatures/humidity) may result in unpredictable results.

Film speed (ISO)

The slower the film speed (ISO 50–100) the better are the sharpness and resolution, and the finer is the grain of the film (Hill & Wolfe, 1993). Most wildlife photographers use ISO 50–200 unless there are really special circumstances that can be shot only with ISO 400. ISO 400 and higher films are usually too grainy for publishable results, but may be necessary to get ID shots of monkeys in low light conditions. Slide film with the same name and ISO may come in two versions: regular and professional. Professional is more expensive and usually means the film should be refrigerated and processed quickly for optimum results.

IMPORTANT THINGS TO REMEMBER FOR STILL PHOTOGRAPHY

When you get a new piece of equipment, test it immediately with slide film. If the lens or camera does not work properly, return it promptly for exchange or refund. Every camera and meter is different and you may need to change the exposure compensation to get good exposures. Whatever you do, do not buy a camera two days before leaving for the field and assume it will work. It may, but finding out how it works before you go will help you to get better pictures. If you already have a PSC take that too. All cameras can fail and having a back-up means that you can still take pictures.

Take the instruction book or a photocopy with you and read it carefully. Modern cameras do many great things if you know how to operate them. Once you have a working knowledge of camera basics, the best way to develop your skills is to practise and look critically at the results. Practise capturing fast-moving arboreal animals on film in difficult lighting conditions by taking photographs at the nearest park.

VIDEOGRAPHY

TYPES OF VIDEO CAMERA

The least expensive video cameras use a single CCD and capture images in exactly the same way as digital still cameras, except that instead of reading out a single image to the microprocessor, they send a 'frame' of video every 1/25 second or just less than 1/30 second (depending on the video format). Somewhat more expensive three-chip cameras assign a chip to each of the three primary colours. The industry divides cameras into three groups: 'consumer' (US$ 400–1500); 'prosumer' (the cameras most often chosen as part of a field researcher's kit, US$ 1500–5000); and 'professional' (above US$ 5000). Consumer cameras have few controllable features, are small, light and universally have one receptor chip. Connecting microphones and other professional devices can be problematic. Most inexpensive video cameras record a fair amount of electronic 'noise' (hiss in the audio and 'snow' in the video). Prosumer cameras offer better connectivity to other recording gear, larger CCDs, more manual control (particularly manual focus and zoom) and, at the top end of the price range, three chips. This class ranges widely in weight (600 g to >3 kg) and size (180 mm to >300 mm).

VIDEO FORMATS

There are three international transmission video formats, National Television System Committee (NTSC), Phase Alternation by Line (PAL) and Système Eléctronique Couleur Avec Mémoire (SECAM). These video standards are generally mutually exclusive of one another, but international or tri-standard playback video cassette recorders (VCRs) and monitors solve this problem. Digital video (DV) format cameras are built for either PAL or NTSC recording, but each can playback video recorded on the other system. However, this is true in-camera only, and not when the camera is attached to a monitor.

TAPE FORMATS

DV and Digital8 (SONY) presently offer the best picture, price and performance. Very good DV cameras are available in a wide range of prices and features. Tape is available in 60-minute mini-tapes. The effective resolution is greater than 500 lines (much better than most televisions). Most importantly, all digital formats can be duplicated to another tape with virtually no loss of quality (unlike earlier analogue video technologies). The recorded information on Digital8 is identical to DV except that it is recorded on 8 mm tape rather than the 6 mm DV standard.

If your budget allows only for the most inexpensive camera get a Video Home System (VHS). Although this suffers from low resolution (240 lines at best) and poor colour fidelity, it is still the price and long-playing champion. It is also the only cassette that you can drop into a home VCR. Avoid cameras employing VHS mini-tapes, because these are often unreliable. In this era, also avoid S-VHS, 8 mm, and Hi8 tape formats, as the advantages they offer over VHS are minor relative to the new generation of digital recording formats.

VIDEO FEATURES USEFUL (AND NOT USEFUL) FOR FIELD RESEARCHERS

Select a video camera based on budget, image quality, size, weight, battery options, and connectivity with other recording equipment. You may want a DV camera that has an internal clock and prints the time on the non-visible portion of the picture (this can be read by an appropriate playback device). Other cameras have an electronic version of a film camera's shutter speed. The image for each frame of video is read from the chip in as long as 1/30 second or as fast as 1/2000 second. The shorter the 'exposure' the less blur in the individual frame. This is an important feature for frame-by-frame analysis of motion.

As for digital still cameras, digital zoom for video is useless and should not be used. Similarly, digital effects such as posterisation, sepia and in-camera dissolves are better added later. Be careful when enabling the visible time and date stamp – it may be useful, even necessary, in some circumstances, but it ruins the video for other applications.

A BASIC VIDEO CAMERA FIELD KIT

A basic field video kit includes the camera (two cameras if video is essential and your budget allows), a neck strap, the camera manual and the following items:

Batteries and charger
Ideally you should carry enough batteries to videotape for the entire period when you have no access to power. In practice this can be difficult, although the advent of lithium ion batteries has improved the situation. Carry at least three of the largest batteries appropriate to your camera. Most modern camera's AC source/battery chargers are designed for international use and can use any voltage and cycles per second available, but you will need the necessary plug adapters. Manufacturers often use the camera as a charger, meaning that you cannot simultaneously shoot with one battery and charge another unless you purchase an additional separate charger. An adaptor designed to charge batteries from a 12 V source such as a car battery is also useful, as are portable solar arrays with 6 or 12 V output. Many cameras have an additional small watch battery that powers the camera's memory and date functions – always carry a spare.

Video tripod and head (camera support)
Very little acceptable video can be taken without a tripod. Particularly when one is using extreme telephoto – the norm for most nature videography – without a tripod, camera shake will render the videotape unwatchable. Unlike still photographic tripods, video support systems are designed so that the

operator can pan, tilt and zoom. Unlike still cameras, video cameras must be levelled; otherwise an apparently straight horizon in the viewfinder will tilt as the camera is panned. Most video tripods have bubble levels to adjust the camera support. Many camera heads are attached to the tripod by means of a 'claw-ball', an adjustable connection that allows the operator to level the camera without adjusting the length of the legs. Tripods are heavy (often >4 kg), so buy a comfortable shoulder strap. It is also important to determine the highest inclination that the tripod head can be adjusted to, since primate videography is often practically straight up. Quality video supports can easily be nearly as expensive as cameras themselves, but inexpensive tripods severely limit creativity and flexibility in the field.

Tape
Buy good-quality tape from a well-known company. A head-cleaning tape should also be part of your basic equipment.

Camera case and useful accessories
Take a good-quality waterproof padded case with a shoulder strap. It should be large enough for the camera, charger, batteries and tape, but small enough to qualify as carry-on luggage. Add lens tissue, a small torch, a multi-tool (e.g. Leatherman™), headphones, resealable plastic bags, duct tape, electrician's tape and a set of small screwdrivers. Lastly, add two indelible marking pens to label cassettes.

Other equipment
Beyond the basic kit there are several other items that will be important for many productions. These include: portable tungsten lights (make sure that the voltage matches that of the host country), telephoto and wide-angle lens adapters, camera protection (custom rain covers and padded slipover cases are available for most 'prosumer' cameras), and lens filters. If audio will be more than incidental to the project, you will also need appropriate microphones: a short shotgun, one or more lavalieres, and a dynamic handheld (does not require battery or camera power) (Chapter 15).

VIDEO PRODUCTION TECHNIQUES

While good still images often tell stories, each picture stands alone. By contrast, video always tells a story, and videotaping technique is therefore as much about story telling as it is about image and sound recording. Plan the story you intend to tell. Even if you plan simply to play back the tape, you must make some editorial choices as you shoot.

Make a habit of describing what you are shooting into the microphone. This may include date, time, location, behaviours observed, the number of individuals in the troop, and any other pertinent information. In short, describe everything that you would otherwise write in a field book. You can often use the set-up time while you are adjusting the camera's position and focus for this. Record a few minutes of 'wild' sound as well, that is to say ambient sound with no human voice. This audio can be edited in later to replace the actual audio, which may include inappropriate sound (e.g. the videographer cursing mosquitoes). When you tape anyone important to your story, make sure that you spell out his or her name and title into the microphone as you begin.

Use a tripod and make sure that the camera is square and level before you start recording. Compose the frame so that, even if there were no primate present, you would still have an attractive, balanced picture. Keep the camera still. With small cameras this usually means composing the shot, hitting 'record', and then taking your hands off the camera. Pan and zoom as little as possible. If you do, pan or zoom from one well-composed picture at the beginning to another well-composed picture at the end. Try several pans or zooms so that you can choose among them later. If possible, shoot both close-ups and shots wide enough to show the animals' habitat and social relationships, and change your position several times so that each subject has a variety of backgrounds. To end a shot, let the animal move out of frame. Keep shooting – tape is cheap.

The most aesthetically pleasing video is made around dawn or dusk, when the sun is low in the sky. With primates it is often necessary to videotape almost straight up. If you have a choice, shoot with the sun behind you. If the sky fills the entire background, the animal will be a silhouette and most details will be lost. For this problem try the settings for 'backlight' or 'spotlight', or adjust the iris on your camera. The contrast ratio (whitest white to blackest black) of video is much lower than that of film, and video exposure settings are consequently much less forgiving. It is usually better to move so that some of the background is leaves, or at least so that the sun's angle creates less contrast.

Always label both tapes and their jackets as soon as they are spent, and, if you are keeping a journal, reference the videotape in it. Playing back videotape in the field is always risky, as sooner or later everyone forgets they have done so and shoots over something important.

KNOW YOUR VIDEO CAMERA

As with still photography, the most important aspect of videotaping in the field is to understand your camera thoroughly. Camcorders have dozens of controls and manuals in the hundreds of pages. There will be a limited set of features that will be an occasional but essential aid in the field when shooting live subjects, when there is no time to experiment. Know what all the arcane symbols that appear and disappear in the viewfinder mean. Try every option and every effect before stepping into the field. This is vital because, with less expensive cameras, there is no standardised terminology and many of the 'special effects' will completely ruin your work if turned on accidentally. Many videographers cover the special effects buttons with cardboard and tape, or even remove the buttons and cover the holes.

PROTECTING YOUR FILM, TAPE AND CAMERA

Electrical devices never operate optimally in extreme conditions and humidity is particularly detrimental. Rapid change in temperature and humidity will usually cause the humidity sensors in a video camera to prevent it from operating, but occasionally the magnetic surface of the tape delaminates and coats or even scratches the recording head. If your camera stops because of humidity, open the cassette door to expose the recording head and sit the camera out in the sun.

Binoculars, cameras and lenses corrode and provide a substrate for fungus in hot and humid tropical conditions. Keep cameras in a waterproof bag with silica gel to dry them out. A big waterproof bag or plastic garbage bag is useful for protecting all your valuables during boat trips, stream crossings, and from rain and leaky roofs. If electricity is available it may be worth building a dry box with small holes and a 50 W bulb to keep equipment dry. If a camera gets wet with fresh water you can dry it out with a low temperature oven (50 °C) or a hair dryer, but salt water will probably render it worthless (Fitzharris, 1996). Where humidity and exposure to water are ongoing hazards, a watertight housing may be necessary. An umbrella can be used to shoot in the rain.

Film and videotape should not be exposed to direct sun, high humidity, or water. Keep it in resealable bags wherever it is coolest. Take the number of rolls or tapes you will need for that day (five to eight film rolls, 2 × 60-minute videotapes). If you are in the field for more than two months, send still film home with someone so that it can be processed rather than have it sit around until you leave. Exposed film that is undeveloped for four months or more may lose contrast and appear underexposed.

TRAVELLING WITH PHOTOGRAPHY AND FILM EQUIPMENT

Small video cameras and photographic equipment are usually ignored by Customs if you have a tourist visa. Register your equipment with your home country's Customs Service to avoid the possibility of being required to pay duty on your own equipment upon return. Check in advance when travelling, as professional gear may require permits from governmental ministries governing film, security and news. What constitutes professional equipment often depends on the judgement of the Customs Agent.

Pack photographic equipment in your carry-on luggage, and prepare to be searched. Have the battery charged and tape in your video camera. You may be asked by security to demonstrate that it is, in fact, a working camera. Videotape is not normally affected by airport X-ray equipment, but X-rays can negatively affect film, especially high speed film.

Before entering a new country, prepare a list of all your equipment including: a description; the manufacturer's name, serial and/or model number; the quantity, weight and cost of each item; and, if possible, photocopies of the purchase receipts. This will expedite completion of the inevitable paperwork. You may be required to pay taxes based on the equipment's value or to post a refundable cash bond equal to the amount of tax payable. An alternative to posting a bond is a 'carnet'. This is essentially a passport and visa for equipment, and entails posting a bond in your country of origin, which is then used as security by the destination country's Customs authority (see 'Useful Internet sites').

Having entered a country you need to consider security, local government and culture. Avoid unnecessary and obvious displays of expensive gear, particularly in urban areas and in crowds. Zoos and reserves commonly levy fees for photography, film and video, and these can vary dramatically depending on perception of your intent. Persons who will be selling their video or photographs are charged, appropriately, more than tourists. Scientists fall between the two, and may be charged as either, neither or both. Often carrying a tripod qualifies you as a professional.

Bear in mind the host government's regulations regarding photography. Taking prohibited video or film can lead to detention, a fine, and/or confiscation of equipment and film. The most likely problem areas include border areas, scenes of civil disorder or other public disturbance, police and/or military personnel or installations, industrial structures (including harbour, rail and airport facilities) and aerial pictures.

Local people and their customs are an integral part of the story, but always respect human subjects and ask permission before you photograph or video. Is payment expected? Are certain places or things, particularly of a cultural or religious nature, subject to photographic or filming restrictions? The most important aspect of photographing and videotaping in another culture is to treat people and their places with respect and err on the side of caution. Giving pictures of themselves to village heads and other leaders may be a simple way to enhance relations between the researcher and the local community (Chapter 1). Carrying photographs of yourself and your family and home to show to local people may also help them understand you better.

RESEARCH AND PHOTOGRAPHY

For the most part you cannot collect good data and take good photographs or video at the same time. You may be able to arrange to collect data on some days and take photographs on others. Unfortunately, it will seem like on the days when you do not have the camera the primates will be closer and in good light, doing amazing things. This may be more psychological than real. The mind forgets how limiting cameras are and remembers what it hopes the camera would have captured.

If you do not have the time, money or ambition to take pictures but you do need photographs, one solution might be to have a dedicated photographer come to your study site to photograph. Many photographers would love the chance. If you choose this route be sure to set guidelines so that you get a photographer who has good equipment, likes fieldwork, and can get good pictures. Make sure that the photographer understands first and foremost that you need your own set of good photographs of subjects you want, that they must pay their own way, plus help to contribute monetarily to the fieldwork. After all, you have spent a long time and done the hard work of habituating or following the study animals so that you can lead the photographer to them to take pictures of them in a short amount of time.

REFERENCES

Fitzharris, T. (1996). *Nature Photography National Audubon Society Guide*. Buffalo, NY: Firefly Books.

Hill, M. & Wolfe, A. (1993). *The Art of Photographing Nature*. New York: Crown Books.

Johnson, D. (2001). *How to do Everything with your Digital Camera*. Berkley, CA: Osborne/McGraw-Hill.

Shaw, J. (2000). *Nature Photography Field Guide*. New York: Amphoto Books, an imprint of Watson Guptill Publications.

Further reading

Compesi, R. &, Sherriffs, R., (1997). *Video Field Production and Editing*. Boston, MA; Allyn and Bacon.

King, J.A. (2000). *Digital Photography for Dummies*, 3rd edition. Foster City, CA: IDG Books Worldwide, Inc.

Zettl, H. (1999). *Television Production Handbook*, 7th edition. San Francisco, CA: Wadsworth Publishing Company.

Useful Internet sites

Carnet : <www.uscib.org> and search for carnet.

PSC, camera trap black and white digital version : <www.cabelas.com>.

17 • Chronobiological aspects of primate research

HANS G. ERKERT

Institute of Zoology, University of Tübingen, Tübingen, Germany

INTRODUCTION

Terrestrial animals live in an environment that undergoes regular variations, ultimately induced by the geophysical conditions prevailing in our solar/earth system. Solar radiation and gravity, in combination with the earth's rotation around its inclined axis and its orbit around the sun, and the moon's revolution around earth, produce marked diurnal, seasonal and lunar as well as tidal periodicities in important physical environmental factors, such as light intensity, ambient temperature, humidity, precipitation, day and night length, and duration of twilight. As a consequence of these periodicities being superimposed on one another, many relevant biotic environmental factors, such as food availability and predator pressure, as well as social contact, communication and competition with conspecifics, and intra- and interspecific competition for food may also vary diurnally, seasonally, lunar periodically or tidally. In this way each animal's environment has a highly complex time structure, is highly repetitive in time and thus highly predictive. Reliable predictivity provides a good substrate for genetically fixed adaptations. Hence, in addition to other general or specific physiological, ecological and/or behavioural adaptations, animals have also evolved endogenous diurnal (circadian), annual (circannual), lunar (circalunar) and/or tidal (circatidal) rhythms.

In non-human primates (in the following referred to only as primates), adaptation to the time structure in their physical and biotic environment is restricted mainly to the development of a circadian timing system involved in the regulation of the pronounced daily (circadian) organisation of physiology and behaviour. Although many primate species show a seasonal pattern in reproduction, strong evidence for the existence of a circannual rhythm underlying this rhythmicity has so far been found only in rhesus macaques (*Macaca mulatta*; Plant *et al.*, 1974), while weak clues to an endogenous annual rhythm have been obtained in squirrel monkeys (*Saimiri sciureus*; DuMond, 1968), ring-tailed lemurs (*Lemur catta*; van Horn, 1975) and mouse lemurs (*Microcebus murinus*; Petter-Rousseaux, 1975). Circatidal rhythms can

Field and Laboratory Methods in Primatology: A Practical Guide, ed. Joanna M. Setchell and Deborah J. Curtis. Published by Cambridge University Press. © Cambridge University Press 2003.

be ruled out in primates, and the unique report on an endogenous lunar periodicity in *Saimiri* by Richter (1968) is not conclusive owing to methodological shortcomings.

Against this background of current knowledge about endogenous timing processes that have evolved in primates, this chapter focuses on diel (24 hour) and circadian rhythmicity. Special emphasis is given to those fundamentals and aspects of comparative chronobiology that should be considered when one is carrying out primatological research in the field. Advice for laboratory research may be inferred from Erkert (1989a, 1997). Ultradian rhythmicity (period length significantly less than 24 hours), which may be found in sleeping behaviour, thermoregulatory processes (Chapter 18) or hormone secretion (Chapter 19), is not dealt with here and neither is ovarian cyclicity – although both are also of interest in chronobiology.

DAILY ORGANISATION OF PHYSIOLOGY AND BEHAVIOUR

Many behavioural activities and most physiological functions of the organism vary regularly over the 24 hour day. In mammals, including primates, several hundred diel rhythms have been established so far. The time course of the different parameters may vary considerably over the 24 hour day. Some functions show a more or less pronounced unimodal pattern, while in others a more bimodal, trimodal or even multi-modal pattern may dominate. Furthermore, the amplitude and the phase of the numerous rhythms, as indicated by the time relation of peaks and/or troughs to certain phases of the external day, may differ. However, the mutual phase relationship of these rhythms is usually relatively constant. In this way the whole system has a relatively stable internal time structure (for reviews, see Moore-Ede *et al.*, 1982; Erkert, 1997) that may be essential for the well-being of the individual and for the organism's adaptation to the challenges of its temporally changing environment.

Depending on the part of the solar day in which the individuals of a given species exhibit most of their behavioural activities, primates are usually assigned to one of the two ecotypes of diurnal or nocturnal animals. Almost all simian species – with the exception of the mostly nocturnal South American owl monkeys (*Aotus* spp.; Erkert, 1999) – are strictly diurnal. Most prosimian species – except a few diurnal Madagascan lemurs such as the indris and sifakas – show a predominantly nocturnal lifestyle (Martin, 1990). True crepuscular species, which limit their daily activity period to dawn and dusk, have not been described in primates. However, some species in the family Lemuridae (Tattersall, 1982; Overdorff & Rasmussen, 1995; Colquhoun, 1998; Donati *et al.*, 2001) and the Argentinian owl monkey *Aotus azarai* (Wright, 1989) have been shown to be active during day and night. This kind of activity has been called cathemeral (Tattersall, 1987) and there is ongoing discussion as to whether it represents an ecological adaptation (Tattersall, 1982;

Engqvist & Richard, 1991; Pereira *et al.*, 1999), is ancestral to the Lemuridae (Tattersall, 1982; Curtis & Rasmussen, 2002), or represents a transitional stage between a primarily nocturnal and diurnal lifestyle (van Schaik & Kappeler, 1996).

THE CIRCADIAN TIMING SYSTEM

Under constant laboratory conditions that do not provide any external 24 hour day time cues, i.e. in an environment with constant lighting (LL) of given illuminance or constant darkness (DD), constant ambient temperature and relative humidity, and without daily social contact with conspecifics or the keeper, representatives of all primate species studied so far continue their diel rhythms of activity, core temperature, feeding and drinking for some time. The period of such free-running rhythms usually deviates systematically from the 24 hours of the solar day and the approximate 24.83 hours of the lunar day. This reveals an endogenous origin and led to the designation 'circadian rhythms'. According to this definition the term 'circadian' always implies an endogenous origin of the rhythm. Therefore, it should only be used if the rhythm has been measured while free-running under constant conditions. Otherwise one should speak of a diel rhythm.

Table 17.1 summarises the range of the spontaneous periods of the free-running circadian rhythms in the few primate species studied under constant environmental conditions. It can be seen that in primates the circadian period varies from about 21 to 26 hours, and that there is no clear relationship either between circadian period length and systematic affiliation or between period and general ecotype.

Lesion experiments carried out in squirrel monkeys have shown that in primates, as in rodents and all other mammals, the main pacemaker regulating circadian rhythmicity is located in the hypothalamic suprachiasmatic nuclei (SCN). These paired nuclei are situated on the base of the third ventricle, just above the optic chiasma (Albers *et al.*, 1984). Details of the cellular structure of the brain's master clock in the SCN, its transmitter and neuropeptide content, afferent and efferent connections, the molecular mechanisms involved in its rhythm generation and their transmission to the various effector systems can be found in Klein *et al.* (1991) and Moore (1999).

ENTRAINMENT OF CIRCADIAN RHYTHMS BY ZEITGEBERS

Since their spontaneous period usually deviates from the 24 hour solar day, circadian systems must be synchronised (entrained) to it by environmental periodicities called zeitgebers. As in other mammals, the most potent zeitgeber cycle that phase-sets a primate's circadian periodicity to the external 24 hour rhythmicity is always the natural (or artificial) light–dark cycle (LD). In some species, large amplitude temperature cycles may also function as a (relatively weak) zeitgeber and entrain circadian rhythmicity under otherwise

Table 17.1. *Circadian spontaneous periods in non-human primates*

Species	Activity[a]	n	Parameter[b]	Illumination intensity range (lx)[c]	Period length (h)		Source
					$\tau \pm$ SD	τ_{min}–τ_{max}	
Prosimiae							
Eulemur fulvus albifr.	Cat.	5	Loc.	0.1–240	24.5 ± 0.6	23.5–25.2	Erkert et al., 1984
Microcebus murinus	No.	30	Loc.	0	22.5 ± 0.6	20.7–23.3	Schilling et al., 1999
				RR	23.0 ± 0.2	n.i.[d]	
Galago senegalensis	No.	5	Loc.	pD–0.1	23.8 ± 0.6	22.8–25.5	Erkert et al., 1984
Otolemur garnettii	No.	5	Loc.	pD–0.1	22.9 ± 0.4	21.3–23.9	Erkert 1989b, and unpublished data
Simiae							
Aotus lemurinus griseimembra	No.	6	Loc.	0.1–360	25.1 ± 0.9	24.2–26.2	Erkert, 1989b
Callithrix jacchus	Di.	10	Loc.	0.2	24.4 ± 0.4	23.9–25.6	Rappold & Erkert, 1994
		14	Loc.	0.1–400	23.3 ± 0.3	22.7–24.0	Erkert, 1989b
		6	Loc.	0.1–430	23.2 ± 0.3	n.i.	Wechselberger, 1995
		8	Loc.	< 0.5	23.3 ± 0.4[e]	n.i.	Glass et al., 2001
Saimiri sciureus	Di.	8	Loc.	0.1–400	25.2 ± 0.4	24.5–26.2	Aschoff & Tokura, 1986
			Feed.	1–600	25.0 ± 0.5	24.3–26.3	Sulzman et al., 1979
			Temp.		24.9 ± 0.6	23.3–25.8	
Macaca mulatta	Di.	4	Loc.	270	24.0 ± 0.3	23.8–24.4	Yellin & Hauty, 1971
		4	Loc.	0–300	24.1 ± 0.3	23.8–24.9	Martinez, 1972
Macaca nemestrina	Di.	3	Loc.	0.003–100	23.1 ± 0.4	22.3–23.8	Tokura & Aschoff, 1978, 1983
Macaca irus	Di.	2	Loc.	0–50	24.2	23.7–24.6	Hawking & Lobban, 1970
Pan troglodytes	Di.	1	Feed.	1–85	24.4	23.7–25.1	Farrer & Ternes, 1969

[a] Cat., cathemeral; No., nocturnal; Di., diurnal.
[b] Loc. = gross locomotor activity; Feed., feeding; Temp., core temperature.
[c] pD, physiological darkness ($\leq 10^{-6}$ lx); 0, constant darkness; RR, constant dim red light.
[d] n.i., not indicated.
[e] Standard error.

constant conditions. This has been shown in pig-tail macaques (*Macaca nemestrina*), squirrel monkeys (Tokura & Aschoff, 1983; Aschoff & Tokura, 1986), and common marmosets (*Callithrix jacchus*) (Pálková *et al.*, 1999). Social entrainment, i.e. synchronisation of circadian rhythmicity by periodic contact with a conspecific (mate), has been demonstrated in *Callithrix* (Erkert & Schardt, 1991), but not in *Saimiri* (Sulzman *et al.*, 1977a). In *Saimiri* entrainment was also achieved by feeding cycles (Sulzman *et al.*, 1977b).

To become effective as a zeitgeber, LD cycles must have a minimum amplitude and the level of luminance prevailing during the light time must exceed a specific threshold. While, for example, the circadian activity rhythm in owl monkeys and mouse lemurs may be entrained by 12:12 hour LDs with only about 0.1 lx (corresponding to full moon luminance) during light time and 0.001 lx, or physiological darkness, during dark time, identical LDs failed to entrain the free-running circadian rhythm in Senegal and Garnet's galagos (*G. senegalensis, Otolemur garnettii*) (Erkert, 1989a). In *Otolemur* the threshold for photic entrainment was between 3–5 and 30–50 lx. It may be hypothesised that such a high threshold for photic entrainment in strictly nocturnal species may be of adaptive value in that it protects their circadian system and its mechanism of photic phase-setting from perturbations by moonlight.

According to current ideas, entrainment of circadian rhythms by a zeitgeber cycle is achieved by a phase-setting mechanism (detailed by Moore-Ede *et al.*, 1982). This mechanism compensates exactly for the time difference between the circadian period and the 24 hour zeitgeber period in such a way that the endogenous rhythmicity adopts a certain phase relationship to the outer day. It is described as the phase angle difference between arbitrarily defined corresponding phases of both rhythms, such as activity onset or end, to sunrise and sunset in diurnal species, and to sunset and sunrise in nocturnal species. Changes in zeitgeber parameters such as amplitude (strength) and form (e.g. light-time:darktime ratio, steepness and duration of rise and fall of light intensity – all of which vary greatly with latitude and season), as well as in certain characteristics of the circadian system, may lead to changes in the phase position of the entrained biological rhythm to its main zeitgeber cycle.

MASKING: DIRECT EFFECTS OF ENVIRONMENTAL FACTORS

Many environmental factors, such as light, ambient temperature, social contact, noise and periodic availability of food, may have a dual effect on the primates' circadian rhythms in both physiology and behaviour. Besides their phase-setting (entraining) zeitgeber effect they often lower or raise, inhibit or enhance the basic level of a parameter as determined by the circadian system. Such rhythm-modulating direct effects of external (and internal) factors, have been called 'masking effects' (Aschoff *et al.*, 1982). They represent a further control mechanism, in addition to the circadian mechanisms, which is of equal importance for the organism's adaptation to varying environmental factors (Erkert & Gröber, 1986; Erkert, 1989a).

More or less pronounced direct effects of light, ambient temperature, relative humidity or social factors may be observed in primates both when living in their natural environment and under controlled conditions in the laboratory. However, a clear separation of the two effects can be achieved only under certain laboratory conditions. As an example, Fig. 17.1 shows the activity rhythm of a male *Otolemur garnettii* in which a large amplitude LD 12 : 12 of 330 : 0.3 lx failed to synchronise the free-running circadian rhythm. It did, however, induce pronounced masking, owing to a strong activity-inhibiting direct effect of the high luminance prevailing during the light time of the LD.

Under natural conditions strong rhythm-masking direct effects occur in several nocturnal primates such as *A. lemurinus* and *Eulemur fulvus*. In both species activity inhibition by light intensities below full moon luminance (about 0.1–0.5 lx) leads to a lunar periodic modulation of the activity pattern and presumably also of the patterns of other circadian functions (Fig. 7.2; Erkert, 1974, 1989a). As in the owl monkeys, and based on the limited data available, nocturnal activity in *Tarsius*, some *Galago* spp. and some cathemeral lemurs parallels mainly the moonlit night hours, while in other cathemeral lemurs and nocturnal prosimians, such as *Microcebus* or *Galago* spp. (H. G. Erkert, unpublished data), moonlight may have a less activity-enhancing or even a more inhibitory direct effect on the activity rhythm (Table 7.2).

Finally, from a chronobiological point of view, it should not be ignored that almost all variations of environmental factors and changes in an organism's physiological status, such as its hormonal, nutrititional, reproductive and/or developmental state, may induce masking effects on circadian or diel rhythmicity, independent of whether they recur regularly or occur stochastically.

ADVICE FOR FIELDWORK

GENERAL CONSIDERATIONS

In all primatological studies to be carried out in the field or in the laboratory it should be borne in mind that most physiological functions and behavioural activities are organised and structured on a circadian basis. Hence it is advisable for physiologists, behaviourists and ecologists not only to make observations or to take measurements or samples at a particular time of day, i.e. at only one fixed circadian phase or at randomly varying circadian phases, but to try to establish the diel pattern(s) of the function(s) of interest. Reliable diel patterns can be obtained only if sampling frequencies are evenly distributed over the 24 hour cycle and if the sampling intervals are not too large. Mere two-point (e.g. day–night or morning–evening) measurements/samplings should be avoided. Although in general the statement 'the shorter the better'

F: L971202.027
C: 27 D: 14 Dec. 1997 09:05 – 4 Mar. 1998 09:00 R: 3–100 (ACTI)

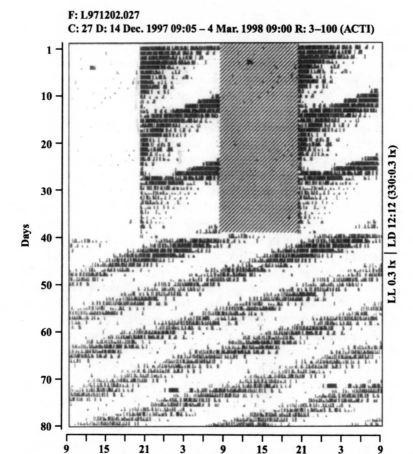

Fig. 17.1. Free-running circadian activity rhythm of a male Garnet's galago (*Otolemur garnettii*) kept in an artificial light–dark cycle (LD 12 : 12 hours) of 330 : 0.3 lx (days 1–40; L: 08 : 00–20:00, hatched area) and subsequently (days 41–80) under constant light (LL) of 0.3 lx (25 ± 1 °C; 60 ± 5%). Owing to the individual's very short spontaneous period τ of only 21:20 h in LL and 22 : 20 h in the LD the zeitgeber period T of 24 hours was outside the circadian system's range of entrainment. The LD therefore failed to entrain its free-running endogenous rhythm. However, owing to a strong activity-inhibiting direct effect of the high luminance during the light time, there was a marked diurnal masking of the circadian rhythm, which immediately disappeared after the LD was switched off on the fortieth day.

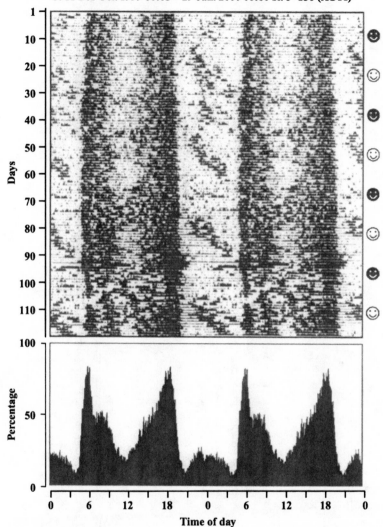

F: L130899.011
C: 11 D:1 Oct. 1999 00:05 – 29 Jan. 2000 00:00 R: 3–150 (ACTI)

Fig. 17.2. Activity rhythm of a wild red-fronted lemur (*Eulemur f. fulvus*) living in its natural habitat and social group at the Kirindy Forest Reserve in Western Madagascar (S 20° 03', E 44° 39'). *Upper diagram*: Double plotted original recordings (5-minute values) carried out over four months (1 Oct. 1999 to 29 Jan. 2000) with an Actiwatch-AW4® accelerometer motion logger. Note the clear lunar periodic variation of the animal's nocturnal activity (☺ full moon, ● new moon), which may be attributed to activity-enhancing or disinhibiting direct effects of moonlight. *Lower diagram*: Mean activity pattern of the animal as averaged over the 120 days of recording. The trimodal pattern is characterised by two very pronounced activity peaks in the early morning and late afternoon and a smaller peak around midnight.

Table 17.2. *Studies assessing direct effects of moonlight on nocturnal activity in primates*

Family	Genus	Moon phase	Effect of moonlight
Cebidae[a]	Aotus	New moon, waxing moon, full moon, waning moon	Increases activity
Cebidae[b]	Aotus		
Galagonidae[c]	Galago	Week centred on new moon, week centred on full moon and waxing or waning half moon	Increases activity
Lemuridae[d]	Eulemur		
Lemuridae[e]	Eulemur	Nocturnal illumination index	No effect
Galagonidae[f]	Galago	Moon absent, half moon, full moon	Increases activity
Lemuridae[g]	Eulemur	Moon absent, half moon, full moon	No effect
Lemuridae[h]	Eulemur	New moon, full moon	Increases activity
Loridae[f]	Loris	New moon, full moon	No effect on activity, but fewer calls during full moon than new moon
Megaladapidae[i]	Lepilemur	New moon, full moon	No effect
Cheirogaleidae[i]	Microcebus		No effect
Tarsiidae[f]	Tarsius	Moon absent, moon present	Increases activity

Source: [a]Erkert, 1974; [b]Wright, 1989; [c]Nash, 1986; [d]Colquhoun, 1998; [e]Curtis et al., 1999; [f]Bearder et al., 2002; [g]Overdorff & Rasmussen, 1995; [h]Donati et al., 2001); [i]Nash, 2000; [j]Niemitz, 1984.
With kind permission of D. Curtis (Thomas & Curtis, 2001).

may be valid, choosing the most promising sampling interval for the determination of the diel pattern in a given function always remains an optimisation task. Sampling periods should be spread out evenly over the animals' entire activity time even when one is establishing time budgets for the various behavioural activities. In long-term studies that require only one trial or measurement per day, week or month, the circadian phase of sampling or at least the time of day (and if necessary also the lunar phase) should be kept constant.

Bearing in mind that certain environmental factors may act as an entraining and/or masking agent for circadian rhythmicity, and that such factors may vary systematically with latitude, season, lunar phase, and/or day time, it is obvious that not only laboratory but also field studies should provide detailed information on all environmental factors which may produce circadian effects or directly influence (mask) diel and circadian rhythms. Among these are, for instance, climate factors such as the time course of ambient temperature and relative humidity (Chapter 5), as well as luminance and all other relevant ecological and social factors.

It is important to determine and describe the exact geographical coordinates (latitude and longitude) of the study area (Chapter 4) and to record details of the dates of beginning and end of the study as well as of day and night length, duration of twilight and the times of sunrise and sunset and their variation during the study period. Whereas day length, often referred to as 'photoperiod', is clearly defined as the span between the true local times of sunrise and sunset, the duration of night is a matter of defining twilight. According to the differentiation between civil, nautical and astronomical twilight we must also differentiate between the civil, nautical and astronomical night. Civil, nautical and astronomical twilight in the evening correspond to the spans lasting from the true local times when the sun's upper margin 'touches' the horizon (about $0°$) to when its centre is $6°$, $12°$ or $18°$ below the horizon, respectively (dusk). In the morning, the span lasts from the true local times at which the centre of the sun crosses $18°$, $12°$ and $6°$ below the horizon, respectively, to when its upper margin crosses the horizon (dawn). Correspondingly, the astronomical, nautical or civil night extends between the times of sun positions (centre of disc) of $-18°$ to $-18°$, of $-12°$ to $-12°$ and of $-6°$ to $-6°$, respectively. In the literature on primate ecology and behaviour 'night' or 'night-time' has often been defined as the period between sunset and sunrise. This may be sufficient in studies on strictly diurnal species that usually awake and become active well after sunrise and start resting and sleeping before sunset. However, for nocturnal or cathemeral primates it may result in incorrect classifications concerning the activity pattern.

When studying primates in the field and evaluating data from such studies, we should be aware that our anthropocentric measurement of (solar) time and the use of certain phases of the 'solar cycle' as reference points (e.g. position of the sun in zenith or at its culmination point corresponds to 12:00

hours local time) may, at least partly, lead to problematic conclusions. We also need to bear in mind that we are diurnal animals, and as such have a diurnally biased view when studying animals with activity rhythms different from our own. For the phase-setting mechanism of an animal's biological clock or circadian system, those times of day at which light intensity rapidly changes, i.e. the periods of dawn and dusk, are much more important than the reference points we use. On cloudless new moon days the luminance prevailing on open areas corresponds to about 10^5 lx around lunchtime, about 10^3 lx at the times of sunrise and sunset, and during the astonomical twilight it falls to and rises from a relatively constant level of about 5×10^{-4} lx throughout the astronomical night. On clear full moon nights the luminance of the moonlit sky amounts to about 3×10^{-1} lx. Thus, according to phase and position of the moon, luminance during the night may vary by a factor of about 1 : 1000. Cloud cover can produce a reduction in light intensity of up to 90% or even 99%, and vegetation can lead to a reduction in luminance by factors of up to the power of 3–4. As an example, Curtis et al. (1999) give concrete data from luminance measurements in a Madagascan seasonally dry, secondary forest, where the amount of light transmitted through the canopy varied between 0.4–1.5% during the wet season and 1–16% during the dry season. The relationships described above show that the best reference phases of the environmental LD-zeitgeber pattern to which the onset and end of the primates' activity times and activity patterns should be related are indeed the times of sunrise and sunset, as well as those of the beginning and end of astronomical twilight. All times must be determined and given as true local time, i.e. corrected according to latitude and longitude of the study site. Summer time should never be used when one is carrying out studies in the field or under semi-natural conditions. This quite artificial and annually changing mode of time measurement may lead to considerable confusion and misinterpretations. Also ensure that the clock on your computer used for data acquisition, programming and reading out data-loggers is never set to summer time. Local times of sunrise and sunset, beginning and end of civil, nautical and astronomical twilight, as well as the dates of the four phases of the moon and the times of moonrise and moonset of the year and period of the study, can either be requested from national institutions and authorities for astronomy or may be extrapolated from tables in The Astronomical Almanac issued every year by the Nautical Almanac Office of the United States, Washington, DC, and Her Majesty's Nautical Almanac Office, London. Alternatively, these data can be obtained from the US Naval Observatory data services or by using a software package designed for this purpose (Thomas & Curtis, 2001).

   In nocturnal and cathemeral and probably also in some diurnal primate species the level of locomotor activity and other behaviours, as well as the body temperature, may depend to different degrees on the luminance prevailing during the night hours. Therefore, it is sometimes indispensable also to consider the lunar cycle, i.e. the lunar phase and the hours of moonlight during sampling nights. In all cases in which strong direct effects of moon-light on behaviour and/or the recorded physiological functions are observed it is advisable to evaluate, i.e. to average, the collected data separately for the four phases of the moon. When doing this one should not only use the data from the day and night in which the respective phase of moon (new moon, waxing moon, full moon, waning moon) occurs, but also those of one or two days before and after it. In this way the effects of accidental variations in luminance, due to varying degrees of cloud cover or unfavourable weather conditions, can be compensated to some degree.

MEASURING LUMINANCE

From a chronobiological point of view, one of the most important environmental factors is the lighting conditions prevailing throughout a study. For fieldwork it is therefore essential to carry out continuous recordings of light intensity. For luminance measurements only devices with a sensor that has a spectral sensitivity that approximately corresponds to the spectral sensitivity of the study species should be chosen. Since most primate eyes have a similar spectral sensitivity to the human eye (Kremers *et al.*, 1999), it seems reasonable to measure light intensities in lux or footcandles (1 fc corresponds to about 10.71 lx). Many photometers (luxmeters) have sensors adapted to the human spectral sensitivity. For measurement the photocell should be directed towards the light source. It is advisable to use a cosine-corrected cell and always direct it vertically towards the sky in an open area. Reduction factors can be calculated and used for an estimation of the light intensities to which the animals are really subjected throughout their activity time by taking simultaneous reference measurements at the various sites preferentially used by the study subjects. Depending on the habitat structure, primates may be subjected to a very broad range of illumination intensities ($<10^{-4}$–10^5 lx). Their eyes have a broad dynamic range and code light intensity logarithmically over a broad range. This should be considered when measuring and describing the luminance prevailing at certain sites and times in a study area. It is best to report the data rounded up or down on a logarithmic scale with only one digit behind the comma (e.g. $2.3 \pm 0.2 \times 10^{-1}$ lx). Ideally, the luxmeters used in field studies on nocturnal or cathemeral species should be as sensitive as possible (minimum threshold of at least 10^{-3} lx) and allow measurements over a broad range of illumination intensities ($10^4/10^5$ lx). For studies on strictly diurnal species less sensitive devices covering a range from about $10^{-1}/10^0$ to $10^4/10^5$ lx may be sufficient. Digital devices and data-loggers are very useful (e.g. Optometer P970; or similar devices, see list of suppliers), but beware of the recently developed small light intensity loggers – only those that have a sensor with an adequate spectral sensitivity and allow measurements over a broad range of light intensities should be used. The STOWAWAY® and HOBO® light intensity loggers, for instance, are biased in that their spectral sensitivity is strongly shifted to the near infrared (about 930 nm) which is invisible to primate eyes and hence ineffective for the circadian phase-setting mechanism and for eliciting masking effects. In this respect light intensity loggers designed for chronobiological research in humans such as the Actiwatch-L® and Actiwatch-L plus® and the Octagonal Basic Motionlogger-L® are more suitable because they measure illumination intensity in lux. However, the range of measurement is limited to 1–4000 lx (resolution 1 lx) in the Motionlogger-L®, 1–36000 lx (resolution 1 lx) in the Actiwatch-L® and 0.01–50000 lx (0.01 lx intervals) in the Actiwatch-L plus®. With a sampling interval of 5 minutes the Actiwatch-L® allows automatic light intensity recordings over about 110 days and the Actiwatch-L® plus about 35 days. Unfortunately these devices and the appropriate software are very expensive and must be specially adapted for use in primatological fieldwork.

RECORDING ACTIVITY

The best and most common, although most time-consuming and strenuous, way to record primate activities in the field is still through observation. This can be done either directly with a good pair of binoculars (7 × –10 ×) in diurnal species and/or with a sensitive night vision device (starlight scope) in nocturnal species, or indirectly by normal or infrared videotaping and later evaluation on the screen. Both options require the following of individuals or groups of interest throughout several entire activity periods and recording of the behaviours of focal animals for fixed successive time intervals (e.g. 5 or 10 minutes). The great advantage of recording primate activities by observation is that the diel patterns of various behavioural activities like travelling, foraging, grooming, play or other social interactions and resting can be established and the individual and/or group time budgets can be determined simultaneously (cf. Alonso & Langguth, 1989). Continuous long-term recordings of locomotor activity, core temperature, heart rate and/or haemodynamic parameters in freely moving primates are best carried out with automated radiotelemetric recording devices (Chapters 10 and 18). However, most of the commercially available, and usually intraperitoneally or subcutaneously implanted, miniature transmitters have a very limited signal range. Hence the application of such miniaturised single- or multi-channel radiotelemetry devices is largely limited to laboratory conditions (cf. Rauth-Widmann et al., 1991; Schnell & Wood, 1993) or outdoor enclosures (Kerl & Rothe, 1996).

A reasonably good and economic method to record primate locomotor activity in laboratory set-ups or in outdoor enclosures is to use passive infrared motion detectors (PIDs) available from suppliers of electronic and/or security equipment. Combined with a program-controlled mutiplexer and counter device (i.e. an adequately programmed computer), such PIDs provide a simple and efficient device for automatic long-term activity recordings, which can be adapted for almost all primate species (Lerchl et al., 1988). However, this method detects only passages of infrared sources in a given area or space. Hence, it allows only activity registrations of singly kept individuals or of the entire group. For some questions it may be advisable to use several PIDs and to focus the sensors on different important sites of the enclosure, such as feeding and resting sites or the entrance to the sleeping box. In this way a more differentiated picture of space use over time and/or on onset and end of daily activity can be obtained. When PIDs are used for automatic activity records the method must be validated initially by direct observation or videotaping in order to find out which kinds of behaviour and locomotion contribute to the activity recorded.

In the field, primate locomotor activity and activity patterns are best recorded continuously over long spans by radio-tracking (Chapter 10). As a measure of locomotor activity either the distance covered or the time moving around per time unit (usually per hour) can be used. Alternatively, changes in the transmitter signal intensity can be recorded and used as an indicator of motor activity.

An alternative method for automatically recording the (loco)motor activity of individual wild-living or captive primates over several weeks or months is to use one of the small motion loggers (accelerometers) designed for chronobiological research in humans. At present two devices can be adapted for use in medium-sized to larger primates (>1 kg): the Actiwatch® and the Octagonal Basic Motionlogger®. Both devices require a computer with an interface and dedicated software to program (initialise) and read out the loggers and to evaluate the recorded data. To protect the mini-loggers against mechanical damage, humidity or water they should be wrapped in thin polythene sheeting and fitted into a small completely closed hard cover (e.g. aluminium) box which can be fixed to a collar around the animal's neck. As an example of actimetry with an Actiwatch® AW-4 motion logger, Fig. 17.2 (upper diagram) shows a four month sector of the activity rhythm in a wild-living *Eulemur f. fulvus* monitored continuously in its habitat in the Kirindy forest reserve near Morondava, eastern Madagascar.

Automatic recordings of body temperature rhythms can be carried out with subcutaneously or intraperitoneally implanted temperature-sensitive dataloggers (Chapter 18). However, owing to a relatively small measuring acuity, these devices can be recommended only for recordings in species with a high amplitude core temperature rhythm.

SUMMARY AND CONCLUSIONS

Adequate timing of behavioural activities and physiological functions is an essential prerequisite for an individual's survival and reproduction. Thus, when one is planning and carrying out primatological field studies or laboratory research, some fundamentals of chronobiology should be considered. (1) The mammalian organism shows a distinct circadian organisation, and most of its physiological functions and behavioural activities undergo more or less pronounced daily (circadian) variations regulated by a genetically fixed endogenous timing system. (2) The spontaneous period produced by the circadian system usually deviates from 24 hours and must therefore be phase-set to the outer solar day by certain environmental periodicities called zeitgebers, the most important of which is the light–dark cycle (LD). (3) The entrained and free-running circadian rhythms may be modulated (masked) more or less intensely by inhibitory or enhancing direct effects of certain environmental and/or internal factors.

These facts lead to the following conclusions for primate research. (1) Whenever possible the whole (average) diel pattern(s) of the behavioural or physiological function(s) of interest should be established, or sampling should at least be carried out at an invariable circadian phase. (2) For an adequate assessment of the prevailing zeitgeber and masking conditions it is essential to give detailed information on the geographical co-ordinates of the study site, the season or vegetation period of data collection, the lighting conditions or times of sunrise, sunset and twilight duration, as well as on the ambient temperature, precipitation, food abundance, predator pressure, social system, etc. (3) Since in most nocturnal prosimians and owl monkeys moonlight may exert more or less pronounced masking effects on the entrained circadian rhythms of locomotor activity and other parameters, in field studies on such species it is indispensable to note moon phases and check for moonlight effects. (4) For luminance measurements only devices (luxmeters) that have a broad sensitivity range and a spectral sensitivity corresponding to that of the study species should be used. (5) Out of general chronobiological interest, field studies on a given primate species and/or population should record and report the timing of onset and end of the activity and/or the sleeping period in relation to sunrise or sunset. (6) Accelerometer data-loggers can be adapted for automated quantitative locomotor activity recordings in the field, in outdoor enclosures or in laboratory set-ups for medium-sized species, since more economical recordings can be made using computer-based passive infrared sensors. However, all automatic recordings first require calibration by comparing the recordings from several days with simultaneous behavioural observations.

REFERENCES

Albers, H. E., Lydic, R., Gander, P. H. & Moore-Ede, M. C. (1984). Role of suprachiasmatic nuclei in the circadian timing system of the squirrel monkey. I. The generation of rhythmicity. *Brain Res.* **300**, 275–84.

Alonso, C. & Langguth, A. (1989). Ecologia e comportamento de *Callithrix jacchus* (Primates: Callitrichidae) numa ilha de floresta atlantica. *Rev. Nordest. Biol.* **6**, 105–37.

Aschoff, J. & Tokura, H. (1986). Circadian activity rhythms in squirrel monkeys: entrainment by temperature cycles. *J. Biol. Rhythms* **1**, 91–9.

Aschoff, J., Daan, S. & Honma, K. I. (1982). Zeitgebers, entrainment, and masking: some unsettled questions. In *Vertebrate Circadian Systems. Structure and Physiology*, ed. J. Aschoff, S. Daan & G. A. Groos, pp. 13–24. Berlin: Springer-Verlag.

Bearder, S. K., Nekaris, K. A. I. & Buzzell, C. A. (2002). Dangers in the night: are some nocturnal primates afraid of the dark? In *Eat or be Eaten. Predator Sensitive Foraging Among Primates*, ed. L. E. Miller, pp. 21–43. Cambridge: Cambridge University Press.

Colquhoun, I. C. (1998). Cathemeral behaviour of *Eulemur macaco macaco* at Ambato Massif, Madagascar. *Folia Primatol.* **69**, 22–34.

Curtis, D. J. & Rasmussen, M. A. (2002). Cathemerality in lemurs. *Evol. Anthropol.* **11**, Suppl. 1, 83–6.

Curtis, D. J., Zaramody, A. & Martin, R. D. (1999). Cathemeral activity in the mongoose lemur, *Eulemur mongoz*. *Am. J. Primatol.* **47**, 279–98.

Donati, G., Lunardi, A., Kappeler, P. M. & Borgognini Tarli, S. M. (2001). Nocturnal activity in the cathemeral red-fronted lemur (*Eulemur fulvus rufus*), with observations during a lunar eclipse. *Am. J. Primatol.* **53**, 69–78.

DuMond, F. V. (1968). The squirrel monkey in a seminatural environment. In *The Squirrel Monkey*, ed. L. A. Rosenblum & R. W. Cooper, pp. 87–145. New York: Academic Press.

Engqvist A. & Richard, A. (1991). Diet as a possible determinant of cathemeral activity patterns in primates. *Folia Primatol.* **57**, 169–72.

Erkert, H. G. (1974). Der Einfluss des Mondlichtes auf die Aktivitätsperiodik nachtaktiver Säugetiere. *Oecologia* **14**, 269–87.

    (1989a). Lighting requirements of nocturnal primates in captivity: a chronobiological approach. *Zoo Biol.* **8**, 179–91.

    (1989b). Characteristics of the circadian activity rhythm in common marmosets (*Callithrix j. jacchus*). *Am. J. Primatol.* **17**, 271–86.

    (1997). Circadian rhythms in the marmoset: their significance for fundamental and applied research. In *Marmosets and Tamarins in Biological and Biomedical Research*, ed. C. H. Pryce, L. Scott & C. H. Schnell, pp. 128–44. Salisbury: DSSD Imagery.

    (1999). Owl monkeys. In *The UFAW Handbook on the Care and Management of Laboratory Animals*, 7th edition, ed. T. Poole & P. English, pp. 574–90. Oxford: Blackwell.

Erkert, H. G. & Gröber, J. (1986). Direct modulation of activity and body temperature of owl monkeys (*Aotus lemurinus griseimembra*) by low light intensities. *Folia Primatol.* **47**, 171–88.

Erkert, H. G. & Schardt, U. (1991). Social entrainment of circadian activity rhythms in common marmosets, *Callithrix j. jacchus* (Primates). *Ethology* **87**, 189–202.

Erkert, H.G., Nagel, B., Schanz, F. & Thiemann-Jäger, A. (1984). Vergleichende Untersuchungen zur Chronobiologie nichtmenschlicher Primaten. *Verh. Dt. Zool. Ges.* **77**, 217.

Farrer, D.N. & Ternes, J.W. (1969). Illumination intensity and behavioral circadian rhythms. In *Circadian Rhythms in Nonhuman Primates*, ed. F.H. Rohles, *Bibl. Primat.* **9**, 1–7. Basel: S. Karger.

Glass, J.D., Tardiff, S.D., Clements, R. & Mrosowsky, N. (2001). Photic and nonphotic circadian phase resetting in a diurnal primate, the common marmoset. *Am. J. Physiol.* **280**, R191–R197.

Hawking F. & Lobban, M.C. (1970). Circadian rhythms in macaca monkeys (physical activity, temperature, urine and microfilarial levels). *J. Interdisc. Cycle Res.* **1**, 267–90.

Kerl, J. & Rothe, H. (1996). Influence of cage size and cage equipment on physiology and behavior of common marmosets (*Callithrix jacchus*). *Lab. Prim. News.* **35**, 10–13.

Klein D.C., Moore, R.Y & Reppert, S.M. (eds.) (1991). *Suprachiasmatic Nucleus. The Mind's Clock.* New York, Oxford: Oxford University Press.

Kremers, J., Silveira, L.C.L., Yamada, E.S. & Lee, B.B. (1999). The ecology and evolution of primate color vision. In *Color Vision: From Genes to Perception*, ed. K.R. Gegenfurter & L.T. Sharpe, pp. 123–42. Cambridge: Cambridge University Press.

Lerchl, A., Küderling, I., Kurre, J. & Fuchs, E. (1988). Locomotor activity registration by passive infrared detection in saddle back tamarins and tree shrews. *Physiol. Behav.* **44**, 281–84.

Martin, R.D. (1990). *Primate Origins and Evolution: A Phylogenetic Reconstruction.* London: Chapman & Hall.

Martinez, J.L. (1972). Effects of selected illumination levels on circadian periodicity in the rhesus monkey (*Macaca mulatta*). *J. Interdisc. Cycle Res.* **3**, 47–59.

Michael, R.P. & Bonsall, R.W. (1977). A 3-year study of an annual rhythm in plasma androgen levels in male rhesus monkeys (*Macaca mulatta*) in a constant laboratory environment. *J. Reprod. Fert.* **49**, 129–31.

Moore, R.Y. (1999). Circadian timing. In *Fundamental Neuroscience*, ed. M.J. Zigmond, F.E. Bloom, S.C. Landis, J.L. Roberts & L.R. Squire, pp. 1189–206. San Diego, CA: Academic Press.

Moore-Ede, M.C., Sulzman, F.M. & Fuller, C.A. (1982). *The Clocks that Time Us: Physiology of the Circadian Timing System.* Cambridge, MA, London: Harvard University Press.

Nash, L.T. (1986). Influence of moonlight levels on travelling and calling patterns in two sympatric species of Galago in Kenya. In *Current Perspectives in Primate Social Dynamics*, ed. D.M. Taub & F.A. King, pp. 357–67. New York: Van Nostrand Reinhold Co.

(2000). Encounter rate estimates on *Lepilemur leucopus* and *Microcebus murinus* at Neza Mahafaly Special Reserve, Southwestern Madagascar. *Lemur News* **5**, 38–40.

Niemitz, C. (1984). *Biology of Tarsiers*. Stuttgart, New York: Gustav Fischer Verlag.

Overdorff, D. J. & Rasmussen, M. A. (1995). Determinants of nighttime activity in "diurnal" lemurid primates. In *Creatures of the Dark: The Nocturnal Prosimians*, ed. L. G. Alterman, G. A. Doyle & K. Izard, pp. 61–74. New York: Plenum Press.

Pálková, M., Sigmund, L. & Erkert, H. G. (1999). Effect of ambient temperature on the circadian activity rhythm in common marmosets, *Callithrix j. jacchus* (Primates). *Chronobiol. Int.* **16**, 149–61.

Pereira, M. E., Strohecker, R. A., Cavigelli, S. A. Hughes, C. L. & Pearson, D. D. (1999). Metabolic strategy and social behavior in Lemuridae. In *New Directions in Lemur Studies*, ed. B. Rakotosamimanana, H. Rasamimanana, J. U. Ganzhorn & S. M. Goodman, pp. 93–118. New York: Kluwer Academic/Plenum Publishers.

Petter-Rousseaux, A. (1975). Activité sexuelle de *Microcebus murinus* (Miller 1777) soumis à des régimes photopériodiques experimenteaux. *Ann. Biol. Anim. Biochem. Biophys.* **15**, 503–8.

Plant, T. M., Zumpe, D., Dauls, M. & Michael, R. P. (1974). An annual rhythm in the plasma testosterone of adult male rhesus monkeys maintained in the laboratory. *J. Endocr.* **62**, 403–4.

Rappold, I. & Erkert, H. G. (1994). Re-entrainment, phase-response and range of entrainment of circadian rhythms in owl monkeys (*Aotus lemurinus g.*) of different age. *Biol. Rhythm Res.* **25**, 133–52.

Rauth-Widmann, B., Thiemann-Jäger, A. & Erkert, H. G. (1991). Significance of nonparametric light effects in entrainment of circadian rhythms in owl monkeys (*Aotus lemurinus griseimembra*) by light–dark cycles. *Chronobiol. Int.* **8**, 251–66.

Richter, C. P. (1968). Inherent twenty-four and lunar clocks of a primate – the squirrel monkey. *Comm. Behav. Biol.* **1**, 305–32.

Schilling, A., Richard, J. P. & Servière, J. (1999). Duration of activity and period of circadian activity–rest rhythm in a photoperiod-dependent primate, *Microcebus murinus*. *C. R. Acad. Sci.* **322**, 759–70.

Schnell, C. R. & Wood, J. M. (1993). Measurement of blood pressure and heart rate by telemetry in conscious, unrestrained marmosets. *Am. J. Physiol.* **264**, H1509–H1516.

Sulzman, F. M., Fuller, C. A. & Moore-Ede, M. C. (1977a). Environmental synchronizers of squirrel monkey circadian rhythms. *J. Appl. Physiol.* **43**, 795–800.

(1977b). Feeding time synchronizes primate circadian rhythms. *Physiol. Behav.* **18**, 775–9.

(1979). Tonic effects of light on the circadian system of the squirrel monkey. *J. Comp. Physiol.* **129**, 43–50.

Tattersall, I. (1982). *The Primates of Madagascar*. New York: Columbia University Press.

(1987). Cathemeral activity in primates: a definition. *Folia Primatol.* **49**, 200–2.

The Astronomical Almanac. Issued each year by The Nautical Almanac Office, U.S.A. and Her Majesty's Nautical Almanac Office, Royal Greenwich Observatory. Washington, DC: US Government Printing Office; London: HMSO.

Thomas, R. M. & Curtis, D. J. (2001). A novel software application for the study of photoperiodic cueing mechanisms underlying circadian and circannual rhythms and lunar-periodic modulations. *Folia Primatol.* **72**, 187.

Tokura, H. & Aschoff, J. (1978). Circadian activity rhythms of the pig-tailed macaque, *Macaca nemestrina*, under constant illumination. *Pflügers Arch.* **376**, 241–3.

(1983). Effects of temperature on the circadian rhythm of pig-tailed macaques *Macaca nemestrina*. *Am. J. Physiol.* **245**, R800–R804.

van Horn, R. N. (1975). Primate breeding season: photoperiodic regulation in captive *Lemur catta*. *Folia Primatol.* **24**, 203–20.

van Schaik, C. P. & Kappeler, P. M. (1996). The social systems of gregarious lemurs: lack of convergence with anthropoids due to evolutionary disequilibrium? *Ethology* **102**, 915–41.

Wechselberger, E. (1995). Characteristika und Mechanismen der Synchronisation der Circadianperiodik des Weissbüscheläffchens, *Callithrix j. jacchus*. Ph.D. thesis, University of Tübingen.

Wright, P. C. (1989). The nocturnal primate niche in the New World. *J. Hum. Evol.* **18**, 635–46.

Yellin, A. M. & Hauty, G. T. (1971). Activity cycles of the rhesus monkey (*Macaca mulatta*) under several experimental conditions, both in isolation and in a group situation. *J. Interdisc. Cycle Res.* **2**, 475–90.

List of suppliers and their Internet sites

Actiwatch-AW4, Actiwatch-L, Actiwatch-L plus®: Cambridge Neurotechnology, Cambridge, UK Website: <camntech.co.uk>.

The Astronomical Almanac: Nautical Almanac Office of the United States, Washington, DC; Her Majesty's Almanac Office, London. Website: <aa.usno.navy.mil/data/>.

Octagonal Basic Motionlogger, Octagonal Basic Motionlogger-L®: Ambulatory Monitoring, Inc., Ardsley, NY, USA. Website: <www.ambulatory-monitoring.com>.

Optometer P970: Gigahertz Optik, Germany. Website: <www.gigahertz-optik.de>.

US Naval Observatory data services: <aa.usno.navy.mil/data/>.

STOWAWAY® and HOBO® light intensity loggers: <www.onsetcomp.com>.

18 • Thermoregulation and energetics

JUTTA SCHMID
Experimental Ecology, University of Ulm, Ulm, Germany

INTRODUCTION

The study of how animals apportion their time and energy (energetics) can provide much insight into physiology, ecology and evolution (Bartholomew, 1982; Schmidt-Nielsen, 1997). Body temperature has a profound effect on the ability of animals to function effectively. Since all animals generate heat internally to some extent, energetics is closely linked to the problem of heat management and thermoregulation. For example, homeothermic or 'warm-blooded' animals (birds and mammals) must produce a great amount of heat in order to maintain a high and constant body temperature in cold as well as in warm surroundings (Schmidt-Nielsen, 1997). Moreover, natural environments can be extremely variable in their thermal attributes and consequently animals show behavioural and physiological adaptations that enable them to cope with these external gradients.

In recent years, there has been a tremendous increase in methodologies and techniques applicable to studies of energy expenditure and thermoregulation. This is particularly true for the study of daily energy requirements and body temperature rhythms of wild animals behaving normally in their natural habitats. In this contribution, I therefore review methods for the study of energetics and mechanisms of temperature regulation in primates (although these can generally be applied to almost all mammalian species). I also briefly mention other possibilities for physiological measurements. All procedures described here require capture of the study animals, and some require invasive surgical intervention. They therefore raise ethical questions when dealing with wild animals and/or endangered species, and require governmental permits and authorisation.

SCIENTIFIC TERMS AND DEFINITIONS

BMR (*basal metabolic rate*): The energy required by an animal when it is resting and not performing any metabolic work to digest food or maintain its body temperature (also called *resting metabolic rate, RMR*).

Field and Laboratory Methods in Primatology: A Practical Guide, ed. Joanna M. Setchell and Deborah J. Curtis. Published by Cambridge University Press. © Cambridge University Press 2003.

RQ   (*respiratory quotient*): The ratio of carbon dioxide produced to oxygen consumed (gives information about the fuel used in metabolism).

DEE   (*daily energy expenditure*): The total amount of energy an animal expends during an entire day (including the BMR, costs of thermoregulation, locomotion, digestion, reproduction, etc.).

FMR   (*field metabolic rate*): DEE of a wild animal.

STUDYING ENERGY EXPENDITURE

INDIRECT CALORIMETRY (RESPIROMETRY)

Basal and maximal energy expenditures, and energy demands while engaged in specific activities, can be measured routinely using indirect calorimetry. This standard procedure involves measuring the heat production of an animal indirectly from its gaseous exchange. The metabolic rate of an animal is measured as either the rate of O_2 consumption or the rate of CO_2 production, and converted to energy expenditure (for the caloric equivalent, see Schmidt-Nielsen, 1997). For respirometry measurements the subjects are restricted to a relatively small chamber. In general, the size of the study animal is limited by the power of the gas analysis system, and the size of the metabolic chamber available. To conduct respirometry studies it is necessary to collaborate with specialists that have a fully equipped laboratory at their disposal.

Respirometry systems involve extensive and sensitive technical equipment and, until a few years ago, measurements could be carried out only in the laboratory. However, as an alternative to the experimental set-up of laboratory studies, metabolic rate can be measured for animals in the field using portable O_2 and CO_2 analyser assemblies (Schmid *et al.*, 2000) and portable electrical supplies. These are available from various manufacturers (see 'List of suppliers'), and can generally measure O_2 and CO_2 simultaneously. This is ideal, because free-living animals in their natural environment have an extensive behavioural repertoire and thus physiological data for wild animals can contribute a great deal to our understanding of rates of energy utilisation. Another advantage of this technique is that if your study animal naturally uses tree holes or burrows as sleeping places, then you can use these sleeping hollows as metabolic chambers, which significantly reduces disturbance to the animal. However, the estimates of O_2 consumption and CO_2 production made using such field systems are not as reliable and accurate as those from laboratory systems.

A general problem of all standard respirometry techniques is that animals cannot behave naturally in the small area of a metabolic chamber (whether you use a natural chamber in the field or an artificial chamber in the laboratory).

TIME–ENERGY BUDGETS

An alternative approach to estimating the daily energy demands of wild animals is to prepare time–energy budgets. In theory, the animal should be watched carefully throughout the day, and records accumulated of the time spent in each activity. To convert this time budget to an energy budget requires knowledge of the energy expenditure associated with each behaviour logged. Various approaches have been employed to do this (e.g. Schartz & Zimmermann, 1971; Weathers et al., 1984). However, time and energy budgeting has general problems as a method for estimating daily energy demands. It is often difficult to record the behaviours of the subject animals in any great detail; assigning realistic values for the energy expenditure associated with the different behavioural classes of the time budget can be problematic; and, finally, obtaining a detailed time budget is very time consuming.

DOUBLY LABELLED WATER

Theory

Energy expenditure and water metabolism can be measured routinely in both the field and the laboratory by means of doubly labelled water ($D_2{}^{18}O$) (Lifson & McClintock, 1966; Speakman, 1997). In principal, this measures the CO_2 production of the animal in a quantitative manner, which is then converted to energy demands. The technique involves labelling an animal's body water with two isotopes, deuterium (^2H, D) and heavy oxygen (^{18}O), and determining the washout rates of both isotopes. D is incorporated into water molecules in the animal and traces water flux through the animal. The ^{18}O equilibrates rapidly between water and CO_2 dissolved in body water, and so washes out faster than D, because it leaves the animal as $C{}^{18}O_2$ as well as $H_2{}^{18}O$. Thus the difference between washout rates of D and ^{18}O represents CO_2 production alone and is a measure of metabolic rate. Doubly labelled water probably provides the most powerful method for the measurement of CO_2 production, and hence energy expenditure, of free-living animals over protracted periods of one to several days, and can be applied just as well to animals kept in captivity.

The doubly labelled water method gives an integrated estimate of energy demands over relatively long periods (days). By combining several different measurements for animals with varying time budgets, the costs of major activity components can be factored out. Although the doubly labelled water method is less accurate than respirometry, the advantages are less disturbance for the animal, no need for electrical supplies, and that the resultant estimate includes passive and active phases of metabolism. Furthermore, it is easy to make simultaneous measurements on several individuals, and to measure individuals in their natural habitat and social situation. There are three major disadvantages to the doubly labelled water approach: first, a complete and successful measurement requires recapture of the injected individual within a certain period of time; secondly, it is very expensive; and, thirdly, the need for capture and manipulation of animals means that it may not be ethical to conduct such research on endangered species and there may also be problems obtaining permission from the local authorities to carry out such research.

Practice

The doubly labelled water technique is relatively complex and difficult to perform and it is therefore indispensable to contact a specialist and to consult the relevant literature for a precise description. Details concerning the efficacy of the technique in its application to various animal species can be found in Speakman (1997). Here, I outline the general procedure and important aspects that need to be taken into account when planning a study.

The practical minimum size limit for animals to be studied using doubly labelled water is a body mass of about 0.9–1.2 g (Speakman, 1997). The upper size limit relates only to the costs and availability of sufficient isotopes to label the animal. Animals selected for the experiment are injected with doubly labelled water ($D_2^{18}O$). The quantity of isotopes injected depends on the body mass of the animal, duration of the experimental measurement, and the enrichment of the injectate. The minimum duration of an experimental measurement is defined by the time required for sufficient divergence of the labels, and for sufficient isotopes to have been washed out of the body. It is important to know the exact quantity of stable isotopes administered and the exact body mass of the animal. For small animals, precise electronic balances (in the laboratory) or less accurate spring balances (in the field) are available. After injection, animals are held captive to allow complete equilibration of the isotopes in the body water. Estimates of time taken for isotopes to reach equilibrium vary from about 15–60 minutes (body mass less than 100 g) to several hours (humans) (Speakman, 1997). An initial sample is then taken. The most desirable sample sources are blood and saliva, since both are direct measures of the instantaneous state of the body water pool. Urine samples are not suitable for two reasons. First, the bladder acts as a water reservoir, which is not in complete exchange with body fluids, and bladder content therefore represents an integrated sample over an unknown time period prior to sample collection (Speakman, 1997). Secondly, it is unlikely to be possible to obtain a urine sample at a specific time of day, as the researcher cannot control the timing of urination!

The best approach for sampling blood from small animals is some form of peripheral venipuncture (methods in Chapter 8). Samples for isotope analyses should be flame sealed into capillary tubes. Capillaries are best protected in sample tubes containing cotton-wool to avoid breakage. They can be stored at room temperature (around 10–20 °C) and should not be left in the cold because they burst if frozen. If samples have to be taken overseas it is best to carry them in your hand luggage.

A second or final sample is taken 24 hours to several days after the first sample (depending on the experimental measurement). It is therefore necessary to recapture the injected animals. The problem is, however, that recapture cannot be guaranteed for any released animal in the field, particularly not within a certain time frame (before the isotopes have been washed out of the body).

Here are some points to consider before using doubly labelled water. Is one measurement for a given individual sufficient to provide reliable data? If not, what are the ethical implications of multiple captures and recaptures on one animal in terms of disturbing the animals and possibly also their place in the hierarchy within their group (Chapters 7 and 8)?

WHICH TECHNIQUE TO USE

To decide which is the most appropriate technique to measure the energy expenditure of animals you must weigh up the advantages and disadvantages of the various alternative procedures. Furthermore, there are several important factors and practical aspects to be considered when choosing a particular methodology:

- Are the experiments to be carried out in the laboratory or in the field?
- Does the chosen technique provide a sufficiently reliable answer to the specific question asked?
- What animal do you wish to study and does its natural activity or anatomy rule out the possibility of using a particular technique?
- What training will be required for the application of a particular technique? How and where will these training requirements be achieved and how much time is required?
- What are the licensing requirements of the proposed procedures and will they be acceptable to the legal authorities of the host country?

STUDYING BODY TEMPERATURE

CORE BODY TEMPERATURE

The core body temperature ($T_b$core) of an animal can be measured routinely by means of implanted temperature-sensitive transmitters. There are two types of transmitter: temperature-sensitive radio transmitters (Mini-Logger®, accuracy ±0.1 deg.C) or data-loggers (Thermochron ibutton®, temperature range −40 °C to +85 °C; accuracy ±0.5 deg.C). The major difference between the two is that data-loggers have internal memories and data are stored automatically. In contrast, signals from radio-transmitters need to be recorded manually or automatically using a telemetry assembly consisting of receiver and antenna (Chapter 10). When data-loggers are used, recorded data are down-loaded to a computer via its serial or parallel interface. Custom-made computer programs set and reset the data-logger, retrieve the data and define sampling methods. The disadvantage of using data-loggers is that you must recapture the same individual to download the data, or the data are lost. If you use radio-transmitters, each transmitter must be calibrated against a mercury thermometer in a water bath before implantation. Both data-loggers and transmitters should be sealed with a paraffin-wax coating before implantation into the animal to minimise the risk of infection or allergic reaction.

Animals should be weighed to the nearest 0.5 g before surgery. Transmitters can be implanted either into the peritoneal cavity or under the skin on the animal's back. The application of subdermal implantation depends on the size and anatomy of the examined species. All operations are carried out under deep anaesthesia (Chapter 8). Animals should be allowed to recover from the surgery for several hours before release. Surgery must be repeated to remove the transmitters when the experiment is finished (the duration of measurement depends on the experimental design and the researcher's time-table).

A non-invasive technique is to measure $T_b$core rectally by using a probe inserted 2–4 cm (depth depends on the species) into the rectum. This method is easy and is frequently used in the laboratory and in the field (Chapter 8).

A further approach to measuring $T_b$core is to use temperature-sensitive transponders. Transponders are especially designed for animal identification and, owing to their small size (length 10 mm), are suitable for use even in the smallest species. The system contains two basic elements: a passive transponder (ID tag) and the reader. The reader emits a flow-frequency magnetic field via its antenna and when a transponder passes within range, its ID code is received. Transponders are individually packaged in a disposable syringe and are injected subdermally (usually on the animal's back) without anaesthesia. When using temperature-sensitive transponders, you get both a reading of the ID of the animal and its body temperature. It is not necessary to calibrate the transponders before usage. One disadvantage of this approach is that the range of the reader is very small and to get a reading one has to be very close to the ID tag (1–2 cm).

SKIN BODY TEMPERATURE

Radio-collars with integrated temperature sensitive sensors (e.g. Biotrack) can be used to monitor skin body temperature ($T_b$skin) (Chapter 10). The radio-collar must be placed so that the temperature sensor is in contact with the animal's skin (usually around the neck). This $T_b$skin is a reasonable estimate of core body temperature, particularly when resting or torpid animals are in a curled-up position with the transmitter pressed against the ventral surface when it is placed around the neck (Audet & Thomas, 1996). Each radio-collar must be calibrated against a precision mercury thermometer in a water bath before application. Transmitter signals are received automatically or manually using receiver and antenna.

Note: During all body temperature measurements, the ambient temperature (T_a) should also be recorded continuously using data-loggers or calibrated thermometers (Chapter 5).

WHICH TECHNIQUE TO USE

The main practical aspect of monitoring body temperature is to decide which transmitter type is the most appropriate technique for the problem at hand. Transmitters differ in frequency, signal range and reception, battery lifespan, type of data recorded and weight. There are practical limits on the minimum size of an animal that can be studied using data-loggers or radio-collars. Generally, the weight of the transmitter used should be below the tolerable weight limit of 5% of the animal's body mass (Gursky, 1998; Chapter 10).

The most significant advantages of data-loggers are reduced disturbance of the study animal (no need to follow the animal in the wild), and that measurements can be made continuously for a long period without 'time out' periods. Furthermore, it is possible to measure large numbers of individuals at the same time, as well as at different study sites. The resultant measurements are truly simultaneous (repetition), which allows control for possible external factors.

In the field, the most obvious technique is to use temperature-sensitive radio-collars and telemetry equipment. The advantage of using this approach is that it is possible to make measurements simultaneously of home range, distance travelled and body temperature (Chapter 10). Even when data are not recorded automatically (stationary telemetry set-ups can record data automatically but they are expensive and time-consuming to install), it is still possible to make measurements of a sufficient number of individuals at one site over a short period of time. Such manually collected data will involve more irregular measurements, but still allow detection of significant effects.

All these methodologies for the measurement of body temperature are applicable to animals both in the field and in the laboratory. However, measurements of the body temperature of free-living animals in their natural environment are significantly more difficult to obtain than for animals in the laboratory. The principal difficulties in the field are the need for surgery and recovery, and the need to remain in close contact with the animal. The latter problem, however, can be avoided by the use of data-loggers. Implanting transmitters or data-loggers is theoretically and practically feasible in the field, but it is advisable to contact a veterinarian or someone who is experienced in surgery (Chapter 8).

OTHER POSSIBILITIES

Schnell & Wood (1993) have described a telemetry system for the continuous measurement of blood pressure and heart rate in conscious marmosets moving freely in their home cages. In this case pressure transmitters are implanted into the peritoneal cavity under anaesthesia and aseptic conditions. A fluid-filled sensor catheter connected to the transmitter is placed in the aorta below the renal artery (pointing upstream). Several biotelemetry receivers are distributed around the cage and signals are received and stored at defined intervals. Other techniques to record data on blood pressure and heart rate are described by Michel *et al.* (1984) and Wood *et al.* (1987).

Note: All techniques to measure blood pressure and heart rate are invasive and should be carried out by a veterinarian or someone who is experienced in surgery (Chapter 8).

LEGAL ASPECTS

Once you have decided to work with animals in the field or in the laboratory, the first consideration should be the legal aspects of applying a specific technique. Applying for permits can take some time, and you need to obtain the authorisations for your study in time to avoid a delay. Unfortunately, there are no global rules and regulations concerning the legal requirements, and if you intend to work abroad legislative requirements are different in different countries. Ethics committees usually review proposals for animal experiments to decide whether the research project is justified and how it should be carried out. If the procedures are classed as animal experiments, then you will need to comply with the appropriate legislation. Contact the ethics committee in the host country for further details concerning animal experimentation and specific requirements.

Permits or licenses will be required for both you and your assistants to capture and mark animals at a proposed field study site. You may also need a special permit for procedures such as injecting an animal, using stable isotopes, taking blood samples and implanting data-loggers or transmitters. Generally, it is necessary to submit a project proposal including information on the study animal, study site and a precise description of the techniques and material that will be used. For laboratory studies, you may also need to meet certain requirements concerning the conditions in which animals will be kept in captivity. Furthermore, you may have to pass training courses to prove you can perform to the standards of the country where you will be working. If you hold a licence to perform procedures in one country, this does not necessarily mean that other countries will accept your accreditation of competence, since standards differ between countries.

The final factor to consider is transportation of blood samples (and perhaps also isotopes) between countries. You will need a permit to export and import blood samples for the doubly labelled water analyses, and if the blood you have sampled is from an animal that is listed on the Appendices 1–3 of the CITES (Convention on International Trade in Endangered Species) regulations, you will need a CITES permit to import your samples.

TIME NEEDED IN THE FIELD AND IN THE LABORATORY

Estimates of time needed in the field and in the laboratory vary from about one week to several weeks, months, or up to a few years. In general, the length of a project depends on the experimental set-up, the questions asked, and the main emphasis of the project. Other time factors are the animal species with which you want to work, its abundance and 'co-operation' (e.g. trapping success, recapture rate; Chapter 7). Furthermore, if the technologies and materials you will be employing are not already established or used routinely, you may have to allow extra time, since establishing and setting up a new methodology is very time consuming.

COSTS

The costs of a study on thermoregulation and energetics fall into several categories: first, personnel to perform the work; secondly, equipment and materials (such as traps, radio-collars, receivers, data-loggers, balances, isotopes); and, thirdly, costs of the analysis. In addition, there are also travel and subsistence costs (inclusive of daily rates for the study site) for field and laboratory excursions. The costs for each study will depend, principally, on the methodologies you wish to apply, the final sample size, the country where the study will be carried out, and your personal access to technical equipment and analysis. Prices for specific equipment change rapidly, and are best found on the Internet or by consulting suppliers. For example, to complete a doubly labelled water study, at present-day prices (2002), O^{18} costs US$ 200 per ml and a 50 g bottle of deuterated water (minimum order) costs about US$ 75. The quantity of isotopes required will depend on the size of the animal and the required final sample size. Isotope analysis costs are currently about US$ 15 per isotope determination. In total, the costs for a two-sample protocol (initial sample and final sample), analysed in duplicate, come to about US$ 180 per animal successfully recaptured ($15 \times 12 = 180$).

REFERENCES

Audet, D. & Thomas, D. W. (1996). Evaluation of the accuracy of body temperature measurement using external transmitters. *Can. J. Zool.* **74**, 1778–81.

Bartholomew, G. A. (1982). Energy metabolism. In *Animal Physiology: Principles and Adaptations*, ed. M. S. Gordon, pp. 46–93. New York: Macmillan.

Gursky, S. (1998). Effects of radio transmitter weight on a small nocturnal primate. *Am. J. Primatol.* **46**, 145–55.

Lifson, N. & McClintock, R. (1966). Theory of use of the turnover rates of body water for measuring energy and material balance. *J. Theor. Biol.* **12**, 46–74.

Michel, J. B., Wood, J. M., Hofbauer, K. G., Corvol, P. & Menard, J. (1984). Blood pressure effects of renin inhibition by human renin antiserum in normotensive marmosets. *Am. J. Physiol.* **246**, F309–F316.

Schartz, R. L. & Zimmermann, J. L. (1971). The time and energy budget of the male dickcissel (*Spiza americana*). *Condor* **73**, 65–76.

Schmid, J., Ruf, T. & Heldmaier, G. (2000). Metabolism and temperature regulation during daily torpor in the smallest primate, the pygmy mouse lemur (*Microcebus myoxinus*) in Madagascar. *J. Comp. Physiol. B* **170**, 59–68.

Schmidt-Nielsen, K. (1997). *Animal Physiology: Adaptation and Environment.* Cambridge: Cambridge University Press.

Schnell, C. R. & Wood, J. M. (1993). Measurement of blood pressure and heart rate by telemetry in conscious, unrestrained marmosets. *Am. J. Physiol.* **264**, H1509–H1516.

Speakman, J. R. (1997). *Doubly Labelled Water. Theory and Practice.* London: Chapman & Hall.

Weathers, W. W., Buttemer, W. A., Hayworth, A. M. & Nagy, K. A. (1984). An evaluation of time-budget estimates of daily energy expenditure in birds. *Auk* **101**, 459–72.

Wood, J. M., Heusser, C., Gulati, N., Forgiarini, P. & Hofbauer, K. G. (1987). Sustained reduction in blood pressure during chronic administration of renin inhibitor to normotensive marmosets. *J. Cardiovasc. Pharmacol.* **7**, Suppl. 10, S96–S98.

List of suppliers and their Internet sites

Data-loggers for measuring body temperature, ambient temperature and humidity

Thermochron ibutton®: Maxim Headquarters, 120 San Gabriel Drive, Sunnyvale, CA 94086, USA. websites: <www.maxim-ic.com>, <www. ibutton.com>.

Radio-transmitters with attached temperature sensors

Biotrack Ltd: 52 Furzebrook Road, Wareham, Dorset BH20 5AX, UK. Tel: +44 (0) 1929 552 992; Fax: +44 (0) 1929 554 948; website: <www.biotrack.co.uk>.

Mini-Logger®: Mini Mitter Co., Inc., 20300 Empire Avenue Building B-3, Bend, OR 97701, USA. Tel: +1 (0) 541 322 7272 or +1 800 6832999; Fax: +1 (0) 541 322 7277; website: <www.minimitter.com>.

Micro-transponders

EURO I.D. Identifikationssysteme GmbH & Co.KG: Mettenicher Strasse 4, D-53919 Weilerswist, Germany. Tel: +49 (0) 2254 94090; Fax: +49 (0) 2254 940970; website: <www.euroid.com>.

Trovan™ Electronic Identification Systems, Ltd: Fax: +49 (0) 221 395893; website: <www.trovan.com>.

Data-loggers

Gemini Data Loggers (UK) Ltd.: Scientific House, Terminus Road, Chichester, West Sussex, PO 19 2UJ, UK. Tel: +44 (0) 1243 783210; Fax: +44 (0) 1243 531948; website: <www.geminidataloggers.com>.

synoTECH™ Sensor und Messtechnis GmbH: Postfach 1109, D-52437 Linnich, Germany. Tel: +49 (0)2462 9919–0; Fax: +49 (0)2462 9919 79; website: <www.synotech.de>.

Telemetry receivers

Telonics: 932 E. Impala Avenue, Mesa, Arizona 85204–6699, USA. Tel: +1 (0) 602 892 4444; Fax: +1 (0) 602 892 9139; website: <www.telonics.com>.

Televilt TVP Positioning AB: Bandygatan 2, SE-71135 Lindenberg, Sweden. Tel: +46 (0) 581 17195; Fax: +46 (0) 581 17196; website: <www. positioning. televilt.se>.

Oxygen analyser assembly

Ametek Precision Instruments Europe GmbH: Rudolf-Diesel-Strasse 16, D-40670 Meerbusch, Germany. Tel: +49 (0) 2159 9136–0; Fax: +49 (0) 2159 9136–39; website: <www.ametek.de>.

Drägerwerk AG: Dräger Medical AG & Co KgaA, Moislinger Allee 53–55, D-23542 Lübeck, Germany. Tel: +49 (0) 451 882 0; Fax: +49 (0) 451 882 2080; website: <www.draeger.com>.

Ahlborn Mess- und Regelungstechnik GmbH: Eichenfeldstrasse 1–3, D-83607 Holzkirchen, Germany. Tel: +49 (0) 8024 3007–0; Fax: +49 (0) 8024 3007–10; website: <www.ahlborn.com>.

19 • Field endocrinology: monitoring hormonal changes in free-ranging primates

J. KEITH HODGES AND MICHAEL HEISTERMANN
Department of Reproductive Biology, German Primate Centre, Göttingen, Germany

INTRODUCTION

Field endocrinology can be considered as the application of non-invasive methodologies to examine behavioural–endocrine interactions in primates living in natural conditions and social settings. In bringing together laboratory and field-based research methods, the discipline is providing new and exciting opportunities for developing a more integrated approach to studies of primate behavioural ecology (in this chapter 'primate' implies non-human primate).

Traditionally, field studies have relied mainly on visual measures, such as behaviour and/or morphology. Although this has generated a great deal of essential information, its limitation is that interpretation of the meaning of an observation is often based on assumptions concerning the physiological context in which it was made. The availability of non-invasive methodologies, based on measurement of hormones in either urine or faeces, now provides us with quantitative measures (of physiological status) by which the significance of these observational field data can be gauged. In turn, this greatly facilitates the testing of hypotheses concerning the adaptive significance of behavioural traits or mating systems and is helping to provide new insights into reproductive processes in an evolutionary context. For a comprehensive review of the topic of field endocrinology and its application to studies in primatology, readers should consult Whitten *et al.* (1998a).

In this chapter, we review the most important aspects of which fieldworkers have to be aware when planning and carrying out endocrinological studies on free-ranging primates.

SAMPLE COLLECTION

URINE

Options for collecting urine samples under field conditions include, (1) the use of sheeting, a tray or bowl to collect urine mid stream (applicable to

Field and Laboratory Methods in Primatology: A Practical Guide, ed. Joanna M. Setchell and Deborah J. Curtis. Published by Cambridge University Press. © Cambridge University Press 2003.

arboreal or semi-arboreal species); (2) aspiration (using a syringe or pipette) or absorption onto filter paper from foliage; or (3) recovery (usually by centrifugal force) from samples of earth. Volumes as small as 0.1 ml may be sufficient (provided further reduction during transport and storage does not occur), and more than 1 ml is generally not necessary.

FAECES

Faecal samples are usually easier to identify and locate than urine and can be collected directly from the ground. A thumb-nail sized amount (about 0.5–2 g) provides more than enough material for analysis. In species where the bolus size is larger, the sample should be homogenised using a gloved hand or improvised spatula on a leaf or other flat surface, before transferring the required amount (not more than 30% of total volume) to the storage container. Avoid the inclusion of large amounts of non-faecal material, such as seeds and non-digested foliage.

GENERAL POINTS

It is essential that only samples of known origin are collected, in other words only from individuals actually seen to be urinating or defecating. This is particularly important for collection of samples from sleeping sites. It is also essential to avoid sample contamination. Cross-contamination (principally, faeces with urine) is the most likely and can partially be reduced in big samples by collecting an aliquot from within the faecal bolus. Other sources of contamination are dilution of urine with water and sample-to-sample contamination through use of the same syringe, spatula for homogenisation, glove, etc. Where it is not possible to use disposable materials, the implement should be rinsed with water and wiped dry; use of detergents should be avoided.

Time of collection is a variable that needs to be controlled for. Diurnal patterns of secretion are particularly pronounced for some hormones (e.g. testosterone, cortisol). Although they are likely to be more evident in urine than in faeces, they may still be seen in the faeces of certain small-bodied species (e.g. callitrichids; Sousa & Ziegler, 1998) in which faecal passage rate is high. Whenever possible, samples should be collected early in the morning, or at least the period of collection should be restricted to roughly the same time of each day. Urine and faeces, like all other body fluids and tissues from primates, are a potential source of pathogens and thus infection. They should be handled with care and appropriate precautions taken to reduce the risk of infection at all stages of collection, storage, transport and the subsequent analysis.

SAMPLE STORAGE

URINE

While most steroids (progesterone is one exception) are relatively robust and stable at ambient temperature for several days, conjugated forms (the predominant form in urine; present in variable amounts in faeces) and creatinine (the protein breakdown marker used to index urinary water content) are less so. Steroid breakdown due to bacterial metabolism can be a potential problem in faeces, where gastrointestinal bacteria are abundant. Thus, in general, the interval from sample collection to preservation should be kept as short as possible. Immediate preservation at the site of collection is preferable, but otherwise the sample should be kept cool (in an insulated cool-box) until properly preserved after return to the camp.

The principal (most commonly applied) and preferable method of storing urine long term is freezing (-10 °C or below). This may limit long-term sample storage at field sites and, since repeated freezing and thawing should be avoided, also requires use of dry ice to maintain samples in a frozen condition throughout all stages of transport (this can be expensive and certain airlines impose restrictions on transportation of dry ice). Alternative methods are (1) storage as an ethanolic solution, (2) the use of sodium azide (0.1%) as a preservative or (3) absorption onto filter paper and storage dry. The former is preferable and enables samples to be stored at ambient temperatures without deterioration for several weeks. The final solution should be within the range 20–50% ethanol, and, since this must be kept constant for all samples, accurately measured volumes of urine must be used. One major disadvantage with the use of sodium azide is that it may interfere with the hormone assay step, if enzyme immunoassays are used. Storage of aqueous or ethanolic solutions on filter paper has been shown to be useful, but can introduce a number of potential analytical errors and needs to be very carefully controlled (cf. Shideler *et al.*, 1995).

FAECES

Faeces are easier to store under field conditions. Using a minimum ratio of 4 : 1 (volume : weight), samples can be stored in >80% ethanol at ambient temperature for periods of up to two years. Ethanol has a further advantage in that vials or tubes containing a known volume of ethanol can be taken into the field, enabling samples to be transferred to the preservation medium immediately after collection (cf. Ziegler *et al.*, 2000). Potential problems with the use of ethanol include evaporation (acceptable in limited amounts) and spillage or leakage. Particular attention must therefore be paid to the type of storage container used. Many are not as leak proof as they are claimed to be (particularly when alcohol is used) and sample spillage during transport will seriously affect the reliability of the results. Where possible, screw-cap vials (10–20 ml range) should be used and tested rigorously (with ethanol) beforehand. As an extra precaution, ensure that sample vials are kept upright at all times. Glass vials are more inert, but are fragile and need careful packing before transport; plastic tubes are safer to transport, although there is a small risk that steroids might stick more to the wall during long storage periods. Partial purification or extraction of the sample in the field before storage is possible (but in our experience, not advantageous) and is described by Whitten *et al.* (1998a).

Containers should be labelled properly, giving animal name/ID number and date (and time of collection if useful). Special sticky paper labels that resist solvents and freeze–thawing processes are preferable and as an additional precaution, apply clear waterproof tape over the labels. Relevant information should also be written directly on the tube, cap, or (preferably) both, using an indelible, waterproof marker (use black only). A separate list of samples should be kept for each animal for cross-checking purposes. Inappropriate labelling can result in loss of information on sample identity (usually during transport), rendering analysis meaningless. This continues to be one of the major causes of loss of data in this type of work, and Murphy's law will guarantee that the unidentifiable samples will always be the ones of particular importance.

SAMPLE PREPARATION

URINE

Steroids in urine are present predominantly in the conjugated form, either as sulphate or glucuronide residues. These can be measured either directly in appropriately diluted urine or by first cleaving the conjugate by hydrolysis followed by use of an assay designed specifically for the parent compound (e.g. measurement can be either as oestrone 3-glucuronide or after hydrolysis as oestrone). For details of methods for enzymatic and non-enzymatic hydrolysis, see Ziegler *et al.* (1996). If hydrolysis is performed, it is usual to carry out an extraction step, whereby steroids are removed into an organic phase (e.g. diethylether), which is reconstituted in aqueous buffer after evaporation for assay. Steroid conjugates are best extracted with aqueous alcohol or (more usually) assayed directly without an extraction step. A separate aliquot (usually 0.02–0.05 ml) of each sample is required for determination of creatinine content either by a creatinine analyser or micro-titre plate method (cf. Bahr *et al.*, 2000).

FAECES

An extraction step is always necessary to measure steroids in faeces. Various methods exist for extracting steroids from frozen–thawed faeces; the more commonly used involving agitating (shaking/vortexing) a known weight of sample with an aqueous solution (40–80%) of methanol or ethanol.

The suspension is allowed to settle (or centrifuged) and a portion of the supernatant is taken for assay or subjected to further purification steps (such as re-extraction or use of Sep-Pack C-18 mini-columns). The final hormone content is expressed per unit mass (grams) of wet faeces or dry faeces after determination of the dry weight of the faecal pellet. Similar procedures are used with samples stored in ethanol. In our experience the most efficient way of dealing with these is to homogenise the sample in the original solvent (e.g. by using a metal spatula), shaking it overnight and, if necessary (depending on the efficiency of the procedure), re-extracting it with additional (e.g. 10 ml) 80% methanol.

Alternatively (and our own preferred method), faecal samples that are kept frozen can be lyophilised directly in the frozen state (also known as freeze-drying) and then pulverised (usually using a pestle and mortar) before methanol extraction (cf. Heistermann et al., 1995a). Although the combined process is relatively time consuming, the advantages are that: (1) freeze-drying compensates for differences in faecal water content; (2) it allows non-faecal material (e.g seeds, stones, undigested fibre) to be easily removed; (3) pulverisation produces a homogeneous powder that can be pre-weighed, aliquoted and then stored at room temperature; and (4) the extraction step is easier and generally does not require further purification steps before assay. With an efficient freeze-drier, up to 120 samples can be processed simultaneously, requiring a total of about 72 hours.

In certain circumstances chromatographic procedures (thin-layer chromatography, high pressure liquid chromatography) may be required in order to detect and/or aid identification of individual steroid components of sample extracts, usually for assay validation purposes (cf. Bahr et al., 2000).

ASSAY METHODOLOGY

Measurements of hormones and their metabolites are usually carried out by immunological procedures utilising hormone- or hormone-group-specific antibodies. Two main types of immunoassay are available. Radioimmunoassays (RIA), which use radioactively labelled hormone as the competitive tracer in the quantification process, and enzyme immunoassays (EIA) in which either enzyme- or biotin-labelled preparations are employed. Being non-isotopic, EIAs avoid the problems associated with use and disposal of radioactivity and also are less costly.

Since all immunoassays are highly sensitive, assay performance must be carefully assessed both during the initial set-up phase and during routine use. There are four main criteria of validation: sensitivity (minimum amount of hormone that can be detected); precision (within- and between-assay repeatability), accuracy (ability to detect the correct amount of hormone in the sample) and specificity. The latter has two components: the degree of specificity of the antibody itself and the possible influence of interfering substances excreted with urine and faeces (matrix effects), which need to be controlled for and removed by incorporating additional sample purification steps if present. The presence of such matrix effects must therefore be carefully determined before any routine use of an assay.

Concerning antibody specificity, highly specific assays may be useful when the identity of the major metabolite is known and when species comparisons are of interest. Since, however, excreted samples (especially faeces) contain a number of metabolites, a specific measurement is often difficult and might be less useful in cases where the antibody detects only metabolites of low abundance. Group-specific assays utilise antibodies that cross-react with several metabolites of related structure. Since knowledge of the relative abundance of individual metabolites is not necessary, these assays have the advantage that they can usually be applied to a wider range of species (see Heistermann et al., 1995b; Schwarzenberger et al., 1997), thus helping to overcome the problems of species specificity in hormone metabolism. When measuring faecal androgens to assess testicular function, however, group-specific antibodies may lead to problems owing to the co-measurement of androgens of adrenal origin.

Available data show that in the majority of primates, the direct measurement of oestrone conjugates (E1C) and pregnanediol-3-glucuronide (PdG) is most useful for monitoring ovarian function and pregnancy using urine analysis (for a review, see Heistermann et al., 1995b). Alternatively, non-specific measurement of C19/C21-progesterone metabolites, such as androsterone or 20α-hydroxyprogesterone, can be used for macaques. Urine also provides a matrix for the measurement of polypeptide and proteohormones (e.g. follicle-stimulating hormone, luteinising hormone, oxytocin), which cannot be measured in faeces.

Information on the measurement of androgens in primate urine is limited. To date, most studies have determined immunoreactive testosterone, although this is a relatively minor component. Our own recent data suggest that measurement of 5-reduced androstanes might reflect testicular endocrine activity better (Möhle et al., 2002). The measurement of urinary cortisol provides a reliable method for monitoring glucocorticoid output in a variety of primate species (cf. Crockett et al., 1993; Robbins & Czekala, 1997; Whitten et al., 1998b), but the more abundant 11,17-dioxoandrostanes (Bahr et al., 2000) and 5ß-reduced androstanolones might be more suitable, at least in some species (M. Heistermann, unpublished data).

The most appropriate assays for the measurement of oestrogen and progesterone metabolites in faeces are shown in Table 19.1. Assays for oestradiol-17ß or oestrone (or a collective measurement of both) are generally useful for monitoring ovarian function and pregnancy, but in some species oestrogen measurements in faeces appear to be unreliable. The majority of the progestin measurements have been based on the use of non-specific progesterone, pregnanediol assays, which measure 5-reduced 20-oxo- or 20-hydroxylated pregnanes, the most abundant progesterone metabolites in the faeces of most

Table 19.1. *Application of faecal oestrogen and progesterone metabolite assays for the assessment of female reproductive status in primates*

Taxa	Species	Oestrogen[a]	Progestin[a]	Reference
Lemuridae	*Eulemur* spp.	E_t	20α-OHP	Curtis et al., 2000
			5-P-3OH	Ostner & Heistermann, 2003;
				P. Gerber et al., unpublished data
	Hapalemur griseus	E_2	5-P-3OH	Gerber et al., unpublished data
Indridae	*Propithecus verreauxii*	E_1C	P_4	Brockman & Whitten, 1996
Loridae	*Nycticebus coucang*	E_2	–[b]	Jurke et al., 1997
Callitrichidae	*Callithrix jacchus*	E_2	Pd, P_4	Heistermann et al., 1993
	Saguinus oedipus	?[c]	P_4, Pd	Ziegler et al., 1996
	Saguinus fuscicollis	E_1C	Pd	Heistermann et al., 1993
	Leontopithecus rosalia	E_t	PdG	French et al., 2002
	Callimico goeldii	E_2	?[c]	Pryce et al., 1994
Cebidae	*Brachyteles arachnoides*	–[b]	P_4	Strier & Ziegler, 1997
	Cebus apella	E_1C	P_4, Pd	Carosi et al., 1999
	Pithecia pithecia	E_1C	PdG	Shideler et al., 1994
	Ateles geoffroyi	E_1C	PdG	Campbell et al., 2001
Cercopithecinae	*Macaca fascicularis*	E_2	PdG	Shideler et al., 1993
	Macaca silenus	E_2	P_4	Matsumuro et al., 1999
	Macaca fuscata	E_1C	5-P-3OH	Heistermann et al., 2001a
			PdG	Fujita et al., 2001
	Cercocebus torquatus	E_2	P_4	Whitten & Russell, 1996
	Papio cynocephalus	E_2	P_4	Wasser et al., 1991
Colobinae	*Semnopithecus entellus*	E_2	20α-OHP	Heistermann et al., 1995a
			PdG	Ziegler et al., 2000
	Pygathrix nemaeus	E_t	5-P-3OH	M. Heistermann et al., unpublished data
Pongidae	*Pan paniscus*	–[b]	P_4, Pd	Heistermann et al., 1996
	Pan troglodytes	?[c]	PdG	T. Deschner et al., unpublished data
	Gorilla gorilla	E_2	Pd	M. Heistermann et al., unpublished data

[a] E_t, total oestrogen; E_2, oestradiol-17β; E_1C, oestrone conjugates; Pd, pregnanediol; PdG, pregnanediol-glucuronide; 20α-OHP, 20α-hydroxyprogesterone; P_4, progesterone; 5-P-3OH, 5α-pregnane-3α-ol-20-one.

[b] Measurement not successful.

[c] No information available.

mammals, including primates (Schwarzenberger *et al.*, 1997). Which assay best reflects female reproductive status depends on the species of interest (Table 19.1), although in general faecal progestin profiles are easier to interpret (compared with oestrogen profiles) particularly with respect to monitoring of ovarian cycles.

Data on the excretion of testicular and adrenal steroids in primate faeces are limited. In some species (ring-tailed lemur (*Lemur catta*), Cavigelli, 1999; muriqui (*Brachyteles arachnoides*), Strier *et al.*, 1999; chimpanzee (*Pan troglodytes*), Whitten *et al.*, 1998b), cortisol assays appear to yield useful information, although cortisol itself is either barely detectable or not present (Bahr *et al.*, 2000). Measurement of 11,17-dioxoandrostanes or 5ß-reduced androstanolones might be more suitable (Wallner *et al.*, 1999; M. Heistermann *et al.*, unpublished data), but further work is needed to identify the most appropriate faecal glucocorticoid assay in primates. Testosterone metabolism is complex and highly species specific (Möhle *et al.*, 2002). In lemur species, measurement of immunoreactive testosterone has yielded informative results on male androgen status (cf. Brockman *et al.*, 1998; von Engelhardt *et al.*, 2000), whereas in the New World marmoset monkeys (*Callithrix* spp.) testicular endocrine function can be monitored by 5-reduced androstanes, but not by testosterone. In Old World monkey species and great apes, a reliable assessment of male testosterone secretion from faecal samples appears to be extremely difficult, probably due to the confounding effect of dehydroepiandrosterone (DHEA), a weak androgen from the adrenal gland, which produces the same metabolites as testosterone itself (Möhle *et al.*, 2002).

GENERAL CONSIDERATIONS

SPECIES VARIATION IN METABOLISM AND EXCRETION

Steroids circulating in the bloodstream undergo a series of metabolic changes before finally being eliminated from the body. The nature of these changes can vary considerably between species, resulting in differences not only in the nature and identity of the metabolites themselves, but also in their preferred route of excretion. This can have important consequences for the selection of an appropriate measurement system and correct interpretation of results obtained (see Perez *et al.*, 1988).

TIME LAG

Whereas circulating hormones more or less reflect real-time changes in endocrine activity, hormones in excreta reflect events that have occurred in the past. In urine the time lag is usually only about 4–8 hours (cf. Bahr *et al.*, 2000), but even so, with once daily sampling, this can be enough to delay detection of an event (e.g. peak, defined rise) by one day. Time lags associated with faecal measurements are longer and more variable, both between and within species, and gut passage times can be affected by a variety of factors including diet, health status and stress level. In most large-bodied species for which data are available, steroids are excreted in faeces 36–48 hours after appearance in circulation (cf. Shideler et al., 1993), although there are exceptions (e.g. 22 hours for testosterone and cortisol in the chimpanzee; Bahr et al., 2000). In contrast, passage time is quicker in animals of smaller size (e.g. 4–8 hours for testosterone and cortisol in the common marmoset). The two main consequences of these time lags (especially for faeces) are: (1) a delay between the occurrence of a specific event and its detection, which needs to be accounted for when faecal measurements are used to determine the timing of events such as ovulation or implantation; and (2) a dampening effect, reducing the amplitude of hormonal changes, and making it more difficult to detect short-lived endocrine responses to acute situations (e.g. stress).

SAMPLING FREQUENCY

Sample frequency is largely determined by the type of information required, although how often samples can actually be collected is influenced by numerous practical considerations. Information on overall physiological condition/status requires less frequent sampling than information on dynamics or timing of events. As a general rule, in order to compensate for intra-individual (sample-to-sample) variation (and also for statistical considerations), it is recommended that sample frequency be so arranged that not less than 6 and preferably 10 samples are collected per condition (i.e. before versus after birth; breeding versus non-breeding season; before and after change in rank; animal 1 versus animal 2).

Weekly samples should be sufficient to follow the course of pregnancy, although increased sampling (twice per week) may be useful during the period leading up to parturition. To detect the presence of ovarian cycles, twice weekly samples are the absolute minimum, whereas defining the duration or interval between two events requires more regular samples collected at a higher frequency (exactly what frequency depends on the acceptable margin of error). Generally, this will be ± 1 days at three or four samples per week and ± 2 days at two samples per week. For the purposes of timing ovulation, daily samples are usually required. It is useful to try to aim for a slightly higher frequency than is perhaps necessary, since it is unlikely that all samples will be collected as planned.

In addition to sample frequency, a variety of other factors can also affect the reliability of the information obtained from hormone assay of field-based samples. Important among these are assay precision and the criteria used to define the events or parameters under investigation. The first refers to the amount of inherent variability in the measurement system and the second is basically a question of how one interprets the results. As no two samples will have identical hormone content, it is essential to differentiate between random variation and patterns of potential significance in relation to the parameters under investigation. One elevated progesterone value does not indicate an ovulatory cycle, but how many do, at what intervals and at what frequency? What is elevated? Ovulation is best timed retrospectively according to the rise in progesterone occurring at the onset of the luteal phase. Here, the question is 'What is a rise?' There are no hard and fixed rules, but, in our experience, increase (and maintenance) above a threshold value determined as the mean plus two standard deviations of preceding (three to five) baseline values provides a statistically significant and useful and informative method for defining a rise in hormone levels (cf. Heistermann et al., 2001b). Appropriate definitions (providing objective criteria for assessment) are essential to meaningful interpretation of data.

REFERENCES

Bahr, N., Palme, R., Möhle, U., Hodges, J. K. & Heistermann, M. (2000). Comparative aspects of the metabolism and excretion of cortisol in three individual non-human primates. *Gen. Comp. Endocrinol.* **117**, 427–38.

Brockman, D. K. & Whitten, P. L. (1996). Reproduction in free-ranging *Propithecus verreauxi*: estrus and the relationship between multiple partner matings and fertilization. *Am. J. Phys. Anthropol.* **100**, 57–69.

Brockman, D. K., Whitten, P. L., Richard, A. F. & Schneider, A. (1998). Reproduction in freeranging male *Propithecus verreauxi*: the hormonal correlates of mating and aggression. *Am. J. Phys. Anthropol.* **105**, 137–51.

Campbell, C. J., Shideler, S. E., Todd, H. E. & Lasley, B. L. (2001). Fecal analysis of ovarian cycles in female black-handed spider monkeys (*Ateles geoffroyi*). *Am. J. Primatol.* **54**, 79–89.

Carosi, M., Heistermann, M. & Visalberghi, E. (1999). Display of proceptive behaviours in relation to urinary and fecal progestin levels over the ovarian cycle in female tufted capuchin monkeys (*Cebus apella*). *Horm. Behav.* **36**, 252–65.

Cavigelli, S. A. (1999). Behavioural patterns associated with faecal cortisol levels in free-ranging female ring-tailed lemurs, *Lemur catta*. *Anim. Behav.* **57**, 935–44.

Crockett, C. M., Bowers, C. L., Sackett, G. P. & Bowden, D. M. (1993). Urinary cortisol responses of long-tailed macaques to five cage sizes, tethering, sedation, and room change. *Am. J. Primatol.* **30**, 55–74.

Curtis, D. J., Zaramody, A., Green, D. I. & Pickard, A. R. (2000). Non-invasive monitoring of reproductive status in wild mongoose lemurs (*Eulemur mongoz*). *Reprod. Fert. Develop.* **12**, 21–9.

French, J. A., De Vleeschouwer, K., Bales, K. & Heistermann, M. (2002). Lion tamarin reproductive biology. In *Lion Tamarins: Biology and Conservation*, ed.

D.G. Kleiman & A.B. Rylands, pp. 133–56. Washington, DC: Smithonian Institution Press.

Fujita, S., Mitsunaga, F., Sugiura, H. & Shimizu, K. (2001). Measurement of urinary and fecal steroid metabolites during the ovarian cycle in captive and wild Japanese macaques, *Macaca fuscata. Am. J. Primatol.* **53**, 167–76.

Heistermann, M., Tari, S. & Hodges, J.K. (1993). Measurement of faecal steroids for monitoring ovarian function in New World primates, Callitrichidae. *J. Reprod. Fert.* **99**, 243–51.

Heistermann, M., Finke, M. & Hodges, J.K. (1995a). Assessment of female reproductive status in captive-housed Hanuman langurs (*Presbytis entellus*) by measurement of urinary and fecal steroid excretion patterns. *Am. J. Primatol.* **37**, 275–84.

Heistermann, M., Möstl, E. & Hodges, J.K. (1995b). Non-invasive endocrine monitoring of female reproductive status: Methods and applications to captive breeding and conservation of exotic species. In *Research and Captive Propagation*, ed. U. Ganslosser, J.K. Hodges & W. Kaumanns, pp. 36–48. Erlangen: Filander Verlag GmbH.

Heistermann, M., Möhle, U., Vervaecke, H., van Elsacker, L. & Hodges, J.K. (1996). Application of urinary and fecal steroid measurements for monitoring ovarian function and pregnancy in the bonobo (*Pan paniscus*) and evaluation of perineal swelling patterns in relation to endocrine events. *Biol. Reprod.* **55**, 844–53.

Heistermann, M., Uhrigshardt, J., Husung, A., Kaumanns, W. & Hodges, J.K. (2001a). Measurement of faecal steroid metabolites in the lion-tailed macaque (*Macaca silenus*): a non-invasive tool for assessing ovarian function. *Primate Rep.* **59**, 27–42.

Heistermann, M., Ziegler, T., van Schaik, C.P., Launhardt, K., Winkler, P. & Hodges, J.K. (2001b). Loss of oestrus, concealed ovulation and paternity confusion in free-ranging Hanuman langurs. *Proc. Roy. Soc. Lond. B* **268**, 2445–51.

Jurke, M.H., Czekala, N.M. & Fitch-Snyder, H. (1997). Non-invasive detection and monitoring of estrus, pregnancy and the postpartum period in pygmy loris (*Nicticebus pygmaeus*) using fecal estrogen metabolites. *Am. J. Primatol.* **41**, 103–115.

Matsumuro, M., Sankai, T., Cho, F., Yoshikawa, Y. & Yoshida, T. (1999). A two-step extraction method to measure fecal steroid hormones in female cynomolgous monkeys (*Macaca fascicularis*). *Am. J. Primatol.* **48**, 291–8.

Möhle, U., Heistermann, M., Palme, R. & Hodges, J.K. (2002). Characterization of urinary and fecal metabolites of testosterone and their measurement for assessing gonadal endocrine function in male nonhuman primates. *Gen. Comp. Endocrinol.* **129**, 135–45.

Ostner, J. & Heistermann, M. (2003). Endocrine characterization of female reproductive status in wild red-fronted lemurs (*Eulemur fulvus rufus*). *Gen. Comp. Endocrinol.* in press.

Perez, L. E., Czekala, N. M., Weisenseel, K. A. & Lasley, B. L. (1988). Excretion of radiolabelled estradiol metabolites in the slow loris (Nycticebus coucang). Am. J. Primatol. 16, 321–30.

Pryce, C. R., Schwarzenberger, F. & Döbeli, M. (1994). Monitoring fecal samples for estrogen excretion across the ovarian cycle in Goeldi's monkey (Callimico goeldii). Zoo Biol. 13, 219–30.

Robbins, M. M. & Czekala, N. M. (1997). A preliminary investigation of urinary testosterone and cortisol levels in wild male mountain gorillas. Am. J. Primatol. 43, 51–64.

Schwarzenberger, F., Palme, R. Bamberg, E. & Möstl, E. (1997). A review of faecal progesterone metabolite analysis for non-invasive monitoring of reproductive function in mammals. Z. Säugetierk. 62, Suppl. II, 214–21.

Shideler, S. E., Ortuño, A. M., Moran, F. M., Moorman, E. A. & Lasley, B. L. (1993). Simple extraction and enzyme immunoassays for estrogen and progesterone metabolites in the feces of Macaca fascicularis during nonconceptive and conceptive ovarian cycles. Biol. Reprod. 48, 1290–8.

Shideler, S. E., Savage, A., Ortuño, A. M., Moorman, E. A. & Lasley, B. L. (1994). Monitoring female reproductive function by measurement of fecal estrogen and progesterone metabolites in the white-faced saki (Pithecia pithecia). Am. J. Primatol. 32, 95–105.

Shideler, S. E., Munro, C. J., Johl, H. K., Taylor, H. W. & Lasley, B. L. (1995). Urine and fecal sample collection on filter paper for ovarian hormone evaluations. Am. J. Primatol. 37, 305–15.

Sousa, M. B. & Ziegler, T. E. (1998). Diurnal variation in the excretion patterns of fecal steroids in common marmosets (Callithrix jacchus). Am. J. Primatol. 46, 105–17.

Strier, K. B. & Ziegler, T. E. (1997). Behavioral and endocrine chracteristics of the ovarian cycle in wild muriqui monkeys, Brachyteles arachnoides. Am. J. Primatol. 32, 31–40.

Strier, K. B., Ziegler, T. E. & Wittwer, D. J. (1999). Seasonal and social correlates of fecal testosterone and cortisol levels in wild male muriquis (Brachyteles arachnoides). Horm. Behav. 35, 125–34.

von Engelhardt, N., Heistermann, M. & Kappeler, P. (2000). Androgen levels and female social dominance in Lemur catta. Proc. Roy. Soc. Lond. B 267, 1533–9.

Wallner, B., Möstl, E., Dittami, J. & Prossinger, H. (1999). Fecal glucocorticoids document stress in female barbary macaques (Macaca sylvanus). Gen. Comp. Endocrinol. 113, 80–6.

Wasser, S. K., Monfort, S. L. & Wildt, D. E. (1991). Rapid extraction of faecal steroids for measuring reproductive cyclicity and early pregnancy in free-ranging yellow baboons (Papio cynocephalus cynocephalus). J. Reprod. Fert. 92, 415–23.

Whitten, P. L. & Russell, E. (1996). Information content of sexual swellings and fecal steroids in sooty mangabeys (Cercocebus torquatus atys). Am. J. Primatol. 40, 67–82.

Whitten, P. L., Brockman, D. K. & Stavisky, R. C. (1998a). Recent advances in non-invasive techniques to monitor hormone-behavior interactions. *Yrbk. Phys. Anthropol.* **41**, 1–23.

Whitten, P. L., Stavisky, R. C., Aureli, F. & Russell, E. (1998b). Response of fecal cortisol to stress in captive chimpanzees (*Pan troglodytes*). *Am. J. Primatol.* **44**, 57–69.

Ziegler, T. E., Scheffler, G., Wittwer, D. J., Schultz-Darken, N. & Snowdon, C. T. (1996). Metabolism of reproductive steroids during the ovarian cycle in two species of callitrichids, *Saguinus oedipus* and *Callithrix jacchus*, and estimation of the ovulatory period from fecal steroids. *Biol. Reprod.* **54**, 91–9.

Ziegler, T. E., Hodges, J. K., Winkler, P. & Heistermann, M. (2000). Hormonal correlates of reproductive seasonality in wild female Hanuman langurs (*Presbytis entellus*). *Am. J. Primatol.* **51**, 119–34.

20 • Collection, storage and analysis of non-invasive genetic material in primate biology

BENOÎT GOOSSENS[1], NICOLA ANTHONY[3], KATHRYN JEFFERY[1], MIREILLE JOHNSON-BAWE[1,2] AND MICHAEL W. BRUFORD[1]
[1]School of Biosciences, Cardiff University, Cardiff, UK
[2]Centre International de Recherches Médicales, Franceville, Gabon
[3]Department of Biological Sciences, University of New Orleans, New Orleans, USA

WHY NON-INVASIVE?

Until recently, genetic studies of wild non-human primate populations (henceforth called primates in this chapter) have lagged behind those of other groups of animals studied by ecologists and evolutionary biologists. During the 1980s when, for example, avian, carnivore and hymenopteran molecular studies dominated the pages of high profile journals, primate studies were notable by their absence, and, to date, with a few exceptions, this remains the case. Two reasons underlie this imbalance.

First, primate populations do not provide the necessary replicates, sample sizes and relatively simple structures of many other groups. Social systems are often complex, generation times are long and habituated groups are extremely precious commodities. Molecular ecological methods were not always able to answer the questions that primatologists wanted to ask.

Secondly, obtaining genetic material before the advent of faecal and hair genotyping was logistically and ethically almost impossible for many populations, especially in groups where habituation over many years might have been severely affected by a large-scale, invasive sampling programme. Such problems particularly prevented many of the long-term studies that characterise primatology from taking up the challenge of adding a genetic dimension, even though there was potentially a great deal of new information to be gained. Non-invasive genetic analysis using new, high precision, molecular tools is therefore an extremely important development in primatology. However, it remains extremely difficult, time-consuming, expensive and prone to error. It is therefore important that the potential user is aware of the limitations, pitfalls, and complex methodologies that await them, and we attempt to outline these issues here. We cover protocols for most currently utilised non-invasive sample types, including faeces, urine, hair and food wadges. We also describe storage protocols for blood and tissue biopsy samples for occasions when animals are captured and handled.

Field and Laboratory Methods in Primatology: A Practical Guide, ed. Joanna M. Setchell and Deborah J. Curtis. Published by Cambridge University Press. © Cambridge University Press 2003.

APPLICATIONS

Molecular genetic information is beginning to have a major impact in primatology, with the promise of pioneering studies in the early–mid 1990s (e.g. Morin *et al.*, 1994) now being realised (e.g. Constable *et al.*, 2001). Here we focus on population and individual level studies, where the interface with ecology and behaviour is at its strongest, but it is noteworthy that molecular phylogenetic studies are continually adding to our knowledge of primate diversity, evolution and hence adaptation. Furthermore, phylogenetic analysis below the species level can be used to study the underlying biogeographical factors that have contributed to the diversity present in primate populations today. This approach has been used to highlight new, evolutionarily distinct populations within well-studied species and to pinpoint potentially important geographical barriers that may delimit genetic divergences across the range of species (e.g. Gonder *et al.*, 1997). A recent development has been the amplification, sequence analysis and evolutionary study of viral genomes from faecal samples in wild primate populations since the advent of efficient kit methods for extracting RNA from this material, which has special relevance for immunodeficiency virus research.

At the population level and below, it is now apparent that molecular genetic analysis can be applied in a myriad of contexts from mating systems and paternity through to population history over demographic timescales. Characterising the demographic relationships among, and measuring gene flow between, populations potentially enables prioritisation of distinct population units and can inform conservationists on their optimal management. The management and maintenance of genetic diversity is an issue for concern for all biologists studying threatened species and is of great relevance in primate populations, where habitat fragmentation, separation and population diminution will have major implications for the long-term viability of many species in the near future. Fragmentation necessarily results in reduced population sizes and may in time result in significant genetic drift (changes in allele frequencies due to demographic stochasticity), inbreeding (reproduction between related individuals) and susceptibility to further stochastic demographic contraction (population bottlenecks). These phenomena can act multiplicatively and may lead to rapid loss of genetic diversity, which can be measured at the genetic level through departure of allele frequencies from expectation under random assortment and through loss of heterozygosity and of alleles at genetic loci relative to larger populations (e.g. Frankham *et al.*, 2002).

EVOLUTIONARY ECOLOGY

The implications of non-invasive genetic analysis for resolving questions pertaining to the behavioural ecology and population structure of primates are clear, and the application of these approaches to long-term studies of habit-uated populations is now a major activity in primatology (e.g. Morin *et al.*, 1994; Altmann *et al.*, 1996; Constable *et al.*, 2001). Non-invasive molecular approaches can allow studies of mating systems (paternity, maternity) and can explore fine-scale patterns of group structure previously thought to be impossible. For some populations it is even possible to gain a picture of the social system, including dispersal, territory range and genetic structure, without knowledge of individuals (or even seeing them), by tracking individual genotypes (and molecular sexing) from faeces and hair samples in space and time. 'Genotype matching' approaches of this nature are still in their infancy and have been successfully carried out in a few non-primates to date, but such studies are underway in several primate populations. Rapid population census (measurement of the fraction of adults breeding, operational sex ratios and family size variances) and conservation assessments of unstudied primate populations could potentially be improved in the future using these methods. Molecular tools can also be used to manage populations *ex situ*, both in primate colonies for research and in zoological parks, and, whereas classical genetic fingerprinting (which requires a blood sample) has traditionally been used, non-invasive tools will undoubtedly become more useful in larger enclosures and less intensively managed species.

MOLECULAR MARKERS

While it is not the objective of this chapter to review the molecular markers used in population studies in any detail, it is worth highlighting the fact that different markers have different properties and relative abundance in the material that is likely to be collected in the field. By far the two most commonly used approaches in primatology today are mitochondrial (mt) DNA sequencing and microsatellite fragment analysis. MtDNA is a small, circular plasmid genome found in the mitochondrion and is found in many copies (up to hundreds) per cell. It is inherited clonally (as a single, linked genetic system) through the maternal line only and its DNA sequences evolve approximately an order of magnitude more rapidly than chromosomal DNA in the nucleus of the cell. Although mtDNA evolves rapidly, by accumulating 'point' mutations in the DNA bases, there are plenty of regions that are sufficiently conserved to provide comparisons and enable markers to be developed across closely related species. The primary usage of mtDNA is in phylogeny and phylogeography – and identifying genetically distinct units for conservation (e.g. Gagneux *et al.*, 2001) – but it can also be used to examine genetic distinctiveness within populations, although its behaviour (as a maternally inherited marker) is strongly affected by patterns of female philopatry and dispersal.

Microsatellites are found mainly in the chromosomes in the nucleus and are present in thousands of copies scattered throughout the genome.

They comprise simple repetitive elements (e.g. repeated sections of two DNA bases – e.g. GAGAGAGA ..., etc.) and their variation is derived mainly through changes in the number of DNA repeats at any given location. They are highly variable, with up to 20 alleles per locus being common – and heterozygosity (possession of two different length alleles) at any given locus, or genome location, can commonly be up to 80%. The application of 10–20 of these loci can provide the researcher with both an individual-specific genotype and an estimate of the genetic similarity between individuals. Microsatellites are commonly used in studies of paternity and social structure and are the tools of choice in behavioural studies (see Sunnucks, 2000). Microsatellite primers can produce polymorphic markers in related species (e.g. across apes and Old World monkeys; Coote & Bruford, 1996), but cross-species utility does have its evolutionary limits, and primatologists studying New World monkeys and strepsirrhines may need to isolate, characterise and develop markers in their study species (e.g. Radespiel *et al.*, 2001), although, as more of these species are studied, more markers will become available for general use. Primer sequences for closely related species may be available in GenBank (see 'Useful Internet sites', at the end of this chapter).

NON-INVASIVE GENOTYPING

DNA sampled non-invasively is present in much lower amounts than in blood and tissue. This poses special problems for the geneticist. The polymerase chain reaction (PCR) is used to amplify DNA from such poor sources of material. PCR involves the enzymatic amplification of DNA through the use of short pieces of DNA ('primers'), which identify and anneal (bind) to a specified complementary sequence either end of the DNA under study. PCR is carried out by denaturing (making single-stranded) template DNA at high temperatures (94 °C), and allowing the DNA to cool to a predefined temperature, at which point annealing of the primers to the template DNA occurs. Once the sequences have annealed, a heat-tolerant DNA polymerase (*Taq* polymerase, isolated from *Thermophilus aquaticus*) can extend this small section of double-stranded DNA and make two copies from the one (template) copy. The DNA template and new copy are both then denatured at high temperature and the process begins again. This PCR reaction is repeated, in practice 20–40 times, to produce millions of copies of the DNA template and, crucially, to give the researcher enough genetic material to analyse and genotype.

The problem with non-invasively collected material is that sometimes the DNA is in very low copy number, often it can be highly fragmented and sometimes PCR may be inhibited by co-extracted compounds present in the material. All these problems make contamination of the PCR reaction with extraneous DNA a real possibility. Consequently DNA extraction has to be (1) highly efficient, (2) not unnecessarily destructive to the integrity of the DNA, (3) able to remove inhibitory material during purification and (4) relatively quick and uncomplicated, avoiding unnecessary steps. Low template DNA copy number and PCR inhibition have led to several phenomena being observed in non-invasive genotyping.

First, PCR product may be extremely difficult to generate and resultant product fragments may not be sufficient for analysis. Here, using more PCR cycles (up to 40 is advisable) or a second round of PCR, using the fragments generated in the first to 'seed' the reaction, may help. However, the more cycles used, the greater the possibility that a copying error may have been introduced by the DNA polymerase, giving a false polymorphism, therefore replicate PCRs are imperative to confirm the results. False data may occur in DNA sequences (artificial point mutations) or in microsatellite fragments (false allele lengths due to DNA polymerase 'slippage' during the PCR reaction). DNA polymerase slippage is a general phenomenon of microsatellite PCRs, but can usually be allowed for by recording only the one (for homozygotes) or two (for heterozygotes) most intensely amplified fragments. False alleles may confuse this procedure, but are usually very weak fragments, and should in any case be replicated.

Secondly, the stochastic non-amplification of one of the two potential alleles at a microsatellite locus can occur ('allelic dropout'). This can occur because of low template copy number or DNA degradation. The latter may be a special problem for loci exhibiting a wide range of allele lengths, because longer alleles may not be amplifiable if their length exceeds the maximum fragment size present in the degraded template DNA. Repeated amplifications using several independent DNA extractions (see below) are a minimum requirement in such studies.

COLLECTING, STORING AND TRANSPORTING MATERIAL FROM THE FIELD

FAECES

Use a stick or sterile plastic or wooden spoon (10–15 cm long) to approximately half-fill a clean 50 ml collection tube with faeces. Do not touch the sampling end of the stick, or the inside of the collection tube and lid, to avoid contamination with your own DNA. Film pots make useful collection tubes, but avoid plastic bags unless the faeces are dry and will be stored in silica gel. Collection tubes do not need to be sterile, although of course this is better. Fresh faeces are best (preferably immediately after defecation), and the outer layer should be sampled preferentially. Faeces degrade and/or disappear quickly in tropical forest, but may remain and desiccate in other environments. The efficiency of DNA extraction from faeces samples of different age varies for different species and populations and requires careful testing before a large-scale study is attempted. Wear disposable plastic or vinyl gloves, and never touch faeces with bare hands to avoid cross-contamination of the samples and potential disease transmission (Chapters 1 and 8). Hepatitis B vaccination is essential for those likely to come into direct contact with primate faeces, particularly ape faeces. Label collection tubes using indelible pen. Information should include Global Positioning System (GPS) co-ordinates (Chapter 4), identity, sex and age of the individual (where known), date, the name of the collector and a sample code. Repeat the sample code on the lid of your tube and duplicate all the information in a field notebook (preferably waterproof).

Return to the field station before filling collection tubes with storage solution. Some of the best preservation methods for faeces are to use silica gel beads to dry the sample, or a DMSO-EDTA-Tris-salt solution (DETs: 20% dimethylsulphoxide (DMSO); 0.25 M ethylenediamine tetra-acetic acid (EDTA), sodium salt; 100 mM Tris-(hydroxymethyl) aminomethane; pH 7.5 and NaCl to saturation). Other viable methods use 70% ethanol, absolute ethanol, freezing at −20 °C or in liquid nitrogen. RNA later® (Ambion, Inc.) is probably the optimal method: it is expensive, but excellent for preservation of RNA and DNA. Surgical spirit (which has the advantage of being available almost everywhere) may be used as a last resort. Add storage solution to your collecting tube until it covers the sample. Once the tube has been capped, seal with Parafilm M® (or similar) to prevent leakage. Ethanol can remove labelling ink and one way around this problem is to include a pencil-written label in the tube, but this runs the risk of contamination. Once samples have been stored, check them periodically for leakage and evaporation. Storage times may affect DNA quality, so it is best to extract samples as soon as possible (within three to six months). Our laboratory and others have, however, successfully extracted good-quality DNA from faeces stored in 70% ethanol for three or more years. Finally, if the samples are to be analysed by a third party, always take a hair sample (about 10 plucked hairs) from yourself to check for possible human-mediated contamination.

URINE

Urine has emerged only recently as a potential source of DNA for non-invasive genetic studies (e.g. Hayakawa & Takenaka, 1999; Valière & Taberlet, 2000). In orang-utans (*Pongo pygmaeus*) and chimpanzees (*Pan troglodytes*), researchers have been able to collect urine from plastic sheets placed under sleeping-nests. Individuals will often urinate from the side of the nest on awakening and fluid released onto plastic sheets can be collected and transferred to storage vials using disposable plastic pipettes. Field researchers have also been able to collect urine opportunistically during field observations or from surrounding vegetation following urination. However, little genetic work has been carried out on samples collected from free-ranging individuals in the field (Valière & Taberlet, 2000). Epithelial cells suspended in the urine are likely to be the primary source of DNA. Fermentation and presumably DNA degradation of these cells may occur rapidly after urination (Hayakawa & Takenaka, 1999) and as large a volume as is possible should be collected and transferred into two volumes of 95% ethanol as rapidly as possible after excretion.

HAIR

Hair can be problematic as a source of DNA, but the following guidelines may help. Plucked hairs are by far the best source of hair DNA for both mitochondrial and microsatellite DNA analysis. In studies of humans and gorillas the majority of plucked hairs have been found to be in the active growth or breakdown phase (anagen or catagen) and their roots are therefore packed with mitotic cells. Single plucked hairs with root material should present no problem for genetic analysis, given adequate storage conditions (but if possible, collect more than 10 hairs per individual). However, for shed hair, while mtDNA is present in the root and hair shaft, nuclear DNA is present only in the root material (Linch et al., 2001). The roots of naturally shed hairs have usually undergone programmed cell death before shedding (telogen phase), and much of the nuclear DNA is degraded. Shed hairs will provide a good source of mtDNA, but a much poorer source of nuclear DNA. However, epithelial tissue may be attached to the root and in the case of freshly shed hairs this can provide a source of undegraded nuclear DNA (Linch et al., 1998).

Plucking hair is usually possible only in captive animals or study groups that are subject to veterinary procedures when the animal is under anaesthetic. Plucking hairs from free-ranging primates is rarely attempted, although Valderrama et al. (1999) describe novel methods for doing so with capuchin monkeys (Cebus olivaceus) and baboons (Papio hamadryas). These include shooting duct tape at the target animal and wrapping food baits with duct tape. Gloves should be worn and multiple hairs pulled from the root. Hairs are best placed in clean paper envelopes, since plastic produces static that make hair manipulation difficult. Ensure that the envelopes are not touched on the inside and are stapled or folded shut (never licked).

In studies of many primate species shed hair collection may not be possible, or may prove extremely difficult, but for nesting species it is possible to locate vacated nests and recover hairs shed by the occupant. For a few species (e.g. gorillas, Gorilla spp.), many nests are located on the ground; however, tree dwellers pose a more practical problem of accessibility. While mtDNA may be amplified from single hair shafts from most nests, regardless of age, we have found that microsatellite success can be greatly affected by nest age and field efforts may be better directed to collection from fresh nests only.

Wear gloves and avoid contamination by collecting hairs with tweezers that are sterilised with a flame between each use. Place hairs into paper envelopes as before. Record the following information: date, GPS co-ordinates, sample ID, nest age, height of nest from ground, collector's initials, and reference to any faecal samples collected nearby. Note whether the nest is solitary or in a group. Nest groups should be assigned an ID with the total number of nests recorded, and each nest numbered within the group – the best storage method for hairs is the simplest: desiccated with self-indicating silica granules, either at room temperature or frozen at −80 °C. While DNA degradation occurs over time when the hair is in a nest, we have not observed this problem with hairs sealed in envelopes, desiccated and stored for up to eight years.

WADGES

Takenaka and colleagues have used buccal cells extracted from discarded food items (wadges) as a source of DNA, having successfully amplified mitochondrial and microsatellite DNA from sugarcane chewed and spat out by free-ranging chimpanzees (e.g. Hashimoto *et al.*, 1996). These wadge samples were collected from the ground using gloves, and about 50 g transferred to a sterile 50 ml polypropylene tube filled with 90% ethanol and 1 mM Na_3EDTA. Storage in this solution avoids bacterial and enzymatic degradation of the DNA. The samples were kept in the field for 10–40 days before being transferred to the laboratory for analysis (Sugiyama *et al.*, 1993).

BLOOD AND TISSUE

For invasively extracted material such as blood and tissue biopsy, storage is equally crucial to prevent DNA degradation and damage. Where freezing is not an option, blood can be stored at room temperature in an equal volume of $1 \times$ STE buffer (0.1 M NaCl, 10 mM Tris, and 1 mM EDTA, pH 8.0) or, less optimally, can be stored in 1 ml of absolute ethanol. DNA remains intact in TNE buffer ($2 \times$ TNE buffer is 10 mM Tris-HCl, 1 M NaCl, 0.1 M EDTA) for several months but it is advisable to use a refrigerator as soon as possible. Tissue biopsies are optimally stored in 25% (weight/volume) DMSO dissolved in 6 M (saturated) NaCl: score the surface of the tissue with a blade and immerse the biopsy fully in the DMSO solution. Tissue stored this way preserves DNA effectively indefinitely. Alternatively Queens Lysis Buffer (QLB) has been used and is also a highly efficient method for preserving DNA (Bruford *et al.*, 1998; Radespiel *et al.*, 2001).

TRANSPORTING SAMPLES

CITES (Convention on International Trade in Endangered Species) and health/agriculture department authorisations may be needed for blood, faecal and hair samples from endangered species, and other permits may also be required for import and export. Pack samples in sealed containers, surrounded by sufficient absorbent material to absorb an accidental spillage. The samples and the absorbent material should be sealed in a leak-proof plastic bag and placed in either a metal container, a strong cardboard box or a polystyrene box inserted into an outer cardboard box. Swab the inner and outer packaging with disinfectant before leaving the country of origin. Envelopes containing hairs should be packed inside either plastic bags or large envelopes, then inside well-sealed, airtight plastic boxes containing silica grains.

LABORATORY METHODS

Contamination with human DNA is a serious problem in non-invasive studies of primates. Carry out faecal extractions in a separate laboratory dedicated exclusively to faecal DNA analysis, and include negative controls (reagents only) in each extraction to monitor for contamination. We also recommend: physical separation of the laboratories where pre- and post-PCR experiments are carried out; avoiding handling concentrated DNA extracts in the pre-PCR room; using pipettes dedicated to non-invasive studies in a laminar-flow hood and aerosol-resistant pipette tips (filter tips); the use of face-masks; and continuously monitoring all reagents for DNA contamination via negative PCR controls.

DNA EXTRACTION

Faeces

Cells containing DNA are not uniformly spread throughout faeces, and two or three extracts should be made per sample. It is also important to choose an extraction method that eliminates the need for laborious centrifuging to remove detergents, salts and enzymes from extracts and involves fewer steps and sample transfers, thereby minimising the potential for contamination. Various kits exist for the extraction of DNA from faeces, all of which come with full instructions. We recommend the QIAamp Stool mini kit (QIAGEN, currently US$ 160 for 50 reactions (i.e. 25 samples); Goossens et al., 2000), which has given the best results in primates (lowland gorillas; orang-utans; Hanuman langurs, *Semnopithecus entellus*; baboons, *Papio cynocephalus*) and other mammals (black rhinoceros, *Diceros bicornis*; forest elephants, *Loxodonta cyclotis*; giant panda, *Ailuropoda melanoleuca*) in our laboratory. Other possibilities include silica-based procedures (Boom et al., 1990), diatomaceous earth protocols (Gerloff et al., 1995), magnetic beads (Flagstad et al., 1999) and Chelex-100® (Walsh et al., 1991).

Urine

Once you have arrived back in the laboratory, urine samples should be centrifuged and the resulting sediment washed with physiological buffer containing 1 mM EDTA. These sediments are then re-suspended in 250 µl of STE buffer, 30 µl of 10% sodium dodecylsulphate (SDS), and are digested with proteinase K (20 µl of a 5 mg/ml stock) for 2 hours at 55 °C. Lysates are then extracted once with 200 µl of phenol-chloroform: IAA (chloroform prepared as a chloroform: isoamyl alcohol mixture of 24:1) and in a final step DNA is precipitated and then redissolved in 20 µl of TE (10 mM Tris, 1 mM EDTA, pH 8.0) buffer.

Hair

The rule here is: the simpler the better. The extraction of hairs in Chelex-100® and proteinase K (Walsh *et al.*, 1991) is a popular method in some laboratories. However, Vigilant (1999) obtained superior results from a simpler method that uses *Taq* polymerase PCR buffer as the extraction buffer (see Allen *et al.*, 1998). We recommend this method and have found that using PCR buffer, water and proteinase K in a small extraction volume works very well for shed hairs.

Wadges

In the laboratory, Takenaka *et al.* (1993) transferred their wadge samples into 200 ml of physiological saline and filtered the resulting solution through two sheets of gauze. The cells were pelleted by centrifugation, filtered through a nylon mesh and centrifuged again. Several methods have been employed to extract DNA from wadges (see Hashimoto *et al.*, 1996), all of which have appeared to be equally successful at recovering DNA. For routine applications, Hashimoto *et al.* (1996) suspended cells in 200 µl of CTAB buffer with 2% cetyltrimethyl-ammonium bromide as a detergent, and digested with 10 µl of 5 mg/ml proteinase K solution for 2 hours at 60 °C. DNA was then extracted with 1 : 1 mix of phenol–chloroform, precipitated in two volumes of ethanol, suspended in 200 µl of TE (5 mM Tris, 0.1 mM EDTA) buffer and enzymatically treated with 2 µl of a 10 mg/ml solution of ribonuclease A for 30 minutes at 37 °C and then with 10 µl of a 5 mg/ml solution of proteinase K for 2 hours at 60 °C. DNA was then extracted once more with phenol–chloroform mix, reprecipitated and dissolved in final volume of 200 µl of TE buffer.

Blood and tissue

Many standard methods exist for extracting DNA from blood and tissue. See Bruford *et al.* (1998) for details of salt extraction and phenol extraction methods.

Other sources

Nuclear DNA has also been amplified from male ejaculates, as well as from female menstrual blood and wounds (Formosan macaque, *Macaca cyclopis*; Chu *et al.*, 1999). In these cases, blood and ejaculates were stored in 1 ml of STE buffer and 1 ml of absolute ethanol. Nuclear DNA was subsequently extracted using a standard phenol–chloroform extraction procedure (e.g. Bruford *et al.*, 1998).

DNA AMPLIFICATION

Recent improvements in *Taq* polymerases, including those that become active only after a 10 minute incubation at 95 °C (e.g. AmpliTaq Gold™, Perkin

Elmer), reduce non-target amplification via a 'hot start' PCR and thereby allow more cycles to occur while maintaining accuracy in the PCR product. A multiple tubes procedure (Taberlet *et al.*, 1996) is the best method to provide reliable genotyping when using samples with very small and unknown DNA quantities, to avoid problems associated with allelic dropout, false alleles and sporadic contamination (see Introduction). This involves repeating PCR experiments three to seven times using aliquots of the same DNA extract and deducing the genotype by analysis of the whole set of results.

We strongly recommend a pilot study using non-invasively collected samples (5–10 samples) collected in the field from the proposed study population. Different storage reagents can then be compared directly by extracting and amplifying the DNA. The removal of dietary inhibitors (Wasser *et al.*, 1997) and different faecal drying methods (Murphy *et al.*, 2002) can also be tested at this point.

ADDITIONAL INFORMATION

Once genotyping systems have been optimised, large-scale PCRs may be carried out in confidence. This is important, since the integrity of genetic data is vital for all studies using non-invasively collected DNA and must be subjected to intense scrutiny prior to publication. Genotyping systems such as the 'manual' methods, which use radioactive and silver or other DNA staining methods, or the semi-automated approaches favoured by modern laboratories owing to their increased sensitivity of detection, are beyond the scope of this chapter. However, the use of multiple independent scoring of alleles and sequences is always desirable and, for semi-automated analyses, base-calling and fragment allele length assignments must always be checked manually by several people. Scoring the genotypes of samples using blind testing procedures is also desirable for these studies.

Once a matrix of samples and confirmed genotypes has been assembled, data analysis can commence. The following softwares are available freely for different applications (see Useful Internet sites).

POPULATION GENETICS

GENEPOP computes exact tests for Hardy–Weinberg equilibrium, population differentiation and genotypic linkage disequilibrium among pairs of loci. It computes estimates of classical population parameters, such as F-statistic and other correlations. It is IBM (DOS) compatible.

GENETIX estimates standard population genetics parameters and produces factorial correspondence plots. It is in French and is IBM compatible.

FSTAT 2.9.3 estimates and tests gene diversities and differentiation statistics for co-dominant genetic markers. It computes estimators of gene diversities and F-statistic, and tests them using randomisation methods. It is IBM compatible.

PARENTAGE AND RELATEDNESS

KINSHIP 1.3.1 performs maximum likelihood tests of pedigree relationships between pairs of individuals in a population using genotype information for single locus, co-dominant genetic markers (such as microsatellites). Kinship also calculates pairwise relatedness statistics. It runs on Macintosh only.

CERVUS 2.0 is useful for large-scale parentage analysis using co-dominant loci. It can analyse allele frequencies, run appropriate simulations and carry out likelihood-based parentage analysis, testing the confidence of each parentage using the results of the simulation. Simulations may also be used to estimate the power of a series of loci for parentage analysis, using real or imaginary allele frequencies. It is IBM compatible.

COSTS AND TIMING

Population genetic studies can be costly. For example using 15 microsatellite markers for 300 faecal samples collected in the field will cost approximately US$ 22 000 (US$ 1000 for collection, US$ 4000 for extraction, US$ 12 000 for amplification, US$ 5000 for DNA typing) at today's prices, and may take up to 24 months once the genotyping systems have been optimised, depending on how difficult the samples prove to be. A paternity analysis for 50 individuals using eight microsatellite markers would cost approximately US$ 8000 and may take up to six months.

REFERENCES

Allen, M., Engstrom, A. S., Meyers, S., Handt, O., Saldeen, T., von Haeseler, A., Pääbo, S. & Gyllensten, U. (1998). Mitochondrial DNA sequencing of shed hairs and saliva on robbery caps: sensitivity and matching probabilities. *J. Forensic Sci.* **43**, 453–64.

Altmann, J., Alberts, S., Coote, T., Dubach, J., Geffen, E., Haines, S. A., Wayne R. K., Muruthi, P. & Bruford, M. W. (1996). Behaviour predicts genetic structure in a wild primate group. *Proc. Natl. Acad. Sci., USA* **93**, 5797–801.

Boom, R., Sol, C. J. A., Salimans, M. M. M., Jansen, C. L., Wertheim-van Dillen, P. M. E. & van der Noordaa, J. (1990). Rapid and simple method for purification of nucleic acids. *J. Clin. Microbiol.* **28**, 495–603.

Bruford, M. W., Hanotte, O. & Burke, T. (1998). Single and multilocus DNA fingerprinting. In *Molecular Genetic Analysis of Populations: A Practical Approach*, 2nd edition, ed. A. R. Hoelzel, pp. 225–69. Oxford: Oxford University Press.

Chu, J.-H., Wu, H.-Y., Yang, Y.-J., Takenaka, O. & Lin, Y.-S. (1999). Polymorphic microsatellite loci and low invasive DNA sampling in *Macaca cyclopis*. *Primates* **40**, 573–80.

Constable, J. L., Ashley, M. V., Goodall, J. & Pusey, A. E. (2001). Noninvasive paternity assignment in Gombe chimpanzees. *Mol. Ecol.* **10**, 1279–300.

Coote, T. & Bruford, M. W. (1996). A set of human microsatellites amplify polymorphic markers in Old World Apes and Monkeys. *J. Hered.* **87**, 406–10.

Flagstad, Ø., Røed, K., Stacy, J. & Jakobsen, K. S. (1999). Reliable non-invasive genotyping based on excremental PCR of nuclear DNA purified with a magnetic bead protocol. *Mol. Ecol.* **8**, 879–83.

Frankham, R., Ballou, J. D. & Briscoe, D. A. (2002). *Introduction to Conservation Genetics.* Cambridge: Cambridge University Press.

Gagneux, P., Gonder, M. K., Goldberg, T. L. & Morin, P. A. (2001). Gene flow in wild chimpanzee populations: what genetic data tell us about chimpanzee movement over space and time. *Proc. Roy. Soc. Ser. B* **356**, 889–97.

Gerloff, U., Schlötterer, C., Rassmann, K., Rambold, I., Hohmann, G., Fruth, B. & Tautz, D. (1995). Amplification of hypervariable simple sequence repeats (microsatellites) from excremental DNA of wild living bonobos (*Pan paniscus*). *Mol. Ecol.* **4**, 515–18.

Gonder, M. K., Oates, J. F., Disotell, T. R., Forstner, M. R. J., Morales, J. C. & D. J. Melnick (1997). A new West African chimpanzee subspecies? *Nature* **388**, 337.

Goossens, B., Chikhi, L., Utami, S. S., de Ruiter, J. R. & Bruford, M. W. (2000). A multi-samples, multi-extracts approach for microsatellite analysis of faecal samples in an arboreal ape. *Cons. Genet.* **1**, 157–62.

Hashimoto, C., Furuichi, T. & Takenaka, O. (1996). Matrilineal kin relationship and social behaviour of wild bonobos (*Pan paniscus*): sequencing the D-loop region of mitochondrial DNA. *Primates* **37**, 305–18.

Hayakawa, S. & Takenaka, O. (1999). Urine as another potential source for template DNA in polymerase chain reaction (PCR). *Am. J. Primatol.* **48**, 299–304.

Linch, C. A., Smith, S. L. & Prahlow, J. A. (1998). Evaluation of the human hair root for DNA typing subsequent to microscopic comparison. *J. Forensic Sci.* **43**, 305–14.

Linch, C. A., Whiting, D. A. & Holland, M. M. (2001). Human hair histogenesis for the mitochondrial DNA forensic scientist. *J. Forensic Sci.* **46**, 844–53.

Morin, P. A., Moore, J. J., Chakraborthy, R., Jin, L., Goodall, J. & Woodruff, D. S. (1994). Kin selection, social structure, gene flow, and the evolution of chimpanzees. *Science* **265**, 1193–201.

Murphy, M. A., Waits, L. P., Kendall, K. C., Wasser, S. K., Higbee, J. A. & Bogden, R. (2002). An evaluation of long-term preservation methods for brown bear (*Ursus arctos*) faecal DNA samples. *Cons. Genet.* **3**, 435–40.

Radespiel, U., Sarikaya, Z., Zimmermann, E. & Bruford, M. W. (2001) Socio-genetic structures in a free-living nocturnal primate population: sex-specific differences in the grey mouse lemur (*Microcebus murinus*). *Behav. Ecol. Sociobiol.* **50**, 493–502.

Sugiyama, Y., Kawamoto, S., Takenaka, O., Kumizaki, K. & Norikatsu W. (1993). Paternity discrimination and inter-group relationships of chimpanzees at Bossou. *Primates* **34**, 545–52.

Sunnucks, P. (2000). Efficient genetic markers for population biology. *Trends Ecol. Evol.* **15**, 199–203.

Taberlet, P., Griffin, S., Goossens, B., Questiau, S., Manceau, V., Escaravage, N., Waits, L. P. & Bouvet, J. (1996). Reliable genotyping of samples with very low DNA quantities using PCR. *Nucl. Acids Res.* **24**, 3189–194.

Takenaka, O., Takashi, H., Kawamoto, S., Arakawa, M. & Takenaka, A. (1993). Polymorphic microsatellite DNA amplification customised for chimpanzee paternity testing. *Primates* **34**, 27–35.

Valderrama, X., Karesh, W. B., Wildman, D. E. & Melnick, D. J. (1999). Non-invasive methods for collecting fresh hair tissue. *Mol. Ecol.* **8**, 1749–52.

Valière, N. & Taberlet, P. (2000). Urine collected in the field as a source of DNA for species and individual identification. *Mol. Ecol.* **9**, 2149–54.

Vigilant, L. (1999). An evaluation of techniques for the extraction and amplification of DNA from naturally shed hairs. *Biol. Chem.* **380**, 1329–31.

Walsh, P. S., Metzger, D. A. & Higuchi, R. (1991). Chelex-100 as a medium for simple extraction of DNA for PCR-based typing from forensic material. *Biotechniques* **10**, 506–13.

Wasser, S. K., Houston, C. S., Koehler, G. M., Cadd, G. G. & Fain, S. R. (1997). Techniques for application of faecal DNA methods to field studies of Ursids. *Mol. Ecol.* **6**, 1091–7.

Useful Internet sites

CERVUS 2.0: <helios.bto.ed.ac.uk/evolgen/cervus/cervus.html>.

FSTAT 2.9.3: <www.unil.ch/izea/softwares/fstat.html>.

Genbank: <www.ncbi.nih.gov/Genbank>.

GENEPOP: <wbiomed.curtin.edu.au/genepop>.

GENETIX: <www.univ-montp2.fr/~genetix/genetix/genetix.htm>.

KINSHIP: <gsoft.smu.edu/Gsoft.html>.

21 • Tips from the bush: an A–Z of suggestions for successful fieldwork

SIMON K . BEARDER[1], K. A. I. NEKARIS[1], DEBORAH J. CURTIS[2],
J. LAWRENCE DEW[3], JULIA N. LLOYD[1] AND JOANNA M.
SETCHELL[2]

[1]*Nocturnal Primate Research Group, Department of Anthropology, School of Social Sciences and Law, Oxford-Brookes University, Oxford, UK*
[2]*Centre for Research in Evolutionary Anthropology, School of Life and Sport Sciences, University of Surrey Roehampton, London, UK*
[3]*Department of Anthropology, University of California, Davis, California, USA*

INTRODUCTION

This guide is designed to provide a light-hearted, yet serious, list of suggestions for improving your comfort and safety in isolated tropical locations where the facilities may be relatively basic. We hope that it will also be useful for research and travel under less arduous conditions. It has been written as both a checklist and a set of instructions, gleaned from conversations with a wide variety of fieldworkers over a number of years and through personal experience. We end with a few wise sayings relating to the environment – to keep you going in times of adversity.

A

Adaptors. Check what sort of electrical sockets to expect at your destination and take the right adaptors.

Ant-proof socks. Have you ever had the problem of army ants invading your trousers so that you have to get undressed to pull them out of your skin? Just in case you do, tuck in your trouser bottoms and use Gortex® over-socks. Ants are unable to negotiate the smooth material and never make it to your nether regions. These socks also keep your feet dry, since water can only pass out. Not very glamorous, but at least you can feel smug while your companions disrobe in a hurry or suffer from rotting feet.

B

Bags. Hip bags of various kinds are invaluable, and cloth bags/shoe bags of different sizes and colours can help to store things and enable you to find them quickly at the bottom of a rucksack. Large polythene bags

Field and Laboratory Methods in Primatology: A Practical Guide, ed. Joanna M. Setchell and Deborah J. Curtis. Published by Cambridge University Press. © Cambridge University Press 2003.

will keep things dry and double as laundry bags (see also Zip-lock or self-seal bags).

Batteries. In addition to rechargeable batteries, take as many long-life batteries with you as you can (not rechargeable). Local batteries may look the same but do not always last. Remember to remove all batteries from your devices before getting on a plane as they can leak and destroy expensive equipment. Batteries that appear to have 'had it' can be 'recharged' a bit by laying them in the sun – sufficient to power your radio once they are no longer useful in a torch. Use rechargeable batteries round the camp and standard batteries for work.

Beeper. A watch that beeps at set intervals (60 seconds, 120 seconds, etc.) is invaluable for collecting behavioural data.

Binoculars. You cannot spend too much on binoculars. Spend time choosing, buy the best you can afford or place this item at the top of your list of presents required. If you take good care of your binoculars they will last you a lifetime. Use lens covers, or blow across the lens before use, to avoid getting debris in your eyes.

Books. A paperback book may be just the thing when sitting under your poncho stranded in the rain or on long trips to town. You can always swap with fellow travellers.

Boots. A good stout pair of boots is of paramount importance (ankle-supporting and waterproof if appropriate). Never scrimp on boots as they get hammered. Many rain forest researchers use cheap rubber boots fitted with high quality arch supporting insoles for working in muddy or flooded areas. Plastic, locally made sandals are popular in humid forests and stout trainers in dryer regions.

Boxes. Plastic storage boxes, particularly those of the size normally used to store shoes, are ideal to take to the field and are gratefully accepted as gifts when you leave. They protect small and sensitive equipment from the elements and insects, and are perfect for storing food. Like shoe bags, they also provide order in a top-loading rucksack.

C

Cameras. You will miss that once-in-a-lifetime shot if you do not carry a compact zoom, idiot-proof, flash camera on your belt, and practice your quick-draw technique. A good flash is invaluable in dense forest as well as fast film. Learn simple maintenance of your camera before you go and remember to buy a spare battery. Automatic camera traps can show you things that you may never normally see (Chapter 16). However, there are also plenty of other ways to record your trip, so don't just take a camera (see Tape-recorder, Video, Chapters 15 and 16).

Clothing. Essential field clothes include: three shirts (one on, one drying and one in the wash), three pairs of socks, two pairs of light-weight field trousers (not the heavy cotton type – quick-drying ones are invaluable and rip-proof ones are available). Undergarments can be kept to a minimum if you get into the habit of washing the pair you have worn each day. Piles of used underwear can attract mice and ants that will demolish them! Take some smarter clothes too, as you will sometimes need to look smart to avoid giving offence when meeting village elders and permit givers, or for social events and special occasions (Chapter 1). Being a fieldworker does not give you the right to be scruffy.

Compass. (See Orientation).

Confidence tricksters. When you arrive in a city where visitors are common you are vulnerable to a group of people who gain a living from tourists by relieving them of their belongings. It pays to be wary of those who seem friendly and charming in their desire to 'help' you. A polite but firm response can, for example, halve your taxi fare. Always check exact prices *before* you travel, and negotiate a fair price. You need not avoid going out, but it is wise to do so with the minimum of possessions so that you do not make an attractive target.

Contraceptives. Remember that these may not be available, or may not be of sufficient quality, in your destination country. The same applies to tampons.

D

Data. Perhaps your most treasured possession – always carry your data with you in your hand luggage when travelling (see Xerox copies).

Data sheets are useful for all sorts of work. It's amazing how easy it is to forget to note down the obvious if you don't have a prompt (see also Lists).

Dehydration. Few visitors realise how much water they need in a hot climate when exercising more than usual. Dehydration can happen quickly in the sweaty tropics. Drinking as much as you can *in advance* is the answer. By the time you get headaches it is too late but it is wise to carry re-hydration salts. Paradoxically, dehydration makes you feel less thirsty, and can kill. Carry enough water, even if it is heavy. Think twice about consuming ice (not just in drinks, but often used to keep food cold, etc.) and salad, because of the risk of stomach problems.

Dictaphone. Very useful for recording data in bad weather, or when you can't write it all down. But don't get carried away – you'll have to transcribe it all later!

Driver's licence. If you intend to drive (or even if it's just a remote possibility), take your driver's licence with you, and get an International Driving Licence before you go. You may also need to apply for a local licence if staying for any length of time.

Drying your equipment in the tropics often means using silica gel (reusable by drying in an oven) and airtight containers. Waterproof bags of all kinds are available from camping shops – or use those boxes.

Air-conditioned rooms can result in condensation inside your cameras and binoculars when you take them outside. Keep them in sealed containers until they have warmed up.

Duct tape is tough canvas tape that is good for temporary repairs to almost anything.

E

Electricity. Be prepared for power cuts and erratic power surges. Make sure you have a long-life battery and a spare if using a computer, and it may pay to invest in a surge protector since many computers have been ruined without one (see also Adaptors).

Email. An excellent idea for your sponsors, friends and family is to send reports whenever you can get to the Internet. Such instant feedback on the joys and woes of fieldwork is an exciting way to keep in touch and to involve all those who have helped you. Email facilities are now available in many cities. You may want to scan important documents and send them to your own email account, as well as to a few trustworthy friends. In this way, you have back-ups in case they are lost.

Embassy. Your country's representative in your study country can help in difficult times and provide useful advice. Make sure you have the address and telephone number of the local Embassy, High Commission or Consulate. It may be useful to report to them on arrival, so that they know who you are and where you are.

Excess baggage is sometimes unavoidable. It's always a good idea to let the airline know in advance, rather than just turning up with it.

Eye-drops. Sounds ridiculous, but if your animals live in trees, and you are spending a lot of time looking upwards, you may need them. They can also be useful if you are anaesthetising your study animals, to protect their corneas from desiccation (Chapter 8).

F

Field guides. When searching for elusive species, local people may point to the nearest likeness without you asking leading questions. For example, the Congolese 'dinosaur', Mokele Mbembe, has recently been identified as a rhino.

Film. Always take more film than you need. Visual aids are invaluable in enabling others to share in your experiences, but remember to ask permission and avoid sensitive subjects. Many remote places that you visit may change drastically in years to come and you may be the only person to have recorded what it was like in the past. Set aside time to take key photographs that will be useful and entertaining when you get home, perhaps years into the future, especially people and close-ups of your animals and their habitats (Chapter 16).

First aid. What you take will depend on your particular needs, but a basic medical kit might include paracetemol, itch-relief cream, antihistamines, starters and stoppers, rehydration salts, antibiotics for skin infections and gut problems (medical advice required), fungicides, betadine solution for wounds (treat all wounds, even small ones). Travel stores will provide you with useful information, and some will sell you all sorts of exciting emergency medical kits but these are useless unless you have received training in how to use them. Don't worry too much about those frightening health stories – unless they are very recent! Books like Schroeder (2000) and Werner *et al.* (2002) are extremely useful in remote places (see also Health).

Flip-flops. Going barefoot is not a good idea, owing to the risk of parasitic infection and other potential injuries.

Food fads. Local foods may be safer than 'exotic' ones when you are eating out, especially if you eat with your hands. Packet soups are very handy and it is good to take favourite reminders of home. Packets or tins of drinking chocolate can give you instant energy. Spices and herbs are useful and dried fruit bars or packets of figs, raisins and bananas provide treats without being too heavy. Prevention is better than cure, so take multivitamins to supplement your diet and eat as well as you can.

Foot rot. Ugh! Remember those Gortex socks and foot powder!

Fridge. Portable fridges and freezers are available and are good for cold beer/soda. More seriously, these can be very useful for sample storage (see Chapters 1, 8, and 19).

Fungicide. In case you forget the Gortex socks and to get rid of ringworm.

G

Gadgets. We all have our favourite items under this heading. How about trying one new gadget each trip to see whether it handles a rigorous field test. If not, discard it. For example, how about a folding monopod for your camera or microphone, or one of those light-sticks that you break to provide lighting around camp, a simple toast maker or an edible candle?

Gifts. Before you leave home try to think of easy-to-carry items that will be appreciated far away. Things that are typical of your country or that are of practical use, such as watches or kitchen gadgets, are ideal (see Chapter 1). If you are joining people who are already in the field, take magazines, sweets, etc. with you – you will immediately make yourself very popular!

Gloves. In the event that you need to handle animals, consider bringing along a pair of gardening gloves. These are often unavailable locally and will protect you from nasty bites. Furthermore, a pair of warm gloves may seem unlikely in the tropics, but can be essential for watching nocturnal animals on chilly nights. Finally, sterile latex gloves are invaluable if you are collecting tissue for genetic analysis and do not wish to contaminate them, and to keep you safe when handling biological samples (see Chapters 1, 8, 19 and 20).

Gossip. As a foreigner, you will be the centre of attention. You may also be the only person from your country that people have ever met. You are an ambassador and a celebrity, so you may have to get used to the idea and be cool about it.

GPS. Global Positioning Systems are a valuable modern tool (Chapter 4), but they may not work under canopy and the batteries may run down. Remember that gadgets can (and often do) fail and it is difficult to get them repaired in remote places. Your old-fashioned compass remains an invaluable aid (see Orientation).

H

Handkerchiefs. The multi-purpose functionality of a handkerchief is unsurpassed. Not only is it useful when you have a cold, and do not wish to pollute the environment with tissues, but it can also serve to wipe the sweat and dust from your brow during hikes or long bus journeys. It can be used to wipe your fingers when no cutlery or water are available, to dust off unsavoury bus or roadside seats and benches, or to bind a small wound. Folded into a square it provides a cushion behind the lens of your headlamp, or for other purposes where padding is needed.

Health. Find out about the major health hazards (e.g. chloroquine-resistant malaria) in your field area, get the best medical advice you can on how to deal with them, and bring the right medicines. Get the recommended vaccinations (see Jabs), and also a general health and dental check several weeks before you leave. Get a tropical health screening when you return home too – some diseases (e.g. schistosomiasis) may be symptomless, yet very destructive in the long term. Take a sterile kit with you, with an assortment of needles and syringes, to avoid the potential of infection at small, poorly funded hospitals. Take a dental kit if you have dodgy teeth. See a health professional in your study country to find out information that may be unfamiliar in your home country. For example, a local health professional will be better placed to advise you on malaria prophylaxis than your own doctor at home. Always check the expiry date when buying medicines.

Hip chains are invaluable for hands-free distance measurement, or simply to make sure that you don't get lost in new territory. Just tie the line to a tree or other fixed object, set the counter to zero and start walking. Go for photodegradable line, which breaks down with sun-exposure.

Hippos, crocodiles, elephants and other large animals. Find out what the best reaction to potentially dangerous animals is (preferably by asking people, rather than experimenting). Apparently, if you are in a close encounter with a hippo you should stick your fingers in its nostrils. It is forced to come up for air and therefore lets you go. To be honest, if you are this close it may advantageous to pray a little as well!

I

Insects bite, and can carry all sorts of diseases (malaria, dengue fever, Japanese B encephalitis, filaria, to name just a few). It may be hot, but light coloured (biting insects are attracted to dark colours), tropical-wear cotton shirts with long sleeves and long trousers will help to prevent insect bites. These are especially necessary in the evenings and at night. For the exposed bits all you need is to carry a propelling stick of Avon Skin-so-Soft and you will always be safe – and smooth. If you don't believe this just try it at home on a summer evening before you go. If it doesn't suit you, and you require something with a bit more kick, take a repellent containing DEET (diethyl-toluamide). This is rarely available in tropical countries. DEET sweats off, but impregnated wristbands might work. It also melts plastic. You can 'bug-proof' clothes with permethrin. Head-nets may be needed in extreme conditions and tubes of anaesthetic cream work well to soothe bites. Apparently taking vitamin B makes you less tasty to mosquitoes.

Insurance that covers medical expenses and repatriation is vital. You may already be insured through your institution or university department, so check before buying a policy. Always carry a card with your name, insurance details, medication taken (e.g. malaria prophylaxis), blood group, allergies and details of emergency contacts and procedures.

J

Jabs. Prevention is better than cure. Find out what vaccinations are recommended. Remember to tell your doctor or travel clinic that you will be working with animals, and ask people who have visited your study area for advice.

Journal. Keeping a journal can keep you sane, and will provide you with a wonderful reminder of events and experiences you will otherwise forget.

K

Kettle. Place a portable kettle in a small bag with a pocket stove (one that uses solid fuel and folds flat), a disposable lighter, plastic containers of tea and coffee, etc. and your cup and water bottle – for a refreshing (and calming) break while on the move.

L

Language training. This is now much easier thanks to the Internet and many 'learn in your car' CDs and tapes are also available, often in more obscure languages. Take a phrase book (if available). Knowing even a handful of words in the local language is invaluable. The fact that you are trying to embrace what is in many ways the essence of a local culture will make people more accepting of you. Better yet, try learning a song in the local language. In many places, if you can sing a song, your hosts will be very pleased, and your fieldwork will progress even more smoothly.

Laptop and palmtop computers are being used more and more in the field for direct data entry, but keep a hard copy for back-up (see Email).

Leeches. If your field site has leeches, socks up to your knees allow you to detect them before they get into your shoes. Other ways to deter these bloodsuckers include spraying your boots with roach killing spray or mashing up tobacco and mixing it with water and rubbing it on your boots; the leeches will fall right off. You might also decide that it is better to wear sandals or flip-flops and pick the leeches off as they come; try rubbing your feet with citronella oil, or better yet with soap, to deter them from attaching their mouth parts to you. If none of these works, salt or burning with a lighter will do the trick. Make sure the leech has released its grip before you pull it off or blood will flow!

Lighters. Take lots. It's amazing how they disappear! Keep a bunch in a self-seal bag or tin and also put one in each bag and pocket for convenience. They also make welcome gifts.

Lists. Keep lists of everything that needs to be done. This is especially important when visiting new places, when there is so much to remember. Such lists will prove invaluable when planning your next trip. For example, you will not be able to remember your cravings for particular foods unless you write them down at the time you are starving!

M

Malaria kills. If you're working in a malarial area, start taking prophylaxis before you leave, so that if side effects occur, they occur at home, and you can switch to an alternative medication. Continue taking your prophylaxis. Know the symptoms to look out for. Self-medication for malaria gains you time to get to a doctor, and is not an alternative to medical help (see Health). Self-test kits (using monoclonal antibodies) are commercially available for falciparum malaria (and are under development for other species) and may be helpful in remote places.

Marking and measuring tapes. Orange 'surveyor's tape' can be used to mark out trails and temporarily label trees (Chapter 11). You can write on the tape with permanent marker pens. Tape measures may be needed for mapping trails or measuring animals (Chapter 9). If you lose your measuring tape, simply use string and calibrate it later.

Money. Transfer funds from your bank to the field. Take travellers' cheques in case problems arise. Find out which currency is the most useful to take with you. Keep receipts for claiming. Credit cards may be a useful back-up in cities but do not assume that they will be accepted, or that cash machines will be available.

Mosquito net. It is best to have your own. Treat your net regularly with permethrin.

N

Notebooks. Of course, for all occasions! Write-in-the-rain notebooks are the answer to wet climates and, to save losing all your hard won data, it is a good idea to write your notes (and letters) in duplicate books (you can buy carbonless ones). Do not forget to store the copies *separately* or send them home! If write-in-the-rain books surpass your budget, store your notebook in a plastic bag, and write in pencil, which never runs, and writes on damp surfaces!

O

Orientation. Try to obtain maps and guides to your study area well before you leave and *always take a compass*. Look out for additional material on arrival and ask as many opinions as you can. The more you understand in advance, the easier it will be to settle down and get things done, or to arrive at a decision on whether it is even safe to go at all. Maps of your study area may be available in the capital city, and guidebooks may list where. What to do if you get lost will depend on your study area. Don't panic, just follow your compass in a straight line until you hit a trail, or walk downhill until you reach a stream, then follow it to the river, etc.

Other people. Many field sites have more than one researcher, plus other staff. You must be able to live and work co-operatively in crowded, difficult conditions, with different characters. This can be one of the best parts of the whole experience but, equally, it can also be the worst! Remember too that other researchers may come after you, and how you act will affect their success as well as your own.

P

Pens. Bring lots and make sure they will write well on damp paper. Forestry Suppliers sell a pen used by astronauts, which is sturdy and has a clip so it can be attached to your notebooks with a string. Pens used and sold locally are likely to be cheap and functional. Indelible markers are necessary for labelling sample tubes and bags and flagging tape, and are less likely to be easily available.

Permits and regulations. Allow plenty of time to organise research permits, permits to export and import biological samples, etc., and respect regulations for the sake of your own research, and that of future researchers.

Pictures of your study animals, home and family are invaluable aids to communication when language is a barrier. People will be interested to see pictures of where *you* are from, and to hear about your culture, just as much as you want to learn about theirs. A few passport photos are always useful for permits.

Pockets. You cannot have too many pockets. Some people swear by those organiser waistcoats worn in the tropics in colonial times and sold today as photographer's jackets, or buy army-style field trousers.

Political situation. Always check before you go (e.g. UK Foreign Office website, US State Department) and ask repeatedly when you get to a new country to learn what to do and what not to do.

Poncho. For those sudden storms (beware of falling trees and branches in hill forest), and sometimes the only thing that will keep insects off.

Postcards. Why not buy postcards when you first arrive in a city and bring adhesive address labels for your friends and loved ones. By posting your cards immediately, they may actually arrive before you get home.

Procrastination. Fieldworker's Procrastination Syndrome is a psychological affliction that can render you useless for weeks, particular if you have fallen for the romance of fieldwork rather than the reality. Symptoms include a compulsive desire to get everything just right before you start work. Preparation becomes the end rather than the means. Others conspire to support your delays by tempting you to do more enjoyable things or pointing out all the dangers and pitfalls. You end up doing little or nothing. You will have to be single-minded and at least a bit 'driven' to succeed.

Q

Quick-cook meals. A few dehydrated camping meals take up little room, and provide an excellent source of familiarity and comfort. Select some of your favourite meals and save them to celebrate special occasions or in cases of emergency. If meals are too large an option, bringing several of your favourite herbs stored in film canisters is a fantastic way to make local food taste more like something from home.

Quick-wipe cleaning tissues or travel wipes can be purchased in a flip-top dispenser and provide an excellent way to keep clean at a field site or when on the move. They are also very effective for cleaning stains from clothing. Just dab onto the dirt spot, rub hard when damp and wipe away with a dry face tissue. Magic!

R

Radio. This may help to keep you sane if you are in the middle of the forest and isolated for a while. Short-wave broadcasts from home help to adjust your perspective when you are feeling run down and lonely!

Rest. (See Y).

Rucksack. If you cannot get your stuff into one manageable rucksack and your overnight bag, *leave it at home*. A single 65 litre rucksack has been used successfully for a one-year trip (see Clothing).

S

Sarongs are often available locally, and have all sorts of uses.

Secateurs. These are quieter than a machete/panga and are useful for botanical sampling too. Slip them into a belt bag or custom-made holster.

Sewing kit. Should be obvious really.

Sleeping bag and sheet sleeping bag. It's so much easier to wash a sleeping bag liner than a sleeping bag. Having your own sheets can also be useful in cheap hotels/rest houses.

Snakes. Love them or hate them, you will feel safer if you carry a crepe bandage large enough to bind a limb (as if for a sprain). This slows the effects of snakebite for long enough to travel for help, saving you from panic, and is much safer than other do-it-yourself remedies.

Soap. Camping stores sell liquid soap that will wash you and your clothes in fresh or salt water, and is biodegradable. They also sell antibacterial gel, which cleans hands without the need for water, and may be useful under some circumstances.

Solar panels can be bought cheaply and hooked up to a car battery to provide a source of power, or to charge your rechargeable batteries, thereby being a bit more environmentally friendly.

Specimen tubes. Take lots, of varying sizes.

Spectacles. Daily disposable contact lenses circumvent steamy glasses. If you wear spectacles, wear a sports elastic to keep them firmly attached to you, and carry a spare pair with you in the field (imagine losing your only pair in a swamp 5 km from camp...).

String, spare straps and bungee cords take up little room but always prove useful as washing lines, guy ropes, etc.

Sun cream and sun hat. High factor sun protection is vital for the fair skinned, particularly if you are near water, sand or snow, or taking doxycycline as a malaria prophylaxis.

Swiss army knife or multi-tool (with scissors, saw and most importantly tweezers). Don't carry this with you on the 'plane – pack it in your hold luggage. A simple dissecting kit in a cloth wallet provides a convenient way to carry most of the other tools that you may require – including a small screwdriver for when your glasses fall apart, or to pick thorns out of your boots!

T

Tape-recorder. A small tape-recorder and good directional microphone will enable you to 'capture' animals that you cannot see. These should be considered as essential as a camera for the study of cryptic (nocturnal and forest-dwelling) mammals and birds, but it may help to you collect valuable data on most species. Calls can often be used for identification and to provide additional information about social interactions and predator detection. Once recorded, they are available for further analysis – perhaps years later (Chapter 15).

Tent. A small tent with a sewn-in groundsheet and mosquito netting is often essential.

Thermometer. This provides a good way to monitor your state of health when things get serious.

Ticks. Don't forget the all-essential tick check on returning from the field each day. These tiny creatures can find their way into the most uncomfortable places, and it is best to remove them with your fingernails or a pair of tweezers as soon as possible. Wear protective clothing and don't wander around with your shirt hanging out. Pepper ticks must be squeezed between your thumbnails until they click! Use Vaseline or burn off the larger varieties to avoid leaving their mouthparts in you. Get someone to groom those places you cannot reach.

Toilet paper. Doubles as tissues and should go everywhere with you.

Torches/flashlights. Head torches are now available in an amazing range of sizes and types with additional red filters and halogen bulbs. Not only will you be able to see animals from their eye-shine, and find your way around more easily than with a hand-held torch, but they are invaluable when removing contact lenses and doing chores around camp. Always carry a torch when away from camp. Invest in LED (light emitting diode) torches that last for ages without new batteries for emergency use; you never know when you may get caught with the light fading.

U

Understanding other cultures. Perhaps the most vital key to success is to be aware of the knowledge and customs of local people (see Language). You will make slow progress if you fail to observe rituals of etiquette and politeness that are normal in an area, so listen, learn, and behave accordingly. It can be an offence, or at least insensitive, to show certain parts of your body in some cultures (e.g. belly, thighs, legs, hair (for women)). It is advisable to spend a few days visiting village elders and important officials at the start, to enable them to understand why you are there, and to listen to their advice. If one line of action fails, try another – do not simply give up! You will eventually find someone who can help. There will often be someone who shares your interests and enthusiasm, who will want to meet you, even if they are difficult to find. When trying to influence people when you are away from home, it is helpful to think of how you would react if the situation were reversed, and they were visiting you (Chapter 1).

Remember to adjust your pace of life and try to be at least as calm and patient as the people you meet. Shopping for supplies can take all day, because you should not ask a stranger a question, pass a friend in the road, or buy something at a stall without following local etiquette and stopping to chat. Local assistants/informers/translators/friends are invaluable for cultural information, and you will learn a huge amount about other people just by sitting down to talk – one of the best things about fieldwork.

V

Video technology is now so good that you may decide to take a small camcorder that also functions as a still camera and tape-recorder. For nocturnal observations, the night shot on a video camera can double as a night vision scope (Chapter 16).

Visas and visa extensions. Check on visa requirements as early as possible, since these may take time to organise. Consult with people who have experience in the country that you are visiting, as well as the appropriate Consulates or Embassies, and ensure that you have all the letters of recommendation and other official documents that you need (see Xerox copies).

W

Water containers are *vital*, especially a personal water bottle to keep with you. Folding containers (polythene or canvas) are convenient and even a condom can serve as a lightweight water carrier! Always take plenty of sterilising tablets (either iodine or chlorine) for emergencies when a water filter is not available. Don't forget to bring flavoured drink powder to cover up that nasty tablet taste.

Water filters. Take one with you or make a siphon water filter using a bucket hung on a branch of a tree, a jerry can and Mill-bank bags for filtering (but in this case remember you will still need to sterilise the filtered water).

X

Xerox copies. Photocopy all your important documents, such as passport, insurance details, permits, airline tickets and driver's licence. Leave a copy at home with your family or friends, and take another copy with you. This will make your life easier when they get lost or stolen. It may be safer when out in the city to carry these copies and leave the originals in a more secure place, along with your other valuables. You may need to get the photocopies officially authorised (normally at the police station).

Y

Why? An often-asked question, especially when you are tired, hungry and have just lost your study animals in the middle of nowhere. You have given up your normal life to struggle around on the side of a mountain in the pouring rain and be eaten alive by all sorts of tiny creatures. It is hard to keep a sense of proportion, so it is very important to take breaks. Do not work so hard that you collapse. Rather, put your health as the number one priority. You will be less efficient or useless if you become run down or sick. To maintain a sustainable workload, plan your working week as if you were at home, with time off for relaxation, socialising and sleep.

Z

Zip-lock or self-seal bags. Your mum may swear by them in the freezer but it is also a good idea to take tons if you are collecting data in humid areas or during the rainy season. They can also be used to store specimens.

SOME CLOSING THOUGHTS

- Wastefulness has become a product of civilisation – but it is not a sign of a civilised person.
- We don't inherit the earth from our forefathers; we borrow it from our children.
- Life is like a tin of sardines – we are all looking for the key.
- Failure is the path of least persistence.
- If you don't enjoy what you have, how could you be happier with more?
- After all is said and done – there is more said than done.

- Enthusiasm breakfasts on obstacles, lunches on objections and dines on competition.
- Many a false step is made by standing still.
- People may doubt what you say but they will always believe what you do.
- Telling others what to do in their own country is like them telling you what to do in yours.

ACKNOWLEDGEMENTS

We thank the students taking the M.Sc. in Primate Conservation at Oxford Brookes University, UK, for comments and additions.

REFERENCES

Schroeder, D.G. (2000). *Staying Healthy in Asia, Africa, and Latin America*, 5th edition. Emeryville, CA: Volunteers In Asia and Moon Travel Handbooks.
Werner D., Thuman, C. & Maxwell, J. (2002). *Where There Is No Doctor: A Village Health Care Handbook*, revised edition. Palo Alto, CA: Hesperian Foundation.

Index

Printed in the United States
143671LV00004B/11/P